T0329540

GLYPHOSATE RESISTANCE IN CROPS AND WEEDS

GLYPHOSATE RESISTANCE IN CROPS AND WEEDS

History, Development, and Management

Edited by

VIJAY K. NANDULA
Mississippi State University
Stoneville, MS

WILEY

JOHN WILEY & SONS, INC., PUBLICATION

For general information on our other products and services or for technical support, please
contact our Customer Care Department within the United States at (800) 762-2974, outside the
United States at (317) 572-3993 or fax (317) 572-4002.

Wiley also publishes its books in a variety of electronic formats. Some content that appears in
print may not be available in electronic formats. For more information about Wiley products,
visit our web site at www.wiley.com.

Library of Congress Cataloging-in-Publication Data:

Glyphosate resistance in crops and weeds: history, development, and management / edited by
Vijay K. Nandula.
　　p. cm.
　Includes index.
　ISBN 978-0-470-41031-8(cloth)
　1. Glyphosate.　2. Herbicide resistance.　3. Herbicide-resistant crops.　4. Plants–Effect of
herbicides on.　I. Nandula, Vijay K.
　SB952.G58G59 2010
　632′.954–dc22

2009054245

10　9　8　7　6　5　4　3　2　1

Dr. Stephen O. Duke for his inspiration and mentoring

and

my wife Aparna, daughter Indu, and son Ajay for their love, support, and understanding.

CONTENTS

PREFACE

Since the discovery of its herbicidal properties in 1970 and commercialization in 1974, glyphosate has been used extensively in both croplands and non-croplands. Because of its lack of selectivity, glyphosate use was initially limited to preplant, postdirected, and postharvest applications for weed control. With the introduction of glyphosate-resistant (GR) crops in the mid-1990s, glyphosate is now widely used for weed control in GR crops without concern for crop injury. GR crops are currently grown in several countries, with particularly strong adoption in the United States, Canada, Argentina, and Brazil. The widespread adoption of GR crops has not only caused weed species shifts in these crops, but it has also resulted in evolution of GR weed biotypes. GR weed populations threaten the sustainability of glyphosate and GR crop technology, thereby jeopardizing derived benefits such as reduced fuel costs and improved soil conservation. To date, 18 weed species have evolved resistance to glyphosate worldwide. This number will most likely increase rapidly in the next few years due to increased selection pressure from glyphosate, better monitoring and detection methods, and better awareness of the problem of glyphosate resistance.

Exciting new technologies such as new generation of GR crops and multiple herbicide-resistant (HR) (including glyphosate resistance) crops are in development or approaching commercialization in the next few years, which will help manage GR weeds and reduce their spread. Modern research techniques such as weed genomics are being employed to study GR weed resistance mechanisms, fitness issues, biology, and ecology. Additional avenues of research being pursued are gene flow, population genetics, multiple resistance, modeling, GR weeds as alternative hosts for other pests, and effects on human and animal health, as well as the impact on conservation tillage. GR

ix

crop technology has revolutionized crop production in the developed world, and the benefits are gradually spilling over to the developing world.

The vast body of complex information being generated on glyphosate resistance, one of the pressing issues faced by growers and land managers, makes it hard to keep current with the topic. To sustain an effective, environmentally safe herbicide such as glyphosate and the GR crop technology well into the future, it is imperative that the issue of GR weeds is comprehensively understood. To this end, an up-to-date source of information on glyphosate resistance is essential for researchers, extension workers, land managers, government personnel, and other decision makers, so the bottom line of growers, and conservation and diversity programs is increased. I earnestly hope that this book will fill this niche.

The book is divided into 16 chapters. Chapter 1 provides an overview of more recent research on the use of the herbicide glyphosate and its environmental, toxicological, and physical aspects. Herbicide resistance is defined in Chapter 2 and several aspects related to it are introduced. Chapter 3 reviews the processes involved in the commercialization of currently grown GR crops as well as the next generation of GR and HR crops, including multiple HR traits.

Chapter 4 is a comprehensive review of GR crop development events as well as multiple HR crops that are currently approaching commercialization. Chapter 5 provides an overview of the biochemical, biological, molecular, and physiological procedures used in laboratory, greenhouse, and field research with glyphosate resistance in plants. Chapter 6 summarizes the current knowledge of biochemical mechanisms of evolved glyphosate resistance in weeds and the molecular basis behind it.

Chapter 7 examines the genetics and inheritance of the mechanisms of glyphosate resistance. A genomic approach is taken in Chapter 8, in order to gain a better understanding of the mechanisms and evolution of glyphosate resistance in weeds using GR horseweed, the first broad-leaved weed that has evolved to be resistant to glyphosate, as a model.

Chapter 9 summarizes the effect of GR corn, cotton, and soybean cropping systems on weed species shifts as well as late-season weed problems in the United States. Chapter 10 describes the history of herbicide resistance, evolution of glyphosate resistance, biology and ecology, and glyphosate resistance management in horseweed. Chapter 11 describes the unprecedented nature and magnitude of difficulty in managing GR Palmer amaranth populations. In Chapter 12, the current situation regarding GR cropping systems and weed management issues in midwestern United States is discussed.

Chapter 13 examines the development and management of GR rigid ryegrass from Australia, the first weed to evolve resistance to glyphosate. Latin America is covered in Chapter 14, which comprehensively reviews the history and current status of glyphosate resistance in weed populations there.

Chapter 15 provides insights from an extension perspective on the management of glyphosate weeds, and Chapter 16 presents an analysis of the effects of GR weeds on management costs. There is some overlap in the content presented among chapters, given the nature of the subject matter.

This book is expected to be useful to students, researchers, regulators, industry, and anyone interested in learning about glyphosate resistance around the world.

Mississippi State University   VIJAY K. NANDULA
Stoneville, MS

ACKNOWLEDGMENTS

I wish to thank Mr. Jonathan Rose, editor, John Wiley & Sons, Inc., for recognizing the need for this project and providing constant support and encouragement. I am deeply indebted to all contributing authors who have come together with the common goal of sharing historic and current information on this important subject of glyphosate resistance. I sincerely express my gratitude to all reviewers who have agreed to review and provide their input toward improving the content of the book in a very timely and efficient manner.

CONTRIBUTORS

Laura G. Abercrombie, Department of Plant Sciences, University of Tennessee, Knoxville, TN

Marion Bleeke, Monsanto Company, 800 N. Lindbergh Blvd., St. Louis, MO

Claire A. CaJacob, Monsanto Company, 700 Chesterfield Village Pkwy W., Chesterfield, MO

Janet E. Carpenter, Consultant; Email: janet.e.carpenter@gmail.com

Linda A. Castle, Pioneer Hi-Bred International, Inc., Verdia Research Campus, 700A Bay Road, Redwood City, CA

R. Eric Cerny, Monsanto Company, 700 Chesterfield Village Pkwy W., Chesterfield, MO

Michael J. Christoffers, Department of Plant Sciences, North Dakota State University, Fargo, ND; Email: Michael.Christoffers@ndsu.edu

A. Stanley Culpepper, Department of Crop and Soil Sciences, University of Georgia, Tifton, GA; Email: stanley@uga.edu

Gerald M. Dill, Monsanto Company, 800 N. Lindbergh Blvd., St. Louis, MO; Email: gerald.m.dill.jr@monsanto.com

Greg A. Elmore, Monsanto Company, 700 Chesterfield Village Pkwy W., Chesterfield, MO

Donna Farmer, Monsanto Company, 800 N. Lindbergh Blvd., St. Louis, MO

Paul C. C. Feng, Monsanto Company, 800 N. Lindbergh Blvd., St. Louis, MO; Email: paul.feng@monsanto.com

Leonard P. Gianessi, Crop Protection Research Institute, CropLife Foundation, 1156 15th Street NW, Suite 400, Washington, DC

Jerry M. Green, Pioneer Hi-Bred International, Inc., Stine-Haskell Research Center Bldg. 210, 1090 Elkton Road, Newark, DE; Email: Jerry.M.Green@pioneer.com

Matthew D. Halfhill, Department of Plant Sciences, University of Tennessee, Knoxville, TN; and Department of Biology, Saint Ambrose University, Davenport, IA

Eric A. Haupfear, Monsanto Company, 800 N. Lindbergh Blvd., St. Louis, MO

Gregory R. Heck, Monsanto Company, 700 Chesterfield Village Pkwy W., Chesterfield, MO

Joy L. Honegger, Monsanto Company, 800 N. Lindbergh Blvd., St. Louis, MO

Jun Hu, Department of Plant Sciences, University of Tennessee, Knoxville, TN; and Institute of Plant Genomics and Biotechnology and Department of Plant Pathology and Microbiology, Texas A&M University, College Station, TX

Jintai Huang, Monsanto Company, 700 Chesterfield Village Pkwy W., Chesterfield, MO

William G. Johnson, Department of Botany and Plant Pathology, Purdue University, West Lafayette, IN

Frank Kohn, Monsanto Company, 800 N. Lindbergh Blvd., St. Louis, MO

Keith Kretzmer, Monsanto Company, 800 N. Lindbergh Blvd., St. Louis, MO

Warren M. Kruger, Monsanto Company, 700 Chesterfield Village Pkwy W., Chesterfield, MO

Christopher L. Main, West Tennessee Research and Education Center, University of Tennessee, Jackson, TN

Carol Mallory-Smith, Department of Crop and Soil Science, Oregon State University, Corvallis, OR

Marianne Malven, Monsanto Company, 700 Chesterfield Village Pkwy W., Chesterfield, MO

Susan J. Martino-Catt, Monsanto Company, 700 Chesterfield Village Pkwy W., Chesterfield, MO

Akbar Mehrsheikh, Monsanto Company, 800 N. Lindbergh Blvd., St. Louis, MO

John A. Miklos, Monsanto Company, 700 Chesterfield Village Pkwy W., Chesterfield, MO

Thomas C. Mueller, Department of Plant Sciences, University of Tennessee, Knoxville, TN

Vijay K. Nandula, Delta Research and Extension Center, Mississippi State University, Stoneville, MS; Email: vnandula@drec.msstate.edu

Jason K. Norsworthy, Department of Crop, Soil, and Environmental Science, University of Arkansas, 1366 West Altheimer Drive, Fayetteville, AR

Micheal D. K. Owen, Department of Agronomy, Iowa State University, Ames, IA

Stephen R. Padgette, Monsanto Company, 700 Chesterfield Village Pkwy W., Chesterfield, MO

Yanhui Peng, Department of Plant Sciences, University of Tennessee, Knoxville, TN

Alejandro Perez-Jones, Monsanto Company, St. Louis, MO

Christopher Preston, School of Agriculture, Food & Wine, University of Adelaide, PMB 1, Glen Osmond SA, Australia; Email: christopher.preston@adelaide.edu.au

Priya Ranjan, Environmental Sciences Division, Oak Ridge National Laboratory, Oak Ridge, TN

Murali R. Rao, Department of Plant Sciences, University of Tennessee, Knoxville, TN

Krishna N. Reddy, USDA-ARS, Southern Weed Science Research Unit, Stoneville, MS; Email: krishna.reddy@ars.usda.gov

R. Douglas Sammons, Monsanto Company, 800 N. Lindbergh Blvd., St. Louis, MO

Dale L. Shaner, USDA-ARS, Fort Collins, CO; Email: dale.shaner@ars.usda.gov

Ken Smith, Southeast Research and Extension Center, Division of Agriculture, University of Arkansas, Monticello, AR; Email: smithken@uamont.edu

Lynn M. Sosnoskie, Department of Crop and Soil Sciences, University of Georgia, Tifton, GA

Lawrence E. Steckel, West Tennessee Research and Education Center, University of Tennessee, Jackson, TN; Email: lsteckel@utk.edu

C. Neal Stewart, Jr., Department of Plant Sciences, University of Tennessee, Knoxville, TN; Email: nealstewart@utk.edu

Patrick J. Tranel, Department of Crop Sciences, University of Illinois, Urbana-Champaign, IL

Bernal E. Valverde, Investigación y Desarrollo en Agricultura Tropical (IDEA Tropical), Alajuela, Costa Rica; and University of Copenhagen, Hojebakkegaard Allé 13, Taastrup 2630, Denmark; Email: ideatrop@ice.co.cr/bev@life.ku.dk

Aruna V. Varanasi, Department of Plant Sciences, North Dakota State University, Fargo, ND

Theodore M. Webster, Crop Protection and Management Research Unit, USDA-Agricultural Research Service, Tifton, GA

Stephen C. Weller, Department of Horticulture and Landscape Architecture, Purdue University, West Lafayette, IN; Email: weller@purdue.edu

D. Wright, Monsanto Company, 800 N. Lindbergh Blvd., St. Louis

Alan C. York, Department of Crop Science, North Carolina State University, Raleigh, NC

Joshua S. Yuan, Institute of Plant Genomics and Biotechnology and Department of Plant Pathology and Microbiology, Texas A&M University, College Station, TX

1

GLYPHOSATE: DISCOVERY, DEVELOPMENT, APPLICATIONS, AND PROPERTIES

GERALD M. DILL, R. DOUGLAS SAMMONS, PAUL C. C. FENG, FRANK KOHN, KEITH KRETZMER, AKBAR MEHRSHEIKH, MARION BLEEKE, JOY L. HONEGGER, DONNA FARMER, DAN WRIGHT, AND ERIC A. HAUPFEAR

1.1 HISTORICAL PERSPECTIVE AND MODE OF ACTION

N-(phosphonomethyl)glycine (glyphosate) is a phosphonomethyl derivative of the amino acid glycine. Glyphosate is a white and odorless crystalline solid comprised of one basic amino function and three ionizable acidic sites (Fig. 1.1). Glyphosate was actually invented in 1950 by a Swiss chemist, Dr. Henri Martin, who worked for the small pharmaceutical company, Cilag (Franz et al. 1997). The product had no pharmaceutical application and was never reported in literature. In 1959, Cilag was acquired by Johnson and Johnson, which sold its research samples, including glyphosate, to Aldrich Chemical. Aldrich sold small amounts of the compound to several companies in the 1960s for undisclosed purposes, but no claims of biological activity were ever reported. In its Inorganic Division, Monsanto was developing compounds as potential water-softening agents and over 100 related aminomethylphosphonic acid (AMPA) analogs were synthesized. When these compounds were tested as herbicides by Dr. Phil Hamm, two showed some herbicidal activity on perennial weeds. However, the unit activity was too low to be a commercial herbicide.

Dr. Hamm enlisted the efforts of Monsanto chemist Dr. John Franz. He repeatedly told Dr. Franz that "he just wanted something five times as strong

Glyphosate Resistance in Crops and Weeds: History, Development, and Management
Edited by Vijay K. Nandula
Copyright © 2010 John Wiley & Sons, Inc.

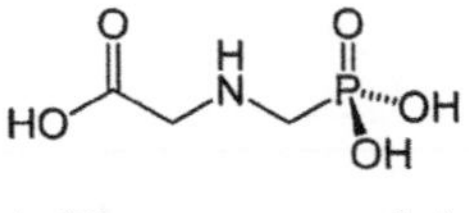

Figure 1.1. The structure of glyphosate.

... that's all." "He convinced me to take a shot at making analogs and derivatives," recalled Dr. Franz. "That didn't yield anything, and I was ready to drop the project. But then I began trying to figure out the peculiarities of those two compounds, and I wondered if they might metabolize differently in the plants than the others ... I began to write out metabolites ... you could write a list of about seven or eight ... it involved completely new chemistry. Glyphosate was the third one I made" (Halter 2007). Glyphosate was first synthesized by Monsanto in May 1970 and was tested in the greenhouse in July of that year. The molecule advanced through the greenhouse screens and field testing system rapidly and was first introduced as Roundup® herbicide by Monsanto Company (St. Louis, MO) (Baird et al. 1971).

Glyphosate inhibits the enzyme 5-enolpyruvylshikimate-3-phosphate synthase (EPSPS) (Amrhein et al. 1980), which is present in plants, fungi, and bacteria, but not in animals (Kishore and Shah 1988). The enzyme catalyzes the transfer of the enolpyruvyl moiety of phosphoenolpyruvate (PEP) to shikimate-3-phosphate (S3P). This is a key step in the synthesis of aromatic amino acids, and ultimately, hormones and other critical plant metabolites. The active site of the EPSPS enzyme in higher plants is very highly conserved (CaJacob et al. 2003). The mechanism of inhibition is also unique in that the binding site for glyphosate is reported to closely overlap with the binding site of PEP (Franz et al. 1997). A diagram of the shikimate pathway and glyphosate's inhibition site is shown in Figure 1.2. No other mode of action for glyphosate has been observed even when very high doses are applied to glyphosate-resistant (GR) soybean and canola (Nandula et al. 2007).

Glyphosate is currently labeled for use in over 130 countries, and current global volume is estimated to be approximately 600 kilotons annually (Research and Markets 2008). The current U.S. glyphosate label of Monsanto Agricultural Herbicides lists over 100 annual broad-leaved and grass species controlled. In addition, over 60 perennial weed species are also included on the label as of the writing of this chapter. It is the broad spectrum perennial weed control that makes glyphosate a very effective product. The ability of the product to translocate to growing meristematic tissues and inhibit an enzymatic process present in plants allows applicators to control underground meristems, corms, rhizomes, and other potential vegetative structures, which regenerate when only upper vegetative material is killed.

Because of its unique properties, glyphosate was initially utilized to control perennial weeds on ditch banks, in right of ways, and fallow fields. However, because it also killed crops, its uses in mainstream agriculture were limited until the use of minimum and no-till practices began to evolve. Spraying

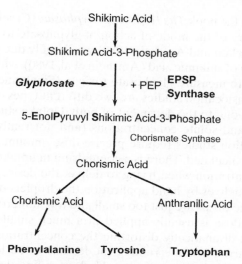

Figure 1.2. The site of inhibition of glyphosate from Dill (2005).

glyphosate to control weeds prior to planting allows growers to substitute chemical weed control with light-duty spray equipment for tillage. This practice saves fuel, preserves soil from erosion, and allows better water permeation into the soil (Dill 2005). Conservation tillage practices have continued to grow with the introduction of GR crops (Dill et al. 2008).

1.2   UPTAKE AND TRANSLOCATION OF GLYPHOSATE

The herbicidal efficacy of glyphosate is strictly dependent on the dose of glyphosate delivered to the symplastic or living portion of the plant. Since glyphosate was first announced (Baird et al. 1971) as a broad spectrum herbicide (and before the evolution of GR weeds), it could be said that all plants could be controlled given delivery of the appropriate dose of glyphosate. The delivery of this efficacious dose has continually been the topic of investigation now for almost 40 years with at least 40 individual weed species now studied in detail to determine the efficiency of uptake and the extent of translocation. The corollary science of pesticide application is an extensive area covering the physics of spray application and the reader is directed to a standard text (Monaco et al. 2002), while here we focus on uptake and translocation.

The first uptake efficiency and translocation studies of ^{14}C-glyphosate (Sprankle et al. 1973) characterized the principal features of glyphosate that we know today: phloem transport and consequent delivery to meristematic growing points in the roots and vegetation. The phloem movement of glyphosate intimately linked the efficiency of translocation to plant health and developmental stage, which are tied to environmental conditions. The early work is well summarized in the book *The Herbicide Glyphosate* (Caseley and Coupland 1985). The discovery of the mode of action of glyphosate to be the inhibition of EPSPS (Steinrucken and Amrhein 1980) was largely due to the very rapid large accumulation of shikimic acid (Amrhein et al. 1980), which now routinely serves as a means to measure glyphosate toxicity (Singh and Shaner 1998).

Uptake and translocation studies are two different types of studies that are often combined as one to the detriment of both. Uptake studies should focus on the drop size and solute concentrations (and not really the total dose), whereas translocation studies require precise dose amounts so that distribution ratios can be calculated. There is a conundrum in uptake studies between volume and concentration when trying to deliver the desired dose. It is virtually impossible to deliver by hand application the droplets dictated by typical carrier volumes; the drops are just too small and too numerous. Consequently, the experimental dose is usually applied in a much smaller volume and/or much larger drop, dramatically distorting the concentration ratios of herbicide:surfactant:carrier volume ruining the lessons to be learned about the efficiency of spray solution penetration. Understanding the impact of spray solution composition on the efficiency of glyphosate penetrating the cuticle to the apoplast and the stepwise entry into the symplast where phloem transport can occur is critical to optimizing herbicide formulation. Normally, the amount of glyphosate "inside" the leaf or not removed by washing is considered the efficiency of uptake. Uptake is dependent on several interdependent factors: droplet size and droplet spread, cuticle composition and thickness, surfactant type and concentration, ionic strength and salt concentration, humidity, and, most importantly, glyphosate concentration. Because of the critical linkages between these factors, the most informative uptake studies are done with a sprayed application using a standard field nozzle and carrier (Feng et al. 2000; Prasad 1989, 1992). However, it is extremely difficult to deliver a precise dose due to the practical problems of leaf intercept of a spray application, and so the leaf intercept efficiency must be included. The interaction of drop size, surfactant, and herbicide concentration does impact the leaf surface cytology and can be correlated to efficiency of uptake (Feng et al. 1998, 2000). The cytotoxic damage caused by the excess surfactant/cuticle surface area provided by a large drop quickly "kills" the loading site for translocation and prematurely stops phloem loading. The exact correlation of drop size and concentration to penetration was determined by using a droplet generator (Prasad and Cadogan 1992). The herbicide concentration in very small droplets did overcome the drop-size factors, and the smaller droplets had minimal negative effect on epidermal cytology (Ryerse et al. 2004), thereby, avoiding the inhibition of transport caused by too much local cell damage (often seen in hand-applied large drops). The concept that small spray droplets do not actually dry but soak into the leaf was shown by coapplication with heavy water (deuterium oxide, D_2O), indicating that the surfactant forms channels to allow the herbicide to penetrate the cuticle as measured by the appearance of D_2O in the leaf (Feng et al. 1999). Spray applications on GR corn then allowed the separation of local droplet-herbicide toxicity from droplet-surfactant injury related to drop size to show that large drops, while being retained less efficiently, were more efficient at loading glyphosate and allowing improved translocation. Consequently, studies that spray ^{14}C-glyphosate provide the best means to mimic field conditions and simultaneously understand the formulated droplet uptake characteristics (Feng and Chiu 2005; Feng et al. 2000, 2003b).

Translocation efficiency is dramatically affected by the self-limitation feature of glyphosate toxicity (vide infra) creating another paradox, optimizing translocation (improving with time) with increasing toxic effect (increasing with time). The negative effects on apical meristems with a small dose of glyphosate are readily accounted for by the observation that individual plant tissues have different sensitivities to glyphosate (Feng et al. 2003a). This toxicity affects the overall glyphosate efficiency and distribution pattern to sink tissues. Dewey (1981) noted that glyphosate easily loaded the phloem, moved from source to sink, and did not usually leave the symplastic assimilant flow. Gougler and Geiger (1981) used a sugar beet model system to demonstrate that glyphosate loads the phloem passively, and this result holds true as no significant active transport of glyphosate has ever been measured. They subsequently showed that reductions in photosynthesis resulted directly in limiting glyphosate translocation (Geiger et al. 1986) and further that glyphosate created a self-limitation of translocation due to its toxicity shutting down photosynthesis and sucrose metabolism (Geiger and Bestman 1990). These observations strongly suggest that the standard practice of overspraying a plant with cold glyphosate at a field rate and then spotting the ^{14}C-glyphosate on a particular leaf to measure translocation is a bad idea. First, the translocation from that source leaf will depend on "its" perception of sink strengths based on its location on the plant. Second, the self-limitation due to whole plant toxicity will prematurely limit translocation. Third, the unknown proportional mixing of cold and ^{14}C-glyphosate precludes learning about the concentration of glyphosate in a tissue. Because translocation studies are more concerned with how "much" glyphosate goes "where" from a source location, then one can simply apply a precise dose to a specific location. The faster the uptake, the better, because the first minute amounts of glyphosate delivered to sinks will begin to initiate the self-limitation, which ultimately stops translocation. Hence, a rapid delivery (but not locally cytotoxic) dose allows more glyphosate to be translocated and reveal the proportional sink strengths from that source location.

The use of GR plants compared with wild-type or a sensitive plant allows the separation of the effects of physiological barriers, like metabolic toxicity from physical barriers such as membranes, cell walls, and cuticles (Feng and Chiu 2005; Feng et al. 2003b). It is not always possible to have a GR plant for this comparison and so that situation can be created by using an ultralow dose of ^{14}C-glyphosate. That is, at some very low dose, the toxicity of glyphosate no longer impacts the uptake and delivery. This concept is particularly useful when characterizing the mechanism of glyphosate resistance in horseweed (Feng et al. 2004). By comparing resistant and sensitive plants below the toxic effect level, the physiological impact of the resistance mechanism on glyphosate translocation and partitioning can be revealed. Studies with GR plants demonstrate restricted translocation in rigid ryegrass (Lorraine-Colwill et al. 2002; Powles and Preston 2006) and horseweed (Feng et al. 2004), but equal translocation in Palmer amaranth (Culpepper et al. 2006; Sammons et al. 2008). Equal translocation requires a modified hypothesis to explain symplastic translocation because apparently, there is no self-limitation of glyphosate delivery. Hetherington et al. (1999) showed increased translocation in GR corn, which is explained by the removal of toxic self-limitation to improve translocation efficiency. Removal of the source perception of toxicity requires a break in the symplastic phloem source–sink connection. The unabated translocation of glyphosate to a sensitive sink tissue would be a simple method of depleting the effective herbicide in the plant by isolating glyphosate in dying sink tissues, mimicking herbivory, and allowing the main plant to resume normal growth. Such a case is described by Patrick and Offler (1996) where an apoplastic step or intervention in phloem delivery insulates the sink from excessive solute concentration or osmotic changes. Studies with GR soybean demonstrate a clear example of self-limitation for apical meristem translocation, but with equal translocation to root tissue from a common source leaf, implicating sink apoplast unloading in soybean root tissue (Sammons et al. 2006). The species of plants using apoplastic unloading is not known and, if common, would change the general understanding we have of source–sink relationships. The facile ability of glyphosate to move from source to sink poses many opportunities to elucidate the regulation of symplastic and apoplastic movement of normal assimilants.

1.3 GLYPHOSATE'S FUNGICIDAL ACTIVITIES

The sensitivity of plant EPSPS enzymes to glyphosate accounts for its efficacy as an herbicide. However, glyphosate is generally recognized as having little to no fungicidal or bactericidal activities. In pure culture, growth of many fungi was inhibited by glyphosate, but only at extremely high concentrations (100 to more than $1000\,mg\,g^{-1}$ for ED_{50}) (Franz et al. 1997). The results of our own *in vitro* screens confirmed that glyphosate has weak activity against many fungi (Table 1.1).

Most GR crops do not metabolize glyphosate and coupled with the use of glyphosate-insensitive CP4 EPSPS results in persistence of glyphosate in crops. Soybean is an exception and has shown slow metabolism of glyphosate to AMPA (Duke et al. 2003; Reddy et al. 2004). GR crops enable the evaluation of disease control effects of glyphosate in the absence of crop injury. We showed in 2005 that glyphosate applied to GR wheat at or below the field use

**TABLE 1.1. Glyphosate Growth Inhibition
(90% Effective Concentration [EC90]) of Important
Agronomic Fungi as Measured by an *In Vitro*
High-Throughput Screen**

Fungi Genus	EC90 (μg g^{-1} or ppm)
Septoria	<100
Pseudocercosporella	<100
Botrytis	<100
Phytophthora	1000
Rhizoctonia	1000
Fusarium	1000
Gaeumannomyces	1000
Puccinia (rust)	5000
Pyricularia	5000

rate of 0.84 kg a.e. ha^{-1} reduced the incidence of leaf and stripe rusts caused by *Puccinia triticina* and *Puccinia striiformis*, respectively (Feng et al. 2005). Laboratory studies showed that disease control was proportional to the spray dose and was correlated to systemic glyphosate concentrations in leaves. Wheat rusts were controlled by tissue glyphosate concentrations at less than 5 ppm, which is 1000 times less than the activity predicted by the *in vitro* screen (Table 1.1). We attributed this difference to the fact that *Puccinia* species are obligate pathogens that may not be amenable to *in vitro* screens. Stripe rust control by glyphosate was confirmed in the field under a natural heavy infestation. Leaf rust control by glyphosate has also been reported by Anderson and Kolmer (2005), and there are reports of activity on other diseases in cropping systems (Sanyal and Shrestha 2008).

Since our initial observation of disease control activities in GR wheat, our attention has shifted to *Phakopsora pachyrhizi*, an obligate pathogen that causes Asian soybean rust (ASR). We reported preliminary data on the activity of glyphosate against ASR in GR soybeans (Feng et al. 2005). Subsequent laboratory studies confirmed that leaf systemic glyphosate was responsible for controlling ASR, and efficacy in the field required application rates of glyphosate at 0.84–1.68 kg ha^{-1} (Feng et al. 2008). Additional laboratory studies using excised soybean trifoliates demonstrated rate-dependent activity of glyphosate against ASR at leaf concentrations ranging from 50 to 200 ppm. Analysis of leaf tissues showed that these concentrations may be reached within 24 h after spray application of glyphosate at 0.84–1.68 kg ha^{-1}.

Field studies conducted in the United States, Brazil, Argentina, and South Africa demonstrated significant reductions in ASR severity and yield loss from the application of glyphosate at rates between 0.84 and 2.5 kg ha^{-1}. These results have been corroborated by independent field studies from several

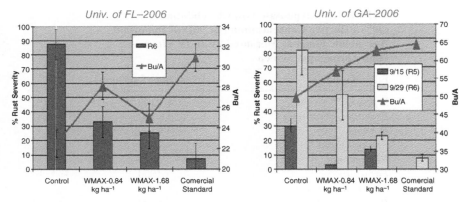

Figure 1.3. Results of field trials conducted by two universities on the effect of glyphosate on percentages of Asian soybean rust severity and yield (Bu/A) in soybeans. Glyphosate (Roundup WeatherMAX®) was applied at 0.84 or 1.68 kg a.e. ha[-1] at R5 or R6 growth stages. The commercial fungicide standard was the labeled rate of pyraclostrobin. WMAX, Roundup WeatherMAX at indicated rates in kg a.e. ha[-1].

universities (R. Kemerait et al. pers. comm.; D. Wright et al. pers. comm.; Harmon et al. 2006). Figure 1.3 shows field results obtained from Universities of Florida and Georgia in 2006. The results showed dose-dependent decrease in ASR severity and preservation of yield from applications of glyphosate at 0.84–1.68 kg ha[-1]. ASR control by glyphosate was less than that of a fungicide control.

We attributed glyphosate's activity to inhibition of fungal EPSPS based on observations that rust control was proportional to glyphosate tissue concentrations and not mediated via induction of pathogenesis-related genes (Feng et al. 2005). Infected plants treated with glyphosate show marked accumulation of shikimic acid, which is a well-established marker for the inhibition of plant EPSPS by glyphosate. Experiments were conducted to determine if shikimate accumulation might also serve as a marker for inhibition of fungal EPSPS. GR soybean leaves do not accumulate shikimate when treated with glyphosate because these plants are engineered with the glyphosate-insensitive CP4 EPSPS (Fig. 1.4). Shikimate levels also remained low when plants were infected with ASR, but without the glyphosate treatment, indicating a low basal level of shikimate in *P. pachyrhizi*. Significant increase in shikimate levels were observed only in infected leaves treated with glyphosate, suggesting that the source of the shikimate is from the fungi. There was an increase in shikimate levels with glyphosate applications from 4 to 10 days after inoculation, and this was coincident with the incubation period of *P. pachyrhizi* in soybeans and also with a reduction in disease severity. These results provided strong evidence that rust control activity of glyphosate is due to inhibition of fungal EPSPS.

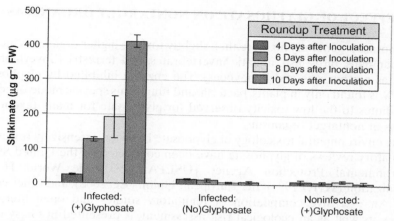

Figure 1.4. Shikimate accumulation in ASR-infected RR soybean leaves after glyphosate treatment. Leaf shikimate levels per gram fresh weight (FW) were measured 2 days after glyphosate treatment ($0.84\,kg\,ha^{-1}$), as a function of glyphosate spray timing (4–10 days after inoculation) on infected plants with glyphosate, infected plants without glyphosate, and noninfected plants with glyphosate treatments. RR soybean plants are resistant to glyphosate and do not accumlate shikimate in response to glyphosate application.

More direct evidence of fungal EPSPS inhibition by glyphosate was obtained by cloning of *P. pachyrhizi* EPSPS. The expression of the *P. pachyrhizi EPSPS* gene complemented the EPSPS-deficient (aroA-) *Escherichia coli* strain thus confirming activity. The growth of the transformed *E. coli* strain was inhibited by glyphosate, demonstrating the sensitivity of *P. pachyrhizi* EPSPS to glyphosate. Enzyme kinetic analysis showed that the *P. pachyrhizi* EPSPS was more sensitive to glyphosate than that of *E. coli* and with a temperature optimum of <37°C. Additional laboratory studies demonstrated a lack of antifungal activity in glyphosate metabolites, which further support the conclusion that glyphosate's antifungal activity is due to direct action on fungal EPSPS.

Similar EPSPS enzymes are found across many classes of plant pathogenic fungi including the Oomycetes, Deuteromycetes, Ascomycetes, and Basidiomycetes. It is therefore reasonable to assume that glyphosate's antifungal activity should be evident in a broader range of fungi. We have shown that glyphosate can suppress disease symptoms and provide yield protection under both greenhouse and field conditions against a range of plant pathogenic fungi. Activity has been demonstrated against powdery mildew (*Microsphaera diffusa*) and *Cercospora* leaf spots (*Cercospora kikuchii* and *Cercospora soja*) in soybeans, against powdery mildew (*Erysiphe pisi*) in peas, and against downy mildew (*Peronospora destructor*) in onions. Our investigations are continuing to determine the potential benefits of disease suppression from the application of glyphosate in GR crops.

1.4 EFFECT OF GLYPHOSATE ON NONTARGET ORGANISMS

Glyphosate is generally no more than slightly toxic to higher organisms including mammals, birds, fish, aquatic invertebrates, and terrestrial invertebrates (such as earthworms and honeybees). The enzyme inhibited by glyphosate, EPSPS, is found only in plants, bacteria, and fungi. This specific mode of action contributes to the low toxicity observed for glyphosate for many taxonomic groups of nontarget organisms.

The environmental toxicology of glyphosate has been extensively reviewed. Regulatory reviews of glyphosate have been conducted by the United States Environmental Protection Agency (USEPA 1993), the World Health Organization (WHO 1994), the European Union (EC 2002), and other countries. An extensive compilation of regulatory studies and open literature studies, as well as an ecological risk assessment, is presented in Giesy et al. (2000). An assessment of risk from overwater application was reported by Solomon and Thompson (2003). A brief review of the ecological effects of glyphosate use in glyphosate tolerant crops is also available (Cerdeira and Duke 2006). The EPA ECOTOX database (http://cfpub.epa.gov/ecotox/) is also a source of regulatory and open literature ecotoxicological studies on glyphosate. Rather than present a comprehensive review of glyphosate effects on nontarget organisms, this section focuses on a few key points regarding ecological toxicology and risk assessment for glyphosate.

Glyphosate toxicity studies have been conducted with a number of different forms of glyphosate. When evaluating the results of glyphosate nontarget organism studies, it is important to note the form of glyphosate that has been tested. Glyphosate has carboxylic acid, phosphonic acid, and amine functionalities (Fig. 1.1). In the protonated acid form, glyphosate is a crystalline solid that is soluble in water at concentrations just over 1% at 25°C. A 1% solution prepared by dissolving crystalline glyphosate without buffering has a pH of 2 (Franz et al. 1997). The pH of glyphosate solutions increases with dilution. The acid form of glyphosate can be neutralized with dilute base to form salts, which are much more soluble in water. In its salt form, glyphosate is soluble at concentrations approaching 50%; these concentrated salt solutions have a pH between 4 and 5. In commercial end-use herbicide products, glyphosate is generally present in the salt form. Counterions used in glyphosate formulations include isopropylamine, potassium, and ammonium.

Commercial products typically also include a surfactant to facilitate the movement of the polar compound glyphosate through the waxy cuticle of plant foliage. While glyphosate and its commercial formulations are generally recognized to pose low toxicity to terrestrial organisms (such as birds, mammals, honeybees, and soil macroorganisms), some commercial formulations have been found to have greater toxicity to aquatic organisms than glyphosate (Folmar et al. 1979) due to the presence of surfactant in the formulation. Table 1.2 compares the toxicity of glyphosate as the acid, as the isopropylamine salt, and as the original Roundup agricultural formulation (MON 2139). Especially for fish, the salt form has less toxicity than the acid

TABLE 1.2. Relative Toxicity of Glyphosate Acid, Glyphosate Isopropylamine Salt, and the Original Glyphosate Formulation, Roundup (MON 2139)

Species	Exposure Duration	LC_{50}/EC_{50}^{a} (mg a.e. L^{-1})		
		Glyphosate Acid	Glyphosate IPA salt	Original Roundup Formulation (MON 2139)[b]
Rainbow trout (*Oncorhynchus mykiss*)	96 h	71.4[b] ST	>460[c] PNT	1.3[d] MT
Bluegill (*Lepomis macrochirus*)	96 h	99.6[b] ST	>460[c] PNT	2.4[d] MT
Daphnia magna	48 h	128[b] PNT	428[c] PNT	3.0[d] MT
		LD_{50} (units as indicated)		
Rat	Single dose	>4275[e] mg a.e. kg bw^{-1} PNT		1550[f] mg a.e. kg bw^{-1} ST
Bobwhite (*Colinus virginianus*)	5 d	>4971[b] mg a.e. kg^{-1} diet ST	—	>1742[f] mg a.e. kg^{-1} diet ST
Honeybee (*Apis mellifera*)	Contact 48 h	>100 μg ae/bee[f] PNT		>31 μg a.e./bee[f] PNT
Earthworm (*Eisenia foetida*)	14 d		>2300 mg a.e. kg^{-1} soil[g] PNT	>1550 mg a.e. kg^{-1} soil[f] PNT

[a]For this comparison, the lowest end points from studies conducted with similar methodology (e.g., fish weight, water chemistry) were employed. EPA toxicity classification (USEPA 2008) is given under the endpoint value except for earthworms where a European toxicity classification is used (Canton et al. 1991). Units for formulation studies have been converted when necessary from mg formulation L^{-1} to mg a.e. L^{-1} for direct comparison of glyphosate concentrations of the acid and salt using the conversion factor 0.31.

[b]Regulatory study reported in USEPA (2008). These values are the values reported for the Analytical Bio-Chemistry Laboratories (ABC, Columbia, MO) studies in Giesy et al. (2000), but with a correction for 83% purity of the test substance.

[c]Values are reported in Giesy et al. (2000) as >1000 mg glyphosate IPA salt L^{-1}; however, review of the study reports indicates this concentration is expressed as the 62% aqueous salt solution rather than a.e. The correction has been made to a.e. using a conversion factor of 0.46.

[d]Folmar et al. (1979). LD_{50} values in this paper are expressed as mg a.e. L^{-1}.

[e]Giesy et al. (2000), with a correction for test substance purity of 85.5%.

[f]Giesy et al. (2000), with a conversion factor of 0.31 applied to convert from formulation units to a.e. units.

[g]Giesy et al. (2000). The LD_{50} value is >3750 mg a.e. kg^{-1} converted from the original study value of 5000 mg kg^{-1} as a 62% IPA salt solution using a salt to acid conversion factor of 0.75; however, since the original test substance was only 62% IPA salt, the original LD_{50} value of 5000 mg kg^{-1} has been corrected to glyphosate acid equivalent using the conversion factor 0.46.

PNT, practically nontoxic; ST, slightly toxic; MT, moderately toxic.

form, which in turn has significantly less toxicity than the original Roundup formulation.

It is also important to note that commercial herbicide products containing glyphosate can contain a number of different surfactants with varying degrees of aquatic toxicity. For example, there are a number of different formulations

TABLE 1.3. Comparative Toxicity of Three Glyphosate Formulations

	LC_{50} (mg formulation L^{-1})		
	Roundup Biactive®	Roundup Transorb®	Roundup Original®
Species	MON 77920	PCP[a]: 25344	PCP: 13644
Green frog (*Rana clamitans*)	>57.7	7.2	6.5

[a]Pest Control Product Registration Number (Canada).

with variations of the Roundup brand name, which exhibit varying degrees of aquatic toxicity (Table 1.3). When reporting results of glyphosate formulation testing, it is very important to provide the complete name of the product tested and any additional information that is available, such as the EPA registration number.

Because there are several forms of glyphosate that can be tested, it is critical that toxicity results clearly indicate whether the values are expressed as glyphosate acid equivalents (a.e.), glyphosate salt (often referred to as active ingredient), or as formulation units. It is also important to note that most concentrated glyphosate formulations have a density greater than 1; therefore, test substance should be measured on a weight basis for accurate conversion between forms based on weight percent units.

The toxicity of glyphosate formulations to amphibians has been a topic of recent investigation by a number of laboratories. Results from amphibian studies by Bidwell and Gorrie and Mann and Bidwell are summarized in Giesy et al. (2000). There have also been a number of more recent investigations regarding the acute toxicity of Roundup formulations to amphibians (e.g., Edginton et al. 2004; Howe et al. 2004; Relyea 2005a, 2005b, 2005c). Altogether, a total of 20 species of amphibians from three continents have been tested for acute toxicity to Roundup formulations. The lowest LC_{50} reported for any of these species for the most sensitive growth stage was 0.88 mg a.e. L^{-1} for *Xenopus laevis* (Edginton et al. 2004). Considering only regulatory studies, the lowest LC_{50} value for a fish species reported for a glyphosate formulation is 5.4 mg formulation L^{-1} (or 1.7 mg a.e. L^{-1}), which is less than two times greater than the lowest amphibian value. Since the United States and the European Union apply a 10- to 100-fold safety factor, respectively, between toxicity values and predicted exposure values, the risk assessments conducted using fish end points are also protective for amphibian species.

Results from monitoring studies can be used to put the reported toxicity values into perspective relative to exposure. Glyphosate concentrations in 51 water bodies in the midwestern United States were measured during three different runoff periods in 2002 (Scribner et al. 2003). The maximum concentration of glyphosate measured in these samples was 8.7 μg a.e. L^{-1} and the ninety-fifth centile concentration ranged from 0.45 to 1.5 μg a.e. L^{-1} for the three sampling periods. A total of 30 sites in southern Ontario, Canada, representing rivers, small streams, and low-flow wetlands were sampled biweekly (April to December) during 2004 and 2005. The maximum concentration measured in these samples was $40.8 \mu g a.e. L^{-1}$. In the wetlands with known amphibian habitat, the upper ninety-ninth centile confidence limit indicates that glyphosate concentrations would typically be below $21 \mu g a.e. L^{-1}$ (Struger et al. 2008). Both of these studies indicate that glyphosate concentrations in the environment are well below concentrations at which toxicity to aquatic animals has been observed in laboratory studies. Consistent with this margin of safety, the EPA recently determined that glyphosate poses no risk of direct effects to the aquatic stage of a threatened aquatic animal (California red-legged frog) (USEPA 2008).

One additional point to consider with respect to the risk assessment for glyphosate formulations is that the tallowamine surfactant often used in these formulations has been demonstrated to rapidly partition out of the water column (Wang et al. 2005). The Wang et al. study, which measured both the disappearance of MON 0818, the surfactant blend in the original Roundup formulation (MON 2139), from the water column and the reduction in toxicity to *Daphnia magna* over time, indicated that the half-life of the surfactant in two sediments was less than 1 day, and the decline in surfactant concentration was correlated with the reduction in toxicity. This rapid partitioning to sediment may also be expected for other surfactants containing long alkyl chains. A number of studies have been conducted that employ extended exposures (16–40 days) in laboratory tests with constant concentrations of glyphosate formulations. Exposures of this duration are not representative of exposures that would occur in the natural environment. Thus, the results of such studies should only be used as an indicator of future investigations to conduct under more realistic exposure scenarios.

The generally low toxicity of glyphosate to nontarget organisms, the rapid disappearance of surfactant from the water column, and the large margin of safety between concentrations of glyphosate in surface water and concentrations at which toxic effects to aquatic animals from glyphosate formulations have been observed, combine to indicate that glyphosate applications in accordance with the label do not pose an unreasonable risk of adverse effects to nontarget organisms.

1.5 PHYSICAL AND ENVIRONMENTAL PROPERTIES OF GLYPHOSATE

Due to its amphoteric nature, glyphosate is readily dissolved in dilute aqueous bases and strong aqueous acids to produce anionic and cationic salts, respectively. The free acid of glyphosate is modestly water soluble ($1.16 g L^{-1}$ at $25°C$), but when converted to monobasic salts, its solubility increases substantially. Due to its limited aqueous solubility, glyphosate is generally formulated as concentrated water solutions of approximately 30–50% in the form of the more soluble monobasic salt (isopropylamine, sodium, potassium, trimethylsulfonium, or ammonium) in a number of commercial herbicidal products. Neither glyphosate acid nor the commercial salts are significantly soluble in common organic solvents. The lack of solubility of glyphosate in nonaqueous solvents has been attributed to the strong intermolecular hydrogen bonding in the molecule (Knuuttila and Knuuttila 1985). The physicochemical properties of glyphosate indicate a favorable environmental profile. For instance, the intermolecular hydrogen bonding results in low volatility of glyphosate (2.59×10^{-5} Pa at 25°C). Glyphosate's low volatility and its high density ($1.75 \, g \, cm^{-3}$) suggest that it is unlikely to evaporate from treated surfaces and move through the air to injure nontarget sources or remain suspended in the air for a long time after application.

With the advent of glyphosate-tolerant crops and the widespread use of glyphosate products in so many different crops (Duke and Powles 2008), glyphosate has been the subject of numerous studies for potential to produce adverse effects. The environmental characteristics of glyphosate have been reviewed by many scientists from the industry (Franz et al. 1997), government regulatory agencies in several countries (USEPA 1993), scientific institutions (Giesy et al. 2000), and international organizations (WHO 1994). A summary of the physical, chemical, and environmental properties of glyphosate from these reviews is shown below.

Chemical decomposition does not contribute to the degradation of glyphosate in the environment because glyphosate is stable to hydrolytic degradations in sterile water in most environmentally relevant pH ranges. Glyphosate is also photolytically stable in sterile water and soil. However, photodegradation can occur in water under certain conditions. Studies using artificial light and solution containing calcium ions reveal slow photodegradation, while studies using natural or simulated sunlight and sterile water show no photodegradation (Franz et al. 1997). Similarly, under intense artificial lights, glyphosate in natural river water degrades via oxidative transformation induced by photochemical excitation of humic acids as reported for other pesticides (Aguer and Richard 1996). Although photodegradation of glyphosate in water can occur, it is not a major pathway of degradation of glyphosate in the environment.

In contrast, glyphosate is readily degraded by microorganisms in soil, nonsterile water, and water/sediment systems. In soil, indigenous microflora degrade glyphosate, under both aerobic and anaerobic conditions. The principle metabolite is AMPA. AMPA is further degraded by soil microflora, although at a slower rate than glyphosate. Studies demonstrate that in soil, up to 79–86% of glyphosate is biodegraded to carbon dioxide during a 6-month period (Franz et al. 1997). The results of over 93 field trials conducted in Europe, Canada, and the United States show that glyphosate dissipates with field half-lives in all cases of less than 1 year, and typically less than 38 days (Giesy et al. 2000). Laboratory and field studies also demonstrate that dissipation times are not affected by the rate of application and that glyphosate and AMPA do not accumulate following multiple applications, either during the same year or over tens of years (Giesy et al. 2000). Biodegradation is also the principle mechanism of degradation of glyphosate in environmental waters under both aerobic and anaerobic conditions. In all cases, the results demonstrate the biodegradation of glyphosate to AMPA and carbon dioxide, and the subsequent biodegradation of AMPA to carbon dioxide.

Glyphosate is only used as a postemergence herbicide, and the potential for root uptake of glyphosate from soils has been reported to be negligible. Lack of glyphosate soil activity is due to its rapid microbial degradation and strong soil-binding properties (Giesy et al. 2000). Glyphosate has been shown to bind tightly to most soils. In laboratory batch equilibrium studies, partition coefficient (K_{oc}) values ranged from 884 to 60,000 for seven soils. Studies have been conducted to investigate the uptake of radiolabeled glyphosate into rotational crops following soil applications to a primary crop. The maximum uptake into plants grown in soil treated with glyphosate was in all cases less than 1% of the total applied. These results demonstrate that glyphosate entry into plants via the root system as a result of applications of glyphosate to the soil is negligible.

1.6 GLYPHOSATE TOXICOLOGY AND APPLICATOR EXPOSURE

Glyphosate and glyphosate-based herbicides are backed by one of the most extensive worldwide human health and safety databases ever compiled for a pesticide product. Before any pesticide product can be registered, distributed, or sold, it is subjected to a rigorous battery of tests to determine that the product does not pose any unreasonable risks to consumers or the environment, when used according to label directions. Governmental regulatory agencies mandate these tests and have experts that review the submitted data for each pesticide. Glyphosate has been thoroughly reviewed and registered by the Canadian Pesticide Management Regulatory Agency (PMRA 1991), the USEPA (1993), the European Commission (EC 2002), and other regulatory agencies around the world. In addition, glyphosate has been reviewed by the WHO (1994), the Joint Meeting of the Food and Agricultural Organization (FAO) Panel of Experts on Pesticide Residues in Food and the Environment and the WHO Core Assessment Group on Pesticide Residues (WHO/FAO 1987, 2004), and third party toxicology experts (Williams et al. 2000).

Comprehensive toxicological studies in laboratory animals have demonstrated that glyphosate has low oral, dermal, and inhalation toxicity and shows no evidence of carcinogenicity, mutagenicity, neurotoxicity, reproductive toxicity, or teratogenicity. In the absence of a carcinogenic potential in animals and the lack of genotoxicity in standard tests, the USEPA (1993) placed glyphosate in its most favorable cancer category, Group E, meaning that there is "evidence of non-carcinogenicity for humans" and the WHO/FAO (2004) concluded that glyphosate was unlikely to pose a carcinogenic risk to humans.

Of almost equal importance to the toxicology data is human pesticide exposure potential. The term "pesticide exposure" may mean different things to different people. If someone had been in a farm field when pesticides were being applied, the person might feel that he or she had been exposed to pesticides. In terms of determining potential risk, however, there is general agreement that exposure should be based on the amount of pesticide that has penetrated into the body, the so-called internal dose (Chester and Hart 1986; Franklin et al. 1986).

Exposure related to the professional use of glyphosate-based formulations, through the monitoring of the single active ingredient, glyphosate, has been the subject of a number of studies. Biomonitoring and passive dosimetry, and exposure modeling are approaches that can be used to estimate applicator exposure to pesticides. Biomonitoring results represent systemic (internal) exposure from all possible routes, whereas the results obtained from passive dosimetry quantify external deposition. There is general agreement that biological measurements as obtained through biomonitoring provide the most relevant information for safety assessments (Chester and Hart 1986; Franklin et al. 1986). There have been six published glyphosate biomonitoring studies (Abdelghani 1995; Acquavella et al. 2004; Centre de Toxicologie du Quebec 1988; Cowell and Steinmetz 1990a, 1990b; Jauhiainen et al. 1991). The authors of each study quantified glyphosate in urine. Urine is an ideal medium for quantifying systemic dose because glyphosate is not metabolized by mammals and is excreted essentially unchanged in urine with a short half-life (Williams et al. 2000).

The most extensive biomonitoring study is the Farm Family Exposure Study (FFES), conducted by investigators at the University of Minnesota with guidance offered by an advisory committee of recognized international experts in exposure assessment (Acquavella et al. 2004). The study monitored farm families. Urine samples were collected the day before glyphosate was to be applied, the day of application, and for 3 days after application. The detection method was capable of detecting 1 part per billion (ppb) glyphosate. In the FFES, 48 farmers applied a Roundup branded herbicide and provided 24-h urine samples the day before, the day of, and for 3 days after the application. Approximately 50% of the applications were on more than 40 ha and application rates were at least 1 kg ha^{-1}. Overall, 40% of the farmers did not have detectable glyphosate in their urine on the day of application. Some farmers did have detectable glyphosate in their urine samples, and the urinary concentrations ranged from <1 to 233 ppb. The maximum systemic dose was estimated to be 0.004 mg kg^{-1}. This would suggest that it is very unlikely for an applicator to get a systemic glyphosate dose that would even approach any level of toxicological concern. For comparison, according to the USEPA (1993), the lowest no observed effect level (NOEL) from glyphosate toxicology studies is considered to be 175 mg kg^{-1} day^{-1}. Regulatory agencies estimate risk to pesticide applicators by using a ratio of the estimated exposure to a relevant NOEL.

This ratio is referred to as the margin of exposure (MOE). Typically, MOEs that are less than 100 will exceed a level of concern for worker risk. Based on estimates of systemic dose, a farmer who did 20 applications per year for 40 years would have a MOE of approximately 1.75 million fold.

In summary, numerous comprehensive toxicological studies in animals conducted over many years clearly demonstrate that there are no significant hazards associated with glyphosate exposure. Glyphosate does not cause cancer, birth defects, mutagenic effects, nervous system effects, or reproductive problems. The comprehensive biomonitoring study of Acquavella et al. (2004) showed that people who regularly work with glyphosate have very low actual internal exposure. Taken together, the results from exposure studies in humans and animal laboratory toxicity studies demonstrate that glyphosate in real-world use conditions would not be expected to pose a health risk to humans when used according to label directions (Williams et al. 2000).

1.7 COMMERCIAL PROCESS CHEMISTRY FOR PREPARING GLYPHOSATE

Many chemical routes for synthesizing glyphosate have been reported (Franz et al. 1997). Such a large number of routes is related to the fact that glyphosate is relatively stable in a variety of reaction environments (i.e., pH, temperature, oxidative, reductive, etc.), thus giving rise to a diversity of synthesis methods.

Although there are many routes reported, only a small fraction of these have yield and other characteristics that make them suitable for commercial operation. Currently, there are two dominant families of chemical pathways for commercial manufacturing of glyphosate: the "alkyl ester" pathways and the "iminodiacetic acid (IDA)" pathways. Each is discussed below.

1.7.1 Alkyl Ester Pathways

A significant number of Chinese manufacturers use a process based on an "alkyl ester" pathway. Although several variations of this pathway exist, commercially, the primary alkyl ester pathway is based on that developed and patented by the Alkaloida Chemical Works of Hungary (Brendel et al. 1984). The "Alkaloida" process uses glycine, dimethylphosphite (DMP), and paraformaldehyde as raw materials.

In the Alkaloida process, the reaction takes place in a nonaqueous medium, where glycine is first added to a mixture of triethylamine and paraformaldehyde (approximately two equivalents) in methanol. Under these conditions, a hydroxymethylglycine intermediate is formed:

GLYCINE   Bis-Hydroxymethylglycine

DMP is then added to the reaction mixture, forming the following phosphonate ester:

Bis-Hydroxymethylglycine   DMP   N-hydroxylmethy-N-phosphonomethylglycine dimethyl phosphonate ester

Concentrated HCl is then added at room temperature, resulting in the removal of the hydroxymethyl group. Then subsequent heating of the solution results in further hydrolysis of the phosphonate ester to produce glyphosate:

N-phosphonomethylglycine dimethyl phosphonate ester   GLYPHOSATE

The various intermediates are not isolated; thus, the reaction system is simple in that the reactions can be carried out in a "single pot." The final solution (containing glyphosate, methanol, etc.) is further processed to isolate glyphosate or a glyphosate solution suitable as a product.

Some of the key advantages to this process (e.g., more stable and neutral pH, lower temperature operation) come from carrying out the reaction in an organic solvent instead of in an aqueous solution and the base choice (Et_3N). These preferred conditions give rise to favorable reaction conditions such that the overall yield of glyphosate is improved.

Process technology developments have led to the recovery and recycling of methanol and Et_3N to the process. Also, attention has been given to developing technologies to recover chloromethane generated during hydrolysis. This captured chloromethane can be sold or used in other processes (e.g., organosilicone production), improving the overall economics of the process.

There are variations of the above process, such as the use of diethyl phosphate (DEP) instead of DMP and other optimized solvents and reaction conditions. Process research continues on the alkyl ester pathways, as a significant amount of China's glyphosate production is based on these processes.

1.7.2   IDA Pathways

The other predominant family of pathways for the commercial production of glyphosate is based on IDA. In general, for these pathways, it is the hydrochloride salt of IDA (IDA·HCl) that participates in a phosphonomethylation reaction via a modified Mannich reaction to form the N-phosphonomethyliminodiacetic acid (PMIDA):

$$\text{HCl·HN} \begin{array}{c} \text{COOH} \\ \text{COOH} \end{array} + H_3PO_3 + HCHO \longrightarrow H_2O_3P-N \begin{array}{c} \text{COOH} \\ \text{COOH} \end{array} + H_2O + HCl.$$

IDA·HCl   PMIDA

One might envision/guess that performing the above phosphonomethylation reaction with glycine rather than IDA would directly generate glyphosate; however, phosphonomethylation of glycine via the Mannich reaction produces glyphosate only in low yield because glyphosate very readily undergoes an additional phosphonomethylation, forming bis-phosphonomethyl glycine. Thus, one can think of the second carboxymethyl group on IDA as a "protecting" group that prevents a second phosphonomethylation from occurring.

Often during the phosphonomethylation reaction to produce PMIDA, both HCl and phosphorous acid are conveniently supplied by feeding PCl_3 to an aqueous solution of IDA. PCl_3 reacts with water accordingly to generate phosphorous acid and hydrochloric acid *in situ*:

$$PCl_3 + 3H_2O \longrightarrow H_3PO_3 + 3HCl.$$

Once PMIDA is formed, it can be isolated, and the protecting group can be removed via oxidation to form glyphosate:

$$H_2O_3P-N \begin{array}{c} \text{COOH} \\ \text{COOH} \end{array} \longrightarrow H_2O_3P-N \begin{array}{c} \text{COOH} \\ \text{H} \end{array}$$

PMIDA   GLYPHOSATE

This oxidation can be achieved by concentrated sulfuric acid, hydrogen peroxide, electrolysis, or oxygen/air over a catalyst.

The production of IDA is often part of the integrated glyphosate process. There are three primary approaches that glyphosate producers use to produce IDA, and they are summarized below.

1.7.2.1 IDA from Iminodiacetonitrile (IDAN)
Caustic is added to IDAN to produce disodium iminodiacetate (DSIDA). Hydrochloric acid is then added to produce IDA:

$$HN \begin{array}{c} \text{CN} \\ \text{CN} \end{array} \xrightarrow{\text{NaOH}} HN \begin{array}{c} COO^-Na^+ \\ COO^-Na^+ \end{array} \xrightarrow{\text{HCl}} HN \begin{array}{c} \text{COOH} \\ \text{COOH} \end{array} + NaCl.$$

IDAN   DSIDA   IDA

Since IDAN is produced from HCN, forming IDA via IDAN is favored in situations where inexpensive or by-product HCN is available.

1.7.2.2 IDA from Diethanol Amine (DEA)
Another means of generating IDA is from DEA. DEA is converted to DSIDA by reacting with caustic

over a catalyst. As above, DSIDA can then be hydrolyzed to IDA or (as shown below) some producers use membrane dialysis to generate IDA and NaOH:

1.7.2.3 IDA from Chloroacetic Acid

In this approach, chloroacetic acid is added to a solution of NH_3 and $Ca(OH)_2$. After the reaction, the solution is then neutralized with HCl to form the hydrochloride salt of IDA:

Of these various strategies for producing IDA, this chloroacetic acid method is the least efficient, as it generates significant quantities of strong acid ($CaCl_2$) waste, leading to lower yields than the IDAN or DEA routes.

1.8   GLYPHOSATE FORMULATION

This section will describe some of the properties of formulations of glyphosate and issues faced in the selection of formulation ingredients. This is meant as a general overview of the subject and not an exhaustive review of the subject area or exhaustive literature review. The formulations discussed will be those principally sold in the United States, not worldwide, although most of the formulations discussed are or have been sold in many countries.

Formulations containing glyphosate have been sold under the trade name of Roundup (Monsanto Company) for more than 30 years. As the original patents on the use of glyphosate as an herbicide and salts of glyphosate expired, other brands such as Touchdown® (Syngenta, Basel, Switzerland), GlyphoMAX® (Dow AgroSciences LLC, Indianapolis, IN), and Gly Star® (Albaugh, Inc., Ankeny, IA), to name only a few, have also come into the market. These commercial mixtures are water solutions of glyphosate salts with most containing a surfactant. Some dry, water-soluble granule or powder formulations have also been sold. Consulting the National Pesticide Information Retrieval System (NPIRS®) (http://ppis.ceris.purdue.edu/) Web site, more than 50 different registered products containing glyphosate are found.

In the design of a glyphosate product formulation, the selection of the type of salt and surfactant has been the principal ingredients studied. The formulation must be stable over the range of temperature extremes that the product will experience in the market place. The formulation must be easily diluted in water and be sprayable without clogging the spray nozzles of application equipment. It must also perform in an efficacious manner as an herbicide. Additional requirements of the formulations are that they have the minimal toxicity to humans and the environment.

1.8.1 Glyphosate Acid and Salt Solubility

The solubility of glyphosate acid is 1.57% in water at 25°C (Vencill 2002). This solubility is too low to be useful for a soluble concentrate commercial product. While it may be possible to formulate the acid as a suspension concentrate, a liquid soluble concentrate product is typically preferred and can have fewer physical stability issues. Hence, the vast majority of commercial products sold to date have been salts of glyphosate. Glyphosate has three acid sites (or exchangeable protons) and one amine available for protonation (Fig. 1.1); thereby, several different types of salts of glyphosate are easily obtained. The simplest forms of salts are produced by reaction of a base with glyphosate acid. As these salts are formed, solubility of the salt can be expressed in terms of the amount of salt soluble, or % ai in solution. This makes the comparison of the amount of glyphosate anion in solution between different salts slightly difficult as the molecular weights of the cations are different. To make comparison easier, the amount of equivalent glyphosate acid dissolved in a salt solution has typically been referred to as "% a.e." The solubility in water of selected salts is shown in Table 1.4 for a variety of glyphosate salts prepared at a 1:1 molar ratio of cation to acid.

One of the first glyphosate formulations sold contained the isopropylamine (IPA) salt. Several other salts have been sold in commercial products since that time, including sodium, tetramethylsulfonium (TMS), potassium, ammonium, monoethanolamine, and dimethylamine salts. An acid salt

TABLE 1.4. Solubility in Water of Various Glyphosate Salts, 1:1 Mole Ratio of Glyphosate:Base (Vencill 2002)[a] and Unpublished Data

Cation	% ai w/w Soluble (20°C)	% a.e. w/w Soluble at 20°C (pH)
H+		1.16 (pH 2.5)[a]
Li+	19	18 (pH 4.5)
Na+	34	30 (pH 3.6)
K+	54	44 (pH 4.2)
TMS+	78.6 (pH 4.06)	54 (pH 4.06)
	50 (pH 4.2)	34 (pH 4.2)
IPA+	63	47 (pH 4.6)
NH₄+	39	35 (pH 4.3)

[a]See references.

NH₄, ammonium; H, hydrogen; IPA, isopropylamine; Li, lithium; K, potassium; Na, sodium; TMS, tetramethylsulfonium.

formulation where the nitrogen atom is protonated using sulfuric acid has also been sold.

Salt solubility is an important factor in preparing a soluble concentrate formulation of glyphosate. The solubility must be high enough such that when the formulation is exposed to extreme low temperatures, the salt will not crystallize and precipitate. Testing of formulations at low temperature expected in the market place is one of the typical hurdles for a formulation to overcome. It is important in these tests that a seed crystal of the salt be added to the formulation since a supersaturated solution can appear to be stable, without a stimulus to crystallize. The seed crystal will give this stimulus and help avoid a false reading.

Most salt formulations of glyphosate contain a "mono" salt of glyphosate or nominally 1 mol of neutralizing cation to 1 mol of glyphosate anion. A way to increase the solubility of a lesser soluble salt is to make a di-cation salt, making use of the second acid site on the glyphosate molecule. This has particularly been used with the ammonium salt as described by Sato et al. (1999). The product Touchdown IQ® (Syngenta) contains this salt. The sesquisodium salt of glyphosate, 1.5 mol of Na per mole of glyphosate, was sold as a water-soluble powder under the product name Polado® (Monsanto Company).

1.8.2 Density of Solutions

Formulations sold in agricultural markets typically describe the active ingredient loading based on the weight of active ingredient per gallon or per liter. The first glyphosate formulation sold as Roundup contained 3 lb of glyphosate a.e. or 4 lb of glyphosate IPA salt per gallon. In metric units, this loading is approximately $360\,g\,L^{-1}$ of glyphosate a.e. or $480\,g\,L^{-1}$ of glyphosate IPA salt. The amount of glyphosate on a weight percent basis in the formulation was 31% glyphosate as the isopropylamine salt or 41.6% IPA salt of glyphosate. It is a simple calculation to obtain the weight per volume for a formulation as shown below in Equations 1.1 and 1.2:

$$\text{Solution specific gravity} \times 1000 \times \%\,w/w = g\,L^{-1}\ \text{active ingredient,} \qquad (1.1)$$

$$\text{Solution specific gravity} \times 8.3283 \times \%\,w/w = lb\,gal^{-1}\ \text{active ingredient.} \qquad (1.2)$$

The specific gravity of a solution is defined as the density of a given solution divided by the density of water at a given temperature. The solution density of glyphosate salt solutions (and hence the active ingredient loading of a formulation) can be affected by the choice of glyphosate salt. Table 1.5 shows the specific gravity of several different solutions of salts of glyphosate. The weight percent of the equivalent amount of glyphosate acid in each solution present as a salt is shown for each salt. This value is abbreviated as % a.e. or percent glyphosate a.e.

TABLE 1.5. Specific Gravity for a Variety of 30% a.e. Glyphosate Salt Solutions and g L^{-1} Loading of Glyphosate (Wright 2003)

Cation	SG 30% a.e.	SG 30% ai
Potassium	1.25	1.20
Ammonium	1.18	1.16
Isopropylamine	1.16	1.11
Ethanolamine	1.24	1.17
Trimethylsulfonium	1.19	1.13

TABLE 1.6. A Partial List of Glyphosate Salts Sold in Commercial Products with a Representative Product Name

Salt Cation	Representative Trade Name	lb gal^{-1} a.e. Glyphosate	% a.e. w/w Glyphosate
Isopropylamine	Roundup® (Monsanto Company)	3	30.4
Tetramethylsulfonium	Touchdown® (Syngenta)	5	39.5
Diammonium	Touchdown IQ® (Syngenta)	3	28.3
Potassium	Roundup WeatherMAX® (Monsanto Company)	4.5	48.8
Dimethylamine	Durango® (Dow AgroSciences LLC)	4	39.7

While the weight per volume or loading of glyphosate possible in a solution of glyphosate salt is determined by the % soluble salt and density of the solution, it is practically limited by the solubility of the salt in water. While two formulations prepared with different salts may contain the same percent glyphosate by weight (% a.e.) the amount of glyphosate expressed in terms of weight per volume can be different. This is demonstrated by comparing the pound per gallon and % a.e. in Touchdown (Syngenta) and Durango® (Dow AgroSciences LLC) that have similar % a.e. concentrations, but the density of the TMS salt solution is much greater than the DMA salt. Some of the salts of glyphosate that have been sold in commercial products are shown in Table 1.6.

1.8.3 Surfactant Selection

The biological efficacy of glyphosate, perhaps more than any other herbicide, can be very dependent on the surfactant in the spray solution. Most of the glyphosate formulations on the market contain a surfactant. When considering a surfactant to include in a formulation, there are two main items to consider: identification of a surfactant that boosts efficacy and identification of a surfactant that is compatible in the formulation. There is a legion of research that has been documented on various surfactants and how they affect the biological performance of glyphosate, much more than could be discussed in a book chapter. The purpose of this section will be to deal with some of the issues to be solved in the selection of a surfactant based on formulation criteria.

Identifying a surfactant that is soluble in concentrated salt solutions can be difficult as many types of surfactants are not soluble in salt solutions and particularly insoluble in glyphosate salt solutions. After the selection of the surfactant, determine if it is soluble in a given concentration of a glyphosate salt solution at room temperature. One measure of the compatibility of the surfactant in the formulation is to measure the cloud point of the solution. As explained by Lange (1999), the "turbidity arises from attractive micelle-micelle interactions. At a higher temperature, phase separation into a water-rich phase and a surfactant-rich phase generally occurs." This can occur at a maximum and a minimum temperature. This is an important consideration when designing a formulation that will experience a wide variety of climatic conditions. If the cloud point is lower than the maximum temperature the product will experience, the formulation may separate into layers that may not be easily reconstituted.

Under most textbook definitions of cloud point, it is described that the higher the ethylene oxide (EO) content of a surfactant, the more soluble it will be in water. This is due to the availability of more oxygen molecules with which water can hydrogen bond. However, in water solutions containing a high amount of salt, as with salt solutions of glyphosate, this is not the case. In fact, most nonionic surfactants such as alkylphenol or alcohol ethoxylates are not soluble to a great extent in solutions containing an appreciable amount of glyphosate salt. The one exception to this rule is alkyl polyglycoside (APG) surfactants (Hill et al. 1996). These nonionic surfactants are highly soluble in salt solutions in general and particularly in glyphosate salt solutions.

Many commercial glyphosate formulations contain the so-called cationic surfactants, or surfactants that can retain a positive charge under acidic conditions. Alkylamine ethoxylates are such surfactants. These surfactants can be compatible in glyphosate salts, but the compatibility is affected by the type of cation carried by the glyphosate salt (Lennon et al. 2006). The compatibility is also affected by the amount of ethoxylation on the alkylamine. With these surfactants, the cloud point does not follow the expected rule of the cloud point of water solutions being higher with higher amounts of EO on the amine. Table 1.7 shows cloud points of formulations containing 30% ae IPA glyphosate salt with ethoxylated cocoamine surfactants at increasing concentration and increasing EO. Note that the cloud point actually decreases with added EO rather than increases as one may find in pure water.

Other types of adjuvants can be used with the application of glyphosate formulations. These adjuvants can be a number of different materials such as

TABLE 1.7. Cloud Point Measurements of Cocoamine Ethoxylate Surfactants in 30% ai Isopropylamine (IPA) Glyphosate Solutions

EO (mol)	% w/w Surfactant	Cloud Point (°C)
5	10	>99
5	15	>99
5	20	>99
10	10	>99
10	15	88
10	20	76
15	10	68
15	15	54
15	20	45

EO, Ethylene oxide.

surfactants, antifoam agents, defoaming agents, drift control materials, and water-conditioning agents. Some of these materials can be included in a formulation without difficulty. Perhaps the most commonly added adjuvant to glyphosate applications as a tank mix ingredient is ammonium sulfate. Adding ammonium sulfate to formulations of glyphosate can be problematic in that you are adding more salt to an already high-salt-containing solution. Particularly in the agricultural formulations, it is desirable to maximize the amount of active ingredient provided in the formulation. Most glyphosate product labels recommend adding 1–2% ammonium sulfate to the spray solution. Adding the amount necessary to obtain this 1–2% concentration to the formulation would greatly reduce the amount of glyphosate in the formulation. Thus, in the U.S. market, few formulations have been sold that contain an appreciable amount of ammonium sulfate.

1.8.4 Dry Granular Formulations

Formulations of glyphosate can be made in the form of water-soluble solids. Both the sodium and ammonium salts have been sold in these types of formulations. The first dry formulation sold in the market was a water-soluble powder, Polado (http://ppis.ceris.purdue.edu/). This product was the sesquisodium salt or 1.5 mol of sodium per mole of glyphosate acid. The monosodium salt of glyphosate was sold as a water-soluble granule in Europe as Roundup Ultragran® (http://ppis.ceris.purdue.edu/). The ammonium salt of glyphosate has been more commercially successful. The ammonium salt as described by Kuchikata et al. (1996) is less hydroscopic than other salts, which offers advantages to the formulator in that it will require less water impermeable packaging. Dry ammonium glyphosate formulations have been sold as Roundup WSD (Monsanto Company) and other commercial labels particularly in South and Central America. A combination of glyphosate, diquat dibromide, and surfactant (Crockett et al. 2006) is also sold as a water-soluble granule under the product name of QuikPro® (Monsanto Company) specifically for the industrial market.

1.8.5   Combination or Package Mix Formulation

Formulations containing more than one active ingredient are commonly referred to as "package mix" formulations. Several products have been sold that contain glyphosate and another herbicide. Typically, this is to place another type of herbicide that offers some benefit to the user such as residual activity or an herbicide with different selectivity. Products sold into the agriculture market have included Bronco® (Monsanto Company) (glyphosate plus alachlor), Landmaster® (Monsanto Company) (glyphosate plus 2,4-D), Fallowmaster® (Monsanto Company) (glyphosate plus dicamba), and Fieldmaster® (Monsanto Company) (acetochlor, atrazine, glyphosate). One benefit of these products is that they offer the convenience of having both active ingredients in the same formulation or container. This can be both a blessing and a curse in that the ratio of active ingredients in the formulation is fixed, which does not allow the user to adjust the ratio of active ingredients based on soil type or species of weeds present in a given field. When preparing a formulation with more than one active, it will typically reduce the concentration in the final formulation for each active ingredient over what could be provided with either active ingredient could have been formulated when provided in separate formulations. In any formulation containing two or more actives, one of the first tests to be conducted is to ensure that one active ingredient does not have a chemical reaction with another that would cause decomposition of one active ingredient. This is particularly true with glyphosate, which can act as a proton donor to aid in the hydrolysis of many actives that contain an ester moiety.

1.8.6   Lawn and Garden Formulations

Sales of glyphosate formulations in the lawn and garden or household consumer market have slightly different requirements when compared with the products sold to farmers in an agricultural market. One principal difference is that the formulations can contain lower concentrations. This is done principally for the convenience of the user. A significant portion of products sold are prediluted or ready-to-use (RTU) formulations. These formulations generally contain the active ingredient as well as surfactants and other additives to potentiate activity in a water solution diluted to a dose that is ready to be applied by the user. These formulations also typically come in a container that is also the applicator, such as a trigger actuated sprayer. Concentrate formulations are also sold to be diluted into pump sprayers. These concentrate formulations can contain as little as 6% and up to 50% glyphosate salt.

In dealing with the consumer expectations, most of the innovations with glyphosate formulations in recent history have been to develop products to provide fast developing symptoms, or yellowing and desiccation of weeds. Arnold et al. (1993) described that pelargonic acid (nominally a C_9 fatty acid or nonanoic acid) can be added to glyphosate formulations to achieve the fast developing symptoms. By controlling the pH of the solution to near neutral, the formulation will be homogeneous; at lower pH values, the fatty acid will separate from the formulation. Faster symptoms in a more concentrated formulation have been obtained by the combination of glyphosate, diquat, and surfactant as described by Crockett et al. (2006). This patent describes that the selection of surfactant is very important so that it allows the glyphosate to get into the plant and the translocation of the glyphosate to occur so that the diquat will not antagonize the biological performance of the systemic herbicide glyphosate. The amount of diquat in the product is only enough to provide yellowing and desiccation of the leaves of the treated weed.

Products that contain both glyphosate and a residual herbicide to keep weeds from germinating in the treated area have been sold. Those products include glyphosate + oxyfluorfen (Ortho®Season-Long®, Scott's Miracle-Gro Company, Marysville, OH), glyphosate + imazapyr (GroundClear®, Scott's Miracle-Gro Company), and glyphosate + imazapic (Roundup Extended Control®, Monsanto Company). Other specialized formulations are also sold such as glyphosate + triclopyr as Roundup Poison Ivy & Tough Brush Control® (Monsanto Company). This combination of actives was developed particularly for use on brushy weeds and vines as described by Wright et al. (2004).

1.9   CONCLUSION

This chapter was meant as an overview of more recent research on the use and environmental, toxicological, and physical aspects of the herbicide glyphosate. Obviously, this compound has been studied extensively over the last 30+ years, and after over three decades of use, glyphosate-based products continue as an important tool for weed control to be used by farmers across the globe. The compound continues to be the leading herbicide used in row crops, orchards, fallow lands, and pastures. Glyphosate's unique and favorable environmental and toxicological properties and its ability to control a broad spectrum of weed species will keep it a key weed management tool.

REFERENCES

Abdelghani, A. A. 1995. Assessment of the exposure of workers applying herbicide mixtures (2,4-D+Roundup, Garlon-3A+Roundup) toxicity and fate of these mixtures in the environment (summary report). State Project #736-14-0067, Louisiana Transportation Research Center.

Acquavella, J. F., B. H. Alexander, J. S. Mandel, C. Gustin, B. Baker, P. Chapman, and M. Bleeke. 2004. Glyphosate biomonitoring for farmer-applicators and their families: results from the Farm Family Exposure Study. *Environmental Health Perspectives* 112:321–326.

Aguer, J. and C. Richard. 1996. Transformation of fenuron induced by photochemical excitation of humic acids. *Pesticide Science* 46:151–155.

Amrhein, N., B. Deus, P. Gehrke, and H. C. Steinrucken. 1980. The site of the inhibition of the shikimate pathway by glyphosate. II. Interference of glyphosate with chorismate formation in vivo and in vitro. *Plant Physiology* 66:830–834.

Anderson, J. A. and J. A. Kolmer. 2005. Rust control in glyphosate tolerant wheat following application of the herbicide glyphosate. *Plant Disease* 89:1136–1142.

Arnold, K. A., A. S. Wideman, R. J. White, M. W. Bugg, and M. N. Cline. 1993. Improved early symptom development with a ready-to-use glyphosate formulation by addition of fatty acid. *Pesticide Science* 38(2–3):270–271.

Baird, D. D., R. P. Upchurch, W. B. Homesley, and J. E. Franz. 1971. Introduction of a new broad spectrum post emergence herbicide class with utility for herbaceous perennial weed control. *Proceedings of the North Central Weed Science Society* 26:64–68.

Brendel, M. H., I. Gulyas, I. Gyoker, K. Zsupan, I. Csorvassy, Z. Salamon, G. Somoguy, I. Szentiralyi, T. Timar, E. C. Biro, I. Fodor, and J. Repasi, inventors; Alkaloida, assignee. 1984. Process for the preparation of N-phosphonomethyl-glycine. U.S. Patent 4,486,359, filed December 4, 1984.

CaJacob, C. A., P. C. C. Feng, G. R. Heck, F. A. Murtaza, R. D. Sammons, and S. R. Padgett. 2003. Engineering resistance to herbicides. In P. Christou and H. Klee, eds. *Handbook of Biotechnology*, Vol. 1. Chichester: John Wiley & Sons, pp. 353–372.

Canton, J. H., J. B. H. J. Linders, R. Luttik, B. J. W. G. Mensink, E. Penman, E. J. van de Plassche, P. M. Sparenburg, and J. Tuinstra. 1991. Catch-up operation on old pesticides. Protection Report No. 678801001, National Institute of Public Health and Environmental, Bilthoven, the Netherlands.

Caseley, J. C. and D. Coupland. 1985. Environmental and plant factors affecting glyphosate uptake, movement and activity. In E. Grossbard and D. Atkinson, eds. *The Herbicide Glyphosate*. London: Butterworths & Co., pp. 92–123.

Centre de Toxicologie du Quebec. 1988. Etude de L'exposition professionnelle des travailleurs forestiers exposes au glyphosate. Centre Hospitalier de l'Université Laval pour le Gouvernement du Québec, ministère de l'Énergie et des Ressources, Direction de la conservation, Service des etudes environnementales, Pub. No. ER89-1110, 65 p. et annexes.

Cerdeira, A. L. and S. O. Duke. 2006. The current status and environmental impacts of glyphosate-resistant crops: a review. *Journal of Environmental Quality* 35(5): 1633–1658.

Chester, G. and T. B. Hart 1986. Biological monitoring of a herbicide applied through backpack and vehicle sprayers. *Toxicology Letters* 33:137–149.

Cowell J. E. and J. R. Steinmetz. 1990a. Assessment of forestry nursery workers exposure to glyphosate during normal operations. Report No. MSL-9655, Monsanto Company, St. Louis, MO.

Cowell J. E. and J. R. Steinmetz. 1990b. Assessment of forest worker exposures to glyphosate during backpack foliar applications of Roundup® Herbicide. Report No. MSL-9656, Monsanto Company, St. Louis, MO.

Crockett, R. P., A. D. Dyszlewski, R. M. Kramer, D. C. Riego, J. J. Sandbrink, D. L. Suttner, D. H. Williamson, and D. R. Wright, inventors. 2006. Herbicidal compositions containing glyphosate and bipyridilium. U.S. Patent 7,008,904, filed March 7, 2006.

Culpepper, S., T. L. Grey, W. K. Vencill, J. M. Kichler, T. M. Webster, S. M. Brown, A. C. York, J. W. Davis, and W. W. Hanna. 2006. Glyphosate-resistant Palmer amaranth (*Amaranthus palmeri*) confirmed in Georgia. *Weed Science* 54:620–626.

Dewey, S. A. 1981. Manipulation of assimilate transport patterns as a method of studying glyphosate translocation in tall morning glory (*Ipomoea purpurea* (L.) Roth). *Dissertation Abstracts International, B* 42:1695–1696.

Dill, G. M. 2005. Glyphosate resistant crops: history, status and future. *Pest Management Science* 61:219–224.

Dill, G., C. CaJacob, and S. Padgette. 2008. Glyphosate-resistant crops: adoption, use and future considerations. *Pest Management Science* 64:326–331.

Duke, S. O. and S. Powles. 2008. Glyphosate: a once-in-a-century herbicide. *Pest Management Science* 64:319–325.

Duke, S. O., A. M. Rimando, P. F. Pace, K. N. Reddy, and R. J. Smeda. 2003. Isoflavone, glyphosate, and aminomethylphosphonic acid levels in seeds of glyphosate-treated, glyphosate-resistant soybean. *Journal of Food and Agricultural Chemistry* 51:340–344.

Edginton, A. N., P. M. Sheridan, G. R. Stephenson, D. G. Thompson, and H. J. Boermans. 2004. Comparative effects of pH and Vision® herbicide on two life stages of four anuran amphibian species. *Environmental Toxicology and Chemistry* 23:815–822.

European Commission (EC). 2002. Review report for the active substance glyphosate. European Commission Document: glyphosate, 6511/VI/99-final. January 21, 2002. http://ec.europa.eu/food/plant/protection/evaluation/existactive/list1_glyphosate_en.pdf (accessed December 12, 2008).

Feng, P. C. C., G. J. Baley, W. P. Clinton, G. J. Bunkers, M. F. Alibhai, T. C. Paulitz, and K. K. Kidwell. 2005. Glyphosate controls rust diseases in glyphosate-resistant wheat and soybean. *Proceedings of the National Academy of Sciences of the United States of America* 102:17290–17295.

Feng, P. C. C. and T. Chiu. 2005. Distribution of [^{14}C]glyphosate in mature glyphosate-resistant cotton from application to a single leaf or over-the-top spray. *Pesticide Biochemistry and Physiology* 82:36–45.

Feng, P. C. C., T. Chiu, and R. D. Sammons. 2003a. Glyphosate efficacy is contributed by its tissue concentration and sensitivity in velvetleaf (*Abutilon theophrasti*). *Pesticide Biochemistry and Physiology* 77:83–91.

Feng, P. C. C., T. Chiu, R. D. Sammons, and J. S. Ryerse. 2003b. Droplet size affects glyphosate retention, absorption, and translocation in corn. *Weed Science* 51: 443–448.

Feng, P. C. C., C. Clark, G. C. Andrade, M. C. Balbi, and P. Caldwell. 2008. The control of Asian rust by glyphosate in glyphosate-resistant soybeans. *Pest Management Science* 64:353–359.

Feng, P. C. C., J. S. Ryerse, C. R. Jones, and R. D. Sammons. 1999. Analysis of surfactant leaf damage using microscopy and its relation to glyphosate or deuterium oxide uptake in velvetleaf (*Abutilon theophrasti*). *Pesticide Science* 55:385–386.

Feng, P. C. C., J. S. Ryerse, and R. D. Sammons. 1998. Correlation of leaf damage with uptake and translocation of glyphosate in velvetleaf (*Abutilon theophrasti*). *Weed Technology* 12:300–307.

Feng, P. C. C., J. J. Sandbrink, and R. D. Sammons. 2000. Retention, uptake, and translocation of ^{14}C-glyphosate from track-spray applications and correlation to rainfastness in velvetleaf (*Abutilon theophrasti*). *Weed Technology* 14:127–132.

Feng, P. C. C., M. Tran, T. Chiu, R. D. Sammons, G. R. Heck, and C. A. CaJacob. 2004. Investigations into glyphosate-resistant horseweed (*Conyza canadensis*): retention, uptake, translocation, and metabolism. *Weed Science* 52:498–505.

Folmar L. C., H. O. Sanders, and A. M. Julin. 1979. Toxicity of the herbicide glyphosate and several of its formulations to fish and aquatic invertebrates. *Archives of Environmental Contamination and Toxicology* 8:269–278.

Franklin C. A., N. I. Muir, and R. P. Moody. 1986. The use of biological monitoring in the estimation of exposure during the application of pesticides. *Toxicology Letters* 33:127–136.

Franz, J. E., M. K. Mao, and J. A. Sikorski. 1997. *Glyphosate: A Unique Global Pesticide.* Washington, DC: American Chemical Society.

Geiger, D. R. and H. D. Bestman. 1990. Self-limitation of herbicide mobility by phytotoxic action. *Weed Science* 38:324–329.

Geiger, D. R., S. W. Kapitan, and M. A. Tucci. 1986. Glyphosate inhibits photosynthesis and allocation of carbon to starch in sugar beet leaves. *Plant Physiology* 82:468–472.

Giesy J. P., S. Dobson, and K. R. Solomon. 2000. Ecotoxicological risk assessment for Roundup® herbicide. *Reviews of Environmental Contamination and Toxicology* 167:35–120.

Gougler, J. A. and D. R. Geiger. 1981. Uptake and distribution of N-phosphonomethylglycine in sugar beet plants. *Plant Physiology* 68:668–672.

Halter, S. 2007. A brief history of Roundup. Proceedings of the First International Symposium on Glyphosate. Agronomical Sciences College of the University of the State of Sao Paulo, Sao Paulo, Brazil, October 15–19.

Harmon, P. F., C. R. Semer, C. L. Harmon, and R. J. McGovern. 2006. Evaluation of fungicides and Roundup for control of Asian soybean rust in Florida, 2006. Plant Disease Management Report No. 1: FC034. http://www.plantmanagementnetwork. org/pub/trial/pdmr/volume1/abstracts/fc034.asp (accessed April 7, 2010).

Hetherington, P. R., T. L. Reynolds, G. Marshall, and R. C. Kirkwood. 1999. The absorption, translocation and distribution of the herbicide glyphosate in maize expressing the CP-4 transgene. *Journal of Experimental Botany* 50:1567–1576.

Hill, K., W. von Rybinski, and G. Stoll, eds. 1996. *Alkyl Polyglycosides.* New York: VCH Publications.

Howe C. M., M. Berrill, P. D. Pauli, C. C. Helbing, K. Werry, and N Veldhoen. 2004. Toxicity of glyphosate-based pesticides to four North American frog species. *Environmental Toxicology and Chemistry* 23:1928–1938.

Jauhiainen A., K. Rasanen, R. Sarantila, J. Nuutinen, and J. Kangas. 1991. Occupational exposure of forest workers to glyphosate during brush saw spraying work. *American Industrial Hygiene Association Journal* 52:61–64.

Kishore, G. M. and D. M. Shah. 1988. Amino acid biosynthesis inhibitors as herbicides. *Annual Review of Biochemistry* 57:627–663.

Knuuttila, P. and H. Knuuttila. 1985. Molecular and crystalline structure of glyphosate. In E. Grossbard and D. Atkinson, eds. *The Herbicide Glyphosate*. London: Butterworths & Co., pp. 18–22.

Kuchikata, M., E. J. Prill, R. O. Richardson, T. Sato, J. M. Surgant, and D. R. Wright, inventors; Monsanto Company, assignee. 1996. Water-soluble herbicide powders or granules containing N-phosphonomethylglycine. PCT Int. Appl. (1990), 45 pp. CODEN: PIXXD2 WO 9007275 A1 19900712 CAN 114:159150 AN 1991:159150 CAPLUS.

Lange, K. R., ed. 1999. *Surfactants: A Practical Handbook*. Cincinnati, OH: Hanser Gardner Publications.

Lennon, P. J., X. Chen, G. B. Arhancet, J. A. Glaenzer, J. N. Gillespie, J. A. Graham, D. Z. Becher, D. R. Wright, H. E. Agbaje, X. C. Xu, W. Abraham, R. J. Brinker, N. R. Pallas, A. S. Wideman, M. D. Mahoney, and S. L. Henke, inventors; Monsanto technology LLC, assignee. 2006. Potassium glyphosate formulations. U.S. Patent 7,049,270, filed May 23, 2006.

Lorraine-Colwill, D. F., S. B. Powles, T. R. Hawkes, P. H. Hollinshead, S. A. J. Warner, and C. Preston. 2002. Investigations into the mechanism of glyphosate resistance in *Lolium rigidum*. *Pesticide Biochemistry and Physiology* 74:62–72.

Monaco, T. J., S. C. Weller, and F. M. Ashton. 2002. *Weed Science: Principles and Practices*, 4th ed. New York: John Wiley & Sons.

Nandula, V. K., K. N. Reddy, A. M. Rimando, S. O. Duke, and D. H. Poston. 2007. Glyphosate-resistant and -susceptible soybean (*Glycine max*) and canola (*Brassica napus*) dose response and metabolism relationships with glyphosate. *Journal of Agricultural and Food Chemistry* 55:3540–3545.

Patrick, J.W. and C. E. Offler. 1996. Post-sieve element transport of photoassimilates in sink regions. *Journal of Experimental Botany* 47:1165–1177.

Pesticide Management Regulatory Agency (PMRA). 1991. Preharvest Use of Glyphosate. Discussion document. Ottawa, Canada: Pesticide Information Division of the Plant Industry Directorate. http://www.hc-sc.gc.ca/cps-spc/pest/part/consultations/_d91-01/index-eng.php (accessed April 5, 2010).

Powles, S. B. and C. Preston. 2006. Evolved glyphosate resistance in plants: biochemical and genetic basis of resistance. *Weed Technology* 20:282–289.

Prasad, R. 1989. Role of some adjuvants in enhancing the efficacy of herbicides on forest species. In P. N. P. Chow, A. M. Hinshalwood, E. Simundsson, and C. A. Grant, eds. *Adjuvants and Agrochemicals. Volume I. Mode of Action and Physiological Activity*. Boca Raton, FL: CRC Press, pp. 159–165.

Prasad, R. 1992. Some factors affecting herbicidal activity of glyphosate in relation to adjuvants and droplet size. In L. E. Bode and D. G. Chasin, eds. *Pesticide Formulations and Application Systems*, Vol. 11. West Conshohocken, PA: American Society for Testing Materials International, pp. 247–257.

Prasad, R. and B. L. Cadogan. 1992. Influence of droplet size and density on phytotoxicity of three herbicides. *Weed Technology* 6:415–423.

Reddy, K. N., A. M. Rimando, and S. O. Duke. 2004. Aminomethylphosphonic acid, a metabolite of glyphosate, causes injury in glyphosate-treated, glyphosate-resistant soybean. *Journal of Agricultural and Food Chemistry* 52:5139–5143.

Relyea, R. A. 2005a. The impact of insecticides and herbicides on the biodiversity and productivity of aquatic communities. *Ecological Applications* 15:618–627.

Relyea R. A. 2005b. The lethal impacts of Roundup® and predatory stress on six species of North American tadpoles. *Archives of Environmental Contamination and Toxicology* 48:351–357.

Relyea R. A. 2005c. The lethal impact of Roundup® on aquatic and terrestrial amphibians. *Ecological Applications* 15:1118–1124.

Research and Markets. 2008. Glyphosate Competitiveness Analysis in China. CCM International Limited. http://www.researchandmarkets.com/reportinfo.asp?cat_id=0&report_id=649031&q=glyphosate&p=1 (accessed April 5, 2010).

Ryerse, J. S., R. A. Downer, R. D. Sammons, and P. C. C. Feng. 2004. Effect of glyphosate spray droplets on leaf cytology in velvetleaf (*Abutilon theophrasti*). *Weed Science* 52:302–309.

Sammons, R. D., R. Eilers, M. Tran, and D. R. Duncan. 2006. Characterization of an exclusion mechanism for glyphosate resistant weeds. Proceedings of the 232nd American Chemical Society (Agrochemical Division) Abstract #189, San Francisco, CA, September 10–14.

Sammons, R. D., A. Herr, D. Gustafson, and M. Starke. 2008. Characterizing glyphosate resistance in *Amaranthus palmeri*. In R. D. van Klinken, V. A. Osten, F. D. Panetta, and J. C. Scanlan, eds. In *Proceedings of the Sixteenth Australian Weeds Conference Proceedings*. Cairns, Australia: The Weed Society of Queensland, p. 116.

Sanyal, D. and A. Shrestha. 2008. Direct effect of herbicides on plant pathogens and disease development in various cropping systems. *Weed Science* 56:155–160.

Sato, T., M. Kuchikata, A. Amano, M. Fujiyama, and D. R. Wright, inventors; Monsanto Company, assignee. 1999. High-loaded ammonium glyphosate formulations. U.S. Patent 5,998,332, filed December 7, 1999.

Scribner, E. A., W. A. Battaglin, J. Dietze, and E. M. Thurman. 2003. Reconnaissance data for glyphosate, other selected herbicides, their degradation products, and antibiotics in 51 streams in nine midwestern states, 2002. Open-File Report 03–217, United States Geological Survey, Lawrence, KS.

Singh, B. K. and D. L. Shaner. 1998. Rapid determination of glyphosate injury to plants and identification of glyphosate-resistant plants. *Weed Technology* 12:527–530.

Solomon K. R. and D. G. Thompson. 2003. Ecological risk assessment for aquatic organisms from over-water uses of glyphosate. *Journal of Toxicology and Environmental Health. Part B* 6:289–324.

Sprankle, P., D. P. Penner, and W. F. Meggitt. 1973. The movement of glyphosate and bentazon in corn, soybean, and several weed species. Abstracts of *Weed Science Society of America* 13:75.

Steinrucken, H. C. and N. Amrhein. 1980. The herbicide glyphosate is a potent inhibitor of 5-enolpyruvylshikimic acid-3-phosphate synthase. *Biochemical and Biophysical Research Communications* 94:1207–1212.

Struger, J., D. Thompson, B. Staznik, P. Martin, T. McDaniel, and C. Marvin. 2008. Occurrence of glyphosate in surface waters of southern Ontario. *Bulletin of Environmental Contamination and Toxicology* 80:378–384.

United States Environmental Protection Agency (USEPA). 1993. Reregistration Eligibility Decision (RED): Glyphosate. Office of Prevention, Pesticides and Toxic Substances. http://www.epa.gov/oppsrrd1/REDs/old_reds/glyphosate.pdf and http://www.epa.gov/oppsrrd1/REDs/factsheets/0178fact.pdf (accessed December 11, 2008).

United States Environmental Protection Agency (USEPA). 2008. Risks of glyphosate use to federally threatened California red-legged frog (*Rana aurora draytonii*). Washington, DC: Environmental Fate and Effects Division, Office of Pesticide Programs, U.S. Environmental Protection Agency Endangered Species Protection Program. October 17, 2008. http://epa.gov/espp/litstatus/effects/redleg-frog/index.html#glyphosate (accessed December 12, 2008).

Vencill, W. K., ed. 2002. *Herbicide Handbook*, 8th ed. Lawrence, KS: Weed Science Society of America, pp. 231–232.

Wang N., J. M. Besser, D. R. Buckler, J. L. Honegger, C. G. Ingersoll, B. T. Johnson, M. L. Kurtzweil, J. MacGregor, and M. J. McKee. 2005. Influence of sediment on the fate and toxicity of a polyethoxylated tallowamine surfactant system (MON 0818) in aquatic microcosms. *Chemosphere* 59:545–551.

World Health Organization (WHO). 1994. Glyphosate: Environmental Health Criteria 159. Geneva, Switzerland: World Health Organization. http://www.inchem.org/documents/ehc/ehc/ehc159.htm (accessed March 23, 2010).

World Health Organization/Food and Agricultural Organization (WHO/FAO). 1987. Pesticide Residues in Food—1986 Evaluations. Part II—Toxicology. FAO Plant Production and Protection Paper 78/2, 1987, nos. 735–755 on INCHEM. http://www.inchem.org/documents/jmpr/jmpmono/v86pr08.htm (accessed April 5, 2010).

World Health Organization/Food and Agricultural Organization (WHO/FAO). 2004. Pesticides Residues in Food—2004. Report of the Joint Meeting of the FAO Panel of Experts on Pesticide Residues in Food and the Environment and the WHO Core Assessment Group on Pesticide Residues (JMPR). FAO Plant Production and Protection Paper 178, World Health Organization and Food and Agriculture Organization of the United Nations, Rome, Italy. http://www.fao.org/ag/AGP/AGPP/Pesticid/JMPR/download/2004/_rep/report2004jmpr.pdf (accessed April 5, 2010).

Williams G. M., R. Kroes, and I. C. Munro. 2000. Safety evaluation and risk assessment of the herbicide Roundup® and its active ingredient, glyphosate, for humans. *Regulatory Toxicology and Pharmacology* 31:117–165.

Wright, D. R., inventor; Monsanto Technology LLC, assignee. 2003. Compact storage and shipping system for glyphosate herbicide. U.S. Patent 6,544,930, filed April 8, 2003.

Wright, D. R., J. J. Sandbrink, and P. G. Ratliff, inventors; Monsanto Technology LLC, assignee. 2004. Herbicidal combinations containing glyphosate and a pyridine derivative. World intellectual property organization WO/2004/093546 (International Application No. PCT/US2004/012368), filed April 11, 2004.

2

HERBICIDE RESISTANCE: DEFINITIONS AND CONCEPTS

Vijay K. Nandula

1.1 INTRODUCTION

Weeds have been in existence since before humans took up cultivation of plants for food, feed, fuel, and fiber. Before the advent of synthetic organic-based herbicides in the 1940s, weeds were controlled for thousands of years by mechanical, cultural, and biological means. 2,4-Dichorophenoxyacetic acid was the first herbicide to be used selectively. Since then, several herbicides belonging to different chemical classes and possessing diverse modes of action have been synthesized and commercialized around the world. Glyphosate, a nonselective herbicide, was discovered in 1970 and marketed in 1974 as Roundup® by the Monsanto Company (St. Louis, MO) for use in both crop and noncrop lands. With the introduction of glyphosate-resistant (GR) crops in the mid 1990s, glyphosate is now widely used for weed control in GR crops without concern for crop injury. Herbicides have vastly contributed to increasing world food production in an efficient, economic, and environmentally sustainable manner. However, repeated application(s) of the same herbicide or a different herbicide with similar mode of action on the same field growing season after growing season has contributed to the widespread occurrence of resistance to herbicides, including glyphosate, in several weed species. The goals of this chapter are to define herbicide resistance and distinguish it from herbicide tolerance, to delineate known mechanisms of herbicide resistance, and to revisit commonly used herbicide resistance terminology.

Glyphosate Resistance in Crops and Weeds: History, Development, and Management
Edited by Vijay K. Nandula
Copyright © 2010 John Wiley & Sons, Inc.

2.2   TOLERANCE AND RESISTANCE

The terms "tolerance" and "resistance" are often used inconsistently by weed scientists and nonweed scientists alike. Even in the weed science community, tolerance and resistance are loosely and often interchangeably used. Also, herbicide manufacturers/seed companies that develop and/or market herbicide-resistant crop cultivars/varieties frequently refer to these as herbicide-tolerant entities. This book will refer to transgenic herbicide-resistant varieties as "resistant" and recognize the definition of herbicide tolerance and resistance established by the Weed Science Society of America (WSSA).

The WSSA defines herbicide tolerance as "the inherent ability of a species to survive and reproduce after herbicide treatment." This implies that there was no selection or genetic manipulation to make the plant tolerant; it is naturally tolerant. Herbicide resistance is defined as "the inherited ability of a plant to survive and reproduce following exposure to a dose of herbicide normally lethal to the wild type. In a plant, resistance may be naturally occurring or induced by such techniques as genetic engineering or selection of variants produced by tissue culture or mutagenesis" (WSSA 1998). This definition of resistance is consistent with a prior description (Powles et al. 1997) where herbicide resistance was defined as the inherited ability of a weed population to survive a herbicide application that is normally lethal to the vast majority of individuals of that species.

A few other classifications of herbicide tolerance and herbicide resistance exist in the literature. The term "tolerance" was used to describe the biochemical and physiological basis of selectivity of herbicides brought about by primary detoxification reaction(s) that provided a given plant species or cultivar/biotype the ability to endure certain herbicide treatments (Devine et al. 1993). Tolerance could be gradual and in a range from low to high as determined by the activity of the detoxifying/metabolizing enzyme. Conversely, "resistance" was considered to be more distinct and originating from an insensitive target-site/binding protein. In other words, tolerance described subtle increases in insensitivity to herbicides, whereas the difference between resistance and susceptibility was more clear-cut. Certain weeds are naturally tolerant to herbicides, which is endowed by morphological, physiological, and genetic plant characteristics (Ware 1994). On the contrary, resistant populations arise due to genetic selection by a herbicide over a period of time typically including several life cycles. The individual resistant variants, however, may occur naturally.

2.3   MECHANISMS OF RESISTANCE

Understanding the processes and means by which weeds withstand labeled herbicide treatments is an important key toward devising effective herbicide resistance management strategies. Currently, five modes of herbicide resistance have been identified in weeds: (1) altered target site due to a mutation at the site of herbicide action resulting in complete or partial lack of inhibition; (2) metabolic deactivation, whereby the herbicide active ingredient is transformed to nonphytotoxic metabolites; (3) reduced absorption and/or translocation that results in restricted movement of lethal levels of herbicide to point/site of action; (4) sequestration/compartmentation by which a herbicide is immobilized away from the site of action in locales such as vacuoles or cell walls; and (5) gene amplification/over-expression of the target site with consequent dilution of the herbicide in relation to the target site. Mechanisms of glyphosate resistance in weeds are discussed in the chapter by Perez-Jones and Mallory-Smith. Nonglyphosate herbicide resistance mechanisms have been extensively dealt with elsewhere in the literature. Newer mechanisms of herbicide (glyphosate) resistance will most likely be discovered in the near future through the employment of advanced techniques such as weed genomics tools.

2.4 DEFINITIONS USED IN HERBICIDE RESISTANCE LITERATURE

Discovery of herbicide resistance in weeds and subsequent research over the past many decades has generated a wealth of information, which in turn, has contributed to a much better understanding of how plants function and respond to the environment in which they thrive. For example, triazine-resistant plants have served as an ideal model system to understand the mode of action of the photosystem II-inhibiting herbicides. The knowledge accumulated from this research has brought forth several concepts and expressions that are frequently used in herbicide resistance discourse. A nonexhaustive compendium of these terms is listed (selected definitions adapted from Raven et al. [1992]).

Accession. A collection of individual plants of a weed species whose characteristics (genetic, physiological, biochemical, or biological) are yet to be determined.

Allele. An alternative form or copy of a gene.

Biotype. A plant selection that has a unique genotypic pedigree.

Cross-Resistance. The expression of a mechanism that endows the ability to withstand herbicides from the same or different chemical classes with similar mode of action (Hall et al. 1994). It can be target-site based or nontarget-site based (reduced uptake, translocation, activation; increased metabolism-deactivation; compartmentation/sequestration).

Dominance. State of an allele whose phenotypic expression is similar both in the homozygous and heterozygous stages.

Ecotype. A biotype that has adapted to a specific growing environment.

Evolution. Progressive change in the gene pool of a given weed (species) population in response to most recent growing conditions (herbicides in this context).

Fitness. Ability of a biotype to survive and reproduce in an environment that may or may not include herbicide treatment.

Genotype. The complement of a plant's complete hereditary information.

Hormesis. Stimulation of growth processes in plants treated with low doses of herbicide(s).

Inheritance. Process of transfer of a genetic trait from one generation to the next.

Mating System. System by which pollen moves from the anthers to the stigma of the same flower or different flowers on the same plant (self-pollination), or to stigma of flowers on a different plant (cross-pollination) of a weed species.

Multiple Resistance. The expression of more than one resistance mechanism endowing the ability to withstand herbicides from different chemical classes (Hall et al. 1994). Multiple-resistant plants may possess two or more distinct resistance mechanisms (Gunsolus 1993).

(Gene) Mutation. An inheritable change to genetic material or the process resulting in such a change.

Negative Cross-Resistance. An expression of mechanism that occurs when a resistant biotype is more susceptible to other classes of herbicides than the susceptible biotype (Gressel 1991).

Population. A group of plants of a single weed species with potential to interbreed and inhabiting a specific geographic area.

Recessive. Condition of an allele whose expression is veiled by a dominant allele in the heterozygous stage.

Selection Pressure. The effectiveness of natural selection in altering the genetic composition of a population over a series of generations (King and Stansfield 2002).

Target Site. A gene or gene product (protein) on which a herbicide is potently inhibitory.

Trait. A genetic characteristic of interest.

2.5 HISTORY OF GLYPHOSATE RESISTANCE DEVELOPMENT IN WEEDS

The first report of any sort implying evolution of glyphosate resistance in a weed species was in rigid ryegrass (*Lolium rigidum* Gaud.) (Pratley et al. 1996). GR rigid ryegrass was confirmed by Powles et al. (1998), who reported a rigid ryegrass population from an orchard in Australia, receiving two or three glyphosate applications per year for 15 years that exhibited 7- to 11-

fold resistance compared with a susceptible population. As of April 8, 2010, 18 weed species have evolved resistance to glyphosate (Heap 2010). In order of the year glyphosate resistance was first documented in the weed species across the world, they are rigid ryegrass, goosegrass (*Eleusine indica* (L.) Gaertn.), horseweed (*Conyza canadensis* (L.) Cronq.), Italian ryegrass (*Lolium perenne* L. ssp. *multiflorum* (Lam.) Husnot), hairy fleabane (*Conyza bonariensis* L.), buckhorn plantain (*Plantago lanceolata* L.), ragweed parthenium (*Parthenium hysterophotus* L.), common ragweed (*Ambrosia artemisiifolia* L.), giant ragweed (*Ambrosia trifida* L.), johnsongrass (*Sorghum halepense* (L.) Pers.), common waterhemp (*Amaranthus rudis* Sauer), Palmer amaranth (*Amaranthus palmeri* S. Wats.), sourgrass (*Digitaria insularis* L. Mez ex Ekman), wild poinsettia (*Euphorbia heterophylla* L.), jungle rice (*Echinocloa colona* (L.) Link), Kochia (*Kochia scoparia* (L.) Schrad.), liverseedgrass (*Urochloa panicoides* Beauvois), and Sumatran fleabane (*Conyza sumatrensis* (Retz.) E. H. Walker). Further details of the above weeds such as year of documentation and worldwide distribution are outlined in Table 2.1. Research investigating these GR weed biotypes and reported in peer-reviewed publications is discussed in other chapters of this book.

2.6 GLYPHOSATE RESISTANCE—AN EVOLUTIONARY TAKE

Glyphosate has recently been deemed as "a once-in-a-century herbicide" (Duke and Powles 2008). Several attributes of glyphosate such as its highly efficacious broad weed spectrum, environmentally and toxicologically safe properties, rapid translocability, unique ability to target the 5-enolpyruvylshikimate-3-phosphate synthase (EPSPS) enzyme with no competing chemistries, the low cost of generic forms of glyphosate, and most importantly, the widespread adoption of GR crops, were put forth to justify this classification. Also, earlier reports maintained that resistance to glyphosate in weeds was unlikely to occur in the field due to inactivity of glyphosate in the soil (resulting in a short period of selection pressure), inefficiency of GR EPSPS forms (due to low affinity for phosphoenol pyruvate, a substrate in the shikimate pathway) (Kishore and Shah 1988), the benign herbicidal properties of glyphosate as outlined above, and the unlikely duplication under normal field conditions of the complex processes involved in the generation of GR crops (Bradshaw et al. 1997).

We have now come to a full circle. Evolved resistance to glyphosate in weeds, which was unheard of until a decade ago when glyphosate was already in commercial use for 20 years, is probably the single most important pest management issue facing GR row crop producers as well as several noncrop land managers. Factors affecting evolution of herbicide resistance in weeds include gene mutation, initial frequency of resistance alleles, inheritance, weed fitness in the presence and absence of herbicide, mating system, gene flow, and farming practices that favor a limited number of dominant weed species

Table 2.1. Distribution of Glyphosate-Resistant Weeds around the World

Weed Species	Year Resistance Was Documented	Country (State)
Rigid ryegrass	1996	Australia (Victoria)
	1997	Australia (New South Wales)
	1998	United States (California)
	1999	Australia (Victoria)
	2000	Australia (South Australia)
	2001	South Africa
	2003	Australia (Western Australia)
	2003	South Africa
	2005	France
	2005	France
	2006	Spain
	2007	Italy
Goosegrass	1997	Malaysia
	2006	Colombia
Horseweed	2000	United States (Delaware)
	2001	United States (Kentucky)
	2001	United States (Tennessee)
	2002	United States (Indiana)
	2002	United States (Maryland)
	2002	United States (Missouri)
	2002	United States (New Jersey)
	2002	United States (Ohio)
	2003	United States (Arkansas)
	2003	United States (Mississippi)
	2003	United States (North Carolina)
	2003	United States (Ohio)
	2003	United States (Pennsylvania)
	2005	Brazil
	2005	United States (California)
	2005	United States (Illinois)
	2005	United States (Kansas)
	2006	China
	2006	Spain
	2007	Czech Republic
	2007	United States (Michigan)
	2007	United States (Mississippi)
Italian ryegrass	2001	Chile
	2002	Chile
	2002	Chile
	2003	Brazil
	2004	United States (Oregon)
	2005	United States (Mississippi)
	2006	Spain
	2007	Argentina

Table 2.1. *Continued*

Weed Species	Year Resistance Was Documented	Country (State)
Buckhorn plantain	2003	South Africa
Hairy fleabane	2003	South Africa
	2004	Spain
	2005	Brazil
	2005	Brazil
	2006	Colombia
	2007	United States (California)
	2009	United States (California)
Common ragweed	2004	United States (Arkansas)
	2004	United States (Missouri)
	2007	United States (Kansas)
Giant ragweed	2004	United States (Ohio)
	2005	United States (Arkansas)
	2005	United States (Indiana)
	2006	United States (Kansas)
	2006	United States (Minnesota)
	2007	United States (Tennessee)
Ragweed parthenium	2004	Columbia
Palmer amaranth	2005	United States (Georgia)
	2005	United States (North Carolina)
	2006	United States (Arkansas)
	2006	United States (Tennessee)
	2006	United States (Tennessee)
	2007	United States (New Mexico)
	2008	United States (Alabama)
	2008	United States (Mississippi)
	2008	United States (Missouri)
Common waterhemp	2005	United States (Missouri)
	2006	United States (Illinois)
	2006	United States (Kansas)
	2007	United States (Minnesota)
Johnsongrass	2005	Argentina
	2006	Argentina
	2007	United States (Arkansas)
Sourgrass	2006	Paraguay
	2008	Brazil
	2008	Paraguay
Wild poinsettia	2006	Brazil
Jungle rice	2007	Australia
Kochia	2007	United States (Kansas)
Liverseedgrass	2008	Australia
Sumatran fleabane	2009	Spain

Source: Heap (2010).

(Jasieniuk et al. 1996; Owen 2001; Thill and Lemerle 2001). Several of the GR weed species listed in Table 2.1 have attributes of these factors that may contribute to glyphosate resistance, including high genetic variability (rigid and Italian ryegrass), prolific seed production (Palmer amaranth and water-hemp), seed dispersal over long distances (horseweed), and cross-pollination (ryegrasses, Palmer amaranth and waterhemp, and weeds belonging to the Asteraceae) to name a few. The biology and management of these weeds is discussed in the following chapters.

2.7 CONCLUSIONS

Glyphosate resistance in weeds is a clear and present economic problem. It is in society's interest to sustain the widely adopted GR crop technology as well as an ideal herbicide such as glyphosate to feed an exponentially growing world population. The benefits of GR crop technology are multifold, with savings in fuel costs coupled with inherent positive effects on the environment, and prevention of top soil loss from erosion arising from zero to low require-ment of tillage operations topping the list of benefits. Also, multiple herbicide-resistant crops, which are currently in development, further warrant sustainability and stewardship of GR crop technology.

REFERENCES

Bradshaw, L. D., S. R. Padgette, S. L. Kimball, and B. H. Wells. 1997. Perspectives on glyphosate resistance. *Weed Technology* 11:189–198.

Devine, M. D., S. O. Duke, and C. Fedtke. 1993. *Physiology of Herbicide Action.* Englewood Cliffs, NJ: PTR Prentice-Hall, p. 96.

Duke, S. O. and S. B. Powles. 2008. Glyphosate: a once-in-a-century herbicide. *Pest Management Science* 64:319–325.

Gressel, J. 1991. Why get resistance? It can be prevented or delayed. In J. C. Caseley, G. W. Cussans, and R. K. Atkin, eds. *Herbicide Resistance in Weeds and Crops.* Oxford, UK: Butterworth-Heinemann, pp. 1–25.

Gunsolus, J. L. 1993. *Herbicide Resistant Weeds.* North Central Regional Extension Publication 468. St. Paul, MN: Minnesota Extension Service, University of Minnesota.

Hall, L. M., J. A. M. Holtum, and S. B. Powles. 1994. Mechanisms responsible for cross resistance and multiple resistance. In J. A. M. Holtum and S. B. Powles, eds. *Herbicide Resistance in Plants: Biology and Biochemistry.* Boca Raton, FL: Lewis Publishers, pp. 243–261.

Heap, I. M. 2010. International Survey of Herbicide-Resistant Weeds. http://www. weedscience.org (accessed April 8, 2010).

Jasieniuk, M., A. L. Brule-Babel, and I. M. Morrison. 1996. The evolution and genetics of herbicide resistance in weeds. *Weed Science* 44:176–193.

King, R. C. and W. D. Stansfield, eds. 2002. *A Dictionary of Genetics*. Oxford, UK: Oxford University Press.

Kishore, G. M. and D. M. Shah. 1988. Amino acid biosynthesis inhibitors as herbicides. *Annual Reviews of Biochemistry* 57:627–663.

Owen, M. D. K. 2001 Importance of weed population shifts and herbicide resistance in the Midwest USA corn belt. In *Proceedings of the Brighton Crop Protection Conference—Weeds*. Farnham, UK: British Crop Protection Council, pp. 407–412.

Powles, S. B., D. F. Lorraine-Colwill, J. J. Dellow, and C. Preston. 1998. Evolved resistance to glyphosate in rigid ryegrass (*Lolium rigidum*) in Australia. *Weed Science* 16:604–607.

Powles, S. B., C. Preston, I. B. Bryan, and A. R. Jutsum. 1997. Herbicide resistance: impact and management. *Advances in Agronomy* 58:57–93.

Pratley, J., P. Baines, P. Eberbach, M. Incerti, and J. Broster. 1996. Glyphosate resistance in annual ryegrass. In J. Virgona and D. Michalk, eds. *Proceedings of the 11th Annual Conference of the Grasslands Society of New South Wales*. Wagga Wagga, Australia: Grasslands Society of NSW, p. 122.

Raven, P. H., R. F. Evert, and S. E. Eichhorn. 1992. *Biology of Plants*, 5th ed. New York: Worth Publishers, pp. 737–762.

Thill, D. C. and D. Lemerle. 2001. World wheat and herbicide resistance. In S. B. Powles and D. L. Shaner, eds. *Herbicide Resistance and World Grains*. New York: CRC Press, pp. 165–194.

Ware, G. W. 1994. *The Pesticide Book*, 4th ed. Fresno, CA: Thomson Publications, p. 200.

Weed Science Society of America (WSSA). 1998. Herbicide resistance and herbicide tolerance defined. *Weed Technology* 12:789.

3

GLYPHOSATE-RESISTANT CROPS: DEVELOPING THE NEXT GENERATION PRODUCTS

Paul C. C. Feng, Claire A. CaJacob, Susan J. Martino-Catt, R. Eric Cerny, Greg A. Elmore, Gregory R. Heck, Jintai Huang, Warren M. Kruger, Marianne Malven, John A. Miklos, and Stephen R. Padgette

3.1 INTRODUCTION

Glyphosate is a broad-spectrum, postemergent herbicide used worldwide in agriculture. The commercialization of glyphosate-resistant (GR) crops in 1996 revolutionized agriculture by enabling the use of glyphosate in crops for weed control (Padgette et al. 1996). Today, GR crops are increasingly adopted in world agriculture accounting for most of the acreages in soybean and cotton, and steadily increasing in corn in the United States (Dill et al. 2008; Gianessi 2008). The rapid adoption of GR crops is attributable to glyphosate's unique properties, which provide efficacious and economical weed control. Although several other herbicide-resistant traits have been commercialized over the years, their adoption has been slow relative to that of glyphosate (CaJacob et al. 2007).

Glyphosate Resistance in Crops and Weeds: History, Development, and Management
Edited by Vijay K. Nandula
Copyright © 2010 John Wiley & Sons, Inc.

3.2   MECHANISMS OF ENGINEERING GLYPHOSATE RESISTANCE

In general, glyphosate resistance can be engineered via deactivation or expression of an insensitive target (Table 3.1); readers are referred to recent reviews for more detailed discussions (CaJacob et al. 2004, 2007; Gressel 2002). The herbicidal activity of glyphosate is due to inhibition of 5-enolpyruvylshikimate-3-phosphate synthase (EPSPS), a key enzyme in the shikimate pathway for the biosynthesis of aromatic amino acids and many secondary metabolites in the phenylpropanoid pathway. The shikimate pathway is found in plants, fungi, and bacteria, but not in animals. Initial attempts to engineer resistance via over-expression of EPSPS did not achieve commercial level of resistance, and experiments were conducted to mutagenize the EPSPS in order to produce a glyphosate-insensitive enzyme. The challenge was to decrease glyphosate binding while maintaining the binding of phosphoenolpyruvate (PEP), which is the natural substrate for EPSPS. The mutagenesis work resulted in the identification of several EPSPS variants that showed reduced glyphosate binding (increased Ki) while maintaining PEP binding (equivalent Km). The initial GR corn product utilized a maize EPSPS with two mutations (TIPS-EPSPS) and since then other glyphosate-insensitive EPSPS variants have been reported (Alibhai et al. 2006; Berg et al. 2008). The glyphosate-insensitive EPSPS that is currently utilized in all GR crops was obtained from a bacterium isolated from the waste stream of a glyphosate manufacturing facility. CP4 EPSPS from an *Agrobacterium* species is highly insensitive to glyphosate showing a Ki-to-Km ratio of 227 as compared with 5.5 and 0.02 for TIPS-EPSPS and maize EPSPS, respectively (CaJacob et al. 2004).

Attempts were also made to engineer GR crops via the deactivation mechanism (Table 3.1). Glyphosate oxidoreductase (GOX) from an *Ochrobactrum* species degrades glyphosate to aminomethylphosphonic acid (AMPA) (Barry et al. 1992). Plants transformed with the *GOX* gene did not achieve commercial level of resistance, partly due to the fact that the AMPA metabolite showed plant toxicity of its own (Reddy et al. 2004). Glyphosate acetyltrans-

TABLE 3.1. **Common Mechanisms for Engineering Glyphosate Resistance in Crops, the Requisite Enzyme, and the Expected Crop Residues**

Resistance Mechanism	Enzyme	Crop Residue
Insensitive target	Glyphosate-insensitive EPSPS (e.g., CP4 EPSPS, TIPS-EPSPS)	Glyphosate
Deactivation	Glyphosate-metabolizing enzyme (e.g., GOX, GAT)	Metabolite (e.g., AMPA, *N*-acetyl glyphosate)

AMPA, aminomethylphosphonic acid; EPSPS, 5-enolpyruvylshikimate-3-phosphate synthase; GAT, glyphosate acetyltransferase; GOX, glyphosate oxidoreductase.

ferase (GAT) is a more recent deactivation enzyme that is being used to develop GR crops (Castle et al. 2004; Siehl et al. 2005).

In theory, glyphosate resistance can be engineered by either the deactivation or the insensitive target mechanisms with one fundamental difference in the fate of glyphosate (Table 3.1). Most crops do not metabolize glyphosate, which persists in plants engineered with the insensitive EPSPS. Recent studies (Anderson and Kolmer 2005; Feng et al. 2005, 2008) have shown that glyphosate is active against fungi and when applied to GR crops provided disease suppression that benefit from the persistence of glyphosate (see further details in Chapter 1). In contrast, the deactivation mechanism converts glyphosate to a new metabolite that is devoid of herbicidal and fungicidal activities. For the latter mechanism, the speed of deactivation relative to glyphosate inhibition of EPSPS is the key to overall resistance. Furthermore, as the glyphosate dose increases, so does the demand on the efficiency of deactivation; however, once deactivated, the risk of injury dissipates. Complications may arise if the metabolite has toxicity of its own or if glyphosate is regenerated via a deconjugation reaction (e.g., deacetylase on *N*-acetyl glyphosate). Another consideration is the substrate specificity of the deactivation enzyme, which ideally is specific to glyphosate and not to plant endogenous metabolites that may alter plant phenotype or composition.

3.3   DEVELOPMENT OF FIRST-GENERATION GR CROPS

The initial approach for the development of GR crops was to over-express the glyphosate-insensitive CP4 EPSPS in all tissues using strong, constitutive viral promoters such as e35S or FMV from cauliflower or figwort mosaic viruses, respectively (Kay et al. 1987). This approach was most successful in soybeans and plants transformed with the *CP4 EPSPS* gene driven by the e35S promoter demonstrated resistance to field use rates of glyphosate; Roundup Ready® (RR; Monsanto, St. Louis, MO) soybean was the first GR crop commercialized in 1996. The first-generation RR cotton was transformed with the *CP4 EPSPS* gene driven by the FMV promoter and was commercialized in 1997. Over-the-top spray application of glyphosate in RR cotton was restricted to plants that had less than four leaves, beyond which postdirected sprays were required. Studies have shown that over-the-top applications beyond the four-leaf stage in RR cotton can result in male sterility and boll drop (Jones and Snipes 1999; Pline et al. 2002). Transformation of corn plants with *CP4 EPSPS* and e35S promoter produced plants that exhibited vegetative resistance to glyphosate but reduced male fertility. The first-generation RR corn utilized the rice actin 1 promoter driving the glyphosate-insensitive *TIPS-EPSPS* and was commercialized in 1998. The first-generation RR canola is unique in utilizing two cassettes with the FMV promoter driving the expression of both *CP4 EPSPS* and *GOX* genes, and was commercialized in 1996. Glyphosate resistance has also been developed for alfalfa and sugar beet using strong viral promoters driving *CP4 EPSPS*. A more detailed discussion of the development of the first-generation products can be found in earlier reviews (CaJacob et al. 2004, 2007).

During the course of developing the first-generation GR crops, we learned that constitutive viral promoters have gaps in their expression patterns that could lead to glyphosate injury in some tissues; as a consequence, the key to achieving higher levels of glyphosate resistance depended on improving the expression pattern of CP4 EPSPS. The strategy for developing our second-generation products is to improve the expression of CP4 EPSPS in those tissues that accumulate glyphosate and therefore are at risk to glyphosate injury. This strategy required a better understanding of glyphosate uptake and translocation in plants.

3.4 IDENTIFICATION OF AT-RISK TISSUES FOR GLYPHOSATE INJURY

Glyphosate undergoes little to no metabolism in most crops. Soybean is an exception and has demonstrated slow metabolism of glyphosate to AMPA (Reddy et al. 2004). Slow or lack of endogenous metabolism coupled with the use of a glyphosate-insensitive CP4 EPSPS translates to persistence of glyphosate and continued translocation from source to sink tissues along the photo-assimilate gradient. Tissues that accumulate glyphosate and have a low CP4 EPSPS expression are considered to be at risk for glyphosate injury. At-risk tissues in a plant vary according to the growth stage, which in turn impacts tissue sink strength and capacity to import glyphosate (CaJacob et al. 2004, 2007).

The identification of at-risk tissues is typically accomplished by examining absorption and translocation of radiolabeled glyphosate. However, meaningful results can only be derived from experimental methods that model realistic field applications. Most published studies employ the "leaf droplet" method whereby ^{14}C-glyphosate is applied as 1-µL drops to a single leaf followed by quantitation of radioactivity to determine absorption and translocation. Our experience is that although the leaf droplet method is easy to use, it is inadequate in modeling field applications where a formulation is diluted to the desired volume, pressurized, atomized through a nozzle, and sprayed over the plants. This is because foliar absorption of glyphosate is affected by numerous variables including the concentrations of glyphosate and surfactant, spray droplet size, and plant coverage. Many of these variables are interdependent; for example, increasing the spray volume increases plant coverage, but reduces glyphosate and surfactant concentrations, which reduce absorption. Furthermore, most commercial spray nozzles produce a wide distribution of droplet sizes, which impact interception, rebound, and canopy penetration that affect foliar retention of spray droplets. Glyphosate translocation is generally in the direction of strong source to sink tissues, and application to a single leaf simply reveals translocation from that leaf and does not model the whole plant spray where foliage of varying source and sink strengths intercept the spray.

Numerous studies have shown that foliar absorption of glyphosate is strongly dependent on the surfactant system (CaJacob et al. 2007; Feng et al. 1998). After penetrating the leaf cuticle and entering the phloem, glyphosate is translocated along the sucrose gradient from source to sink tissues. Once in the cell, glyphosate must enter the chloroplast in order to reach the EPSPS target; many questions still remain as to how a water-soluble molecule such as glyphosate is able to cross so many membrane barriers. Reports have also shown that glyphosate can be compartmentalized in the cell resulting in reduced translocation, which have been proposed as a mechanism for glyphosate resistance in weeds (Feng et al. 2004; Preston and Wakelin 2008). Plant growth stage and spray dose are two factors that affect the concentration of glyphosate in at-risk tissues. A young sink leaf at one growth stage may be a mature source leaf at a later growth stage. As a result, the timing of glyphosate application will impact which sink tissues are at risk. Studies have also shown that glyphosate concentration in sink tissues is linearly proportional to the spray dose; therefore, an accurate assessment of at-risk tissues requires spray application at field use rates.

Given the complexities of field spray applications, we concluded that the only reliable model is to spray the plants with formulations augmented with radiolabeled glyphosate in a track sprayer. Over the years, the track sprayer has been used to measure absorption and translocation from over-the-top spray application of ^{14}C-glyphosate in crops and weeds (CaJacob et al. 2004, 2007; Feng and Chiu 2005; Feng et al. 2003). Results from studies in corn showed that low expression of CP4 EPSPS in tapetum and immature pollen may result in injury from a spray at the V8 stage, but not at the V3 stage. The reason is because these tissues are not yet developed at V3 and show little to no glyphosate import, but these same tissues become strong sinks at V8 with exponentially higher glyphosate import. Similar results in *Arabidopsis*, cotton, and soybean suggest that male reproductive tissues are particularly sensitive to glyphosate, and as these tissues approach the reproductive stage, the increased sink strength results in increased glyphosate import.

3.5   DEVELOPMENT OF SECOND-GENERATION GR CROPS

The strategy for the development of second-generation products was driven by the desire to enhance the protection of at-risk tissues from glyphosate injury. The realization of the spatial and temporal nature of at-risk tissues guided us to develop promoters that could match the expression pattern of CP4 EPSPS to the risk of injury. This was accomplished in two ways, by the use of multiple expression cassettes with different promoters or with chimeric promoters containing enhancer elements to increase the specificity of expression of constitutive promoters.

3.5.1 GR Soybeans

The first-generation RR soybean event 40-3-2, which used the cauliflower mosaic virus 35S promoter to drive *CP4 EPSPS* expression, delivered excellent glyphosate resistance and has been widely adopted in soybean productions around the world (Padgette et al. 1995). Based on the experiences from other GR plants, we examined the possibility of improving performance in soybeans. The second-generation product employs enhancer elements from the figwort mosaic virus 35S promoter upstream of the *Arabidopsis thaliana* promoter from the TSF1 gene (35S/TSF1) in combination with the *cis*-acting TSF1 intron to drive an optimized synthetic version of the *CP4 EPSPS* gene (Axelos et al. 1989; CaJacob et al. 2007; Richins et al. 1987). The second-generation soybean product was commercialized in 2009 as Roundup Ready 2 Yield® (RR2Y; Monsanto), which will be the platform for stacking future soybean traits.

The RR2Y expression cassette was transformed into soybean using an *Agrobacterium*-mediated system (Klee and Rogers 1987). Transformed events were screened for glyphosate resistance in greenhouse and field trials. In addition, extensive molecular analysis was performed to understand the transgene locus, copy number, 5′ and 3′ insert junctions, and stability over multiple generations. Analysis of leaf DNA from R4 to R7 generations confirmed transgene stability across generations. The characterization of the *in planta* CP4 EPSPS protein employed N-terminal sequencing by matrix assisted laser desorption ionization–time of flight (MALDI-TOF) mass spectrometry. Western blot analysis with antibody detection identified the full-length and properly processed 44.0-kDa protein, which showed equivalence to the *Escherichia coli*-produced reference standard. Compositional analysis of grain, forage, and processed fractions from RR2Y soybean MON 89788 showed equivalence to that of conventional soybeans (Lundry et al. 2008).

During the development of RR2Y soybean, the genomic region of the transgene locus was analyzed and used as a criterion for event selection and deployment. While biotechnology is focused on the identification, expression, and stability of the transgene, molecular breeding examines the impact of the deployment of the transgene into the germplasm. Over the years, molecular breeding has advanced from selection for economically important traits based on phenotypes to the use of molecular genetics to identify genomic regions with valuable traits. The genomic region at the transgene insertion site may possess unique agronomic values from a breeding perspective and has potential to augment the performance of the transgene for overall improved phenotypic effect. A further benefit is that valuable genomic regions that contain the transgene are selected for breeding into the germplasm of a crop.

Using molecular breeding tools, we identified and advanced the RR2Y event with a preferred breeding value by employing a proprietary set of DNA markers to classify the soybean genome into regions or haplotype windows. Based on the genomic location of the transgenic events, we estimated the agronomic values (e.g., yield, maturity, plant height, and lodging) and potential impacts on breeding populations. The extensive use of gene mapping has allowed us to identify specific DNA regions in soybean that have a positive impact on yield. The *CP4 EPSPS* gene in the RR2Y event is situated in one of these DNA regions that provide higher yield potential. The result is the identification of an RR2Y event with not only excellent glyphosate resistance but also additional agronomic values.

Near isogenic lines of RR and RR2Y soybeans in maturity group 3.2 were compared in field trials for resistance to glyphosate and other agronomic properties including yield. After backcrossing, these lines were estimated to have 94% genetic similarity with the primary difference being the transgene construct and its location in the soybean genome. Mapping studies showed that the *CP4 EPSPS* gene was inserted in the D1b chromosome in RR soybeans, but in the D1a chromosome in RR2Y soybeans. During the last 4 years, these two soybean lines have undergone extensive side-by-side testing in 73 field trials across six states. Both soybean lines showed comparable agronomic properties and resistance to glyphosate; however, the RR2Y line has consistently yielded 7–11% higher than the RR line (Fig. 3.1) with an average increase of 9% during the 4-year period. Early observations indicate that RR2Y soybean plants produce more seeds per plant resulting from a higher percentage of pods containing three or more seeds than the RR soybean plants. Calculations show that just a few more seeds per plant can significantly increase soybean yield on a per acre basis. RR2Y soybeans were launched in

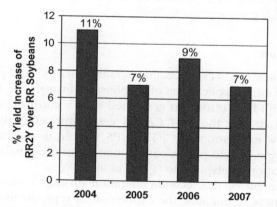

Figure 3.1. Percentages of yield increase in near-isogenic lines of Roundup Ready 2 Yield® (RR2Y) over Roundup Ready® (RR) soybeans in maturity group 3 evaluated in 73 field tests across six states from 2004 to 2007.

a controlled commercial release in the U.S. in the spring of 2009, followed by a broader launch in 2010.

3.5.2   GR Corn

Initial attempts to generate glyphosate resistance in corn utilized the *CP4 EPSPS* gene driven by the e35S promoter (Kay et al. 1987). However, this expression cassette did not achieve adequate reproductive resistance. When sprayed with glyphosate beyond the first few juvenile leaf stages, plants exhibited male sterility due to a notable deficit in expression of the CP4 EPSPS within tapetum and developing microspores (Heck et al. 2005). This lack of male reproductive resistance using a strong e35S promoter spurred interest in complementing its expression pattern with a second cassette, which utilized the rice actin 1 promoter that enhanced expression in male reproductive tissues (McElroy et al. 1990). The actin promoter had been successfully implemented in the first-generation RR corn (GA21 event) driving a mutant maize TIPS-EPSPS albeit with multiple, tandemly integrated cassettes (Sidhu et al. 2000). RR corn was commercialized in 1998.

For the second-generation RR corn 2 product (NK603 event), a combination of tandem CP4 EPSPS cassettes driven by e35S and rice actin 1 promoters, respectively, was used to engineer full resistance in all cell types. This strategy was confirmed by immunolocalization studies showing complementary accumulation of CP4 EPSPS in developing microspores from the rice actin cassette over that of e35S cassette (Heck et al. 2005). The RR corn 2 demonstrates superior vegetative and reproductive resistance to glyphosate and expanded crop safety. The RR corn 2 permits sequential applications of Roundup at $0.84\,kg\,a.e.ha^{-1}$ with a maximum single in-crop application of $1.26\,kg\,a.e.ha^{-1}$, thus providing a wider window of application and higher rate flexibility for tough weeds. RR corn 2 was commercialized in 2001 and has since grown to about 70% of U.S. corn acreage in 2008.

3.5.3   GR Cotton

RR cotton event MON1445 was commercialized in 1997 and provided resistance to over-the-top applications of glyphosate through the four-leaf stage (Nida et al. 1996). RR cotton was generated from a construct containing the FMV promoter driving the *CP4 EPSPS gene*. Immunolocalization studies showed that low expression of CP4 EPSPS protein in pollen mother cells, tapetum, microspores, and pollen contributed to reduced male fertility from off-label application of glyphosate (Chen et al. 2006).

The development of the second-generation product focused, in part, on utilizing promoters to enhance the expression in reproductive tissues. Dual cassette vectors, similar in strategy to that of RR corn 2, were constructed utilizing a strong viral promoter and a plant promoter to provide reproductive tolerance. The Roundup Ready Flex® (RRF, Monsanto) cotton event

MON 88913 contains two chimeric promoters constructed by combining elements from the 35S viral promoters with the TSF1 (FMV/TSF1) and ACT8 (35S/ACT8) promoters each driving the expression of the *CP4 EPSPS* gene in chloroplasts (Cerny et al. in press). Events generated from such a vector showed improved vegetative and reproductive resistance to glyphosate relative to the first-generation product, and with higher expression of CP4 EPSPS protein in reproductive tissues including tapetum and developing microspores (Chen et al. 2006). The RRF event contains a single, complete copy of the T-DNA and produces the identical CP4 EPSPS protein as that in RR cotton MON 1445. RRF shows enhanced vegetative and reproductive resistance to glyphosate and permits over-the-top application of glyphosate from emergence up to 7 days before harvest thus greatly expanding the application window and flexibility in managing weeds. RRF was commercialized in 2006 and has experienced rapid adoption accounting for ~50% of U.S. cotton acreage in 2008.

3.5.4 GR Canola

RR canola event RT73 was commercialized in 1996 and was selected from plants expressing both the *CP4 EPSPS* and *GOX* genes under the constitutive FMV promoter. RR canola is labeled for two sequential applications of glyphosate at 0.4 kg a.e. ha^{-1} or a single application at 0.6 kg a.e. ha^{-1}.

The second-generation RR2 canola is under development to enhance the level of glyphosate resistance and to expand the window of application. The transformation cassette in RR2 canola utilizes a novel chimeric promoter to enhance chloroplastic expression of CP4 EPSPS in male reproductive tissues. Transformants were screened for glyphosate resistance in the greenhouse and also for pollen viability as an indicator of male reproductive fertility. Glyphosate was applied at various rates and growth stages, and pollen grains from the first position flower in the most mature branch were collected and stained for viability (Alexander 1969). For RR RT73 canola, pollen viability declined from 98% to 0% after glyphosate applications from 0.8 to 1.6 kg a.e. ha^{-1} at the 4- and 10-leaf stages, whereas pollen from RR2 plants remained viable even at the rate of 3.6 kg a.e. ha^{-1}. The female reproductive tissues in RT73 remained viable after glyphosate spray and produced normal seed set from cross-pollination with wild-type (WT) pollen. RR2 leaf chlorosis ratings at 3 and 6 days after glyphosate treatments showed no injury from rates as high as 7.2 kg a.e. ha^{-1}, whereas RT73 plants exhibited significant stunting, chlorosis and lodging, and, in addition, poor anther and pollen formation.

Figure 3.2 shows seed production per plant from glyphosate treatment in RR RT73 versus RR2 canola plants. The results showed that while RT73 plants showed decreased seed production in response to increased glyphosate rate, the RR2 plants were unaffected by rates as high as 3.6 kg a.e. ha^{-1}. The second-generation RR2 canola product displays enhanced glyphosate resistance and a wider window of application for expanded weed control options.

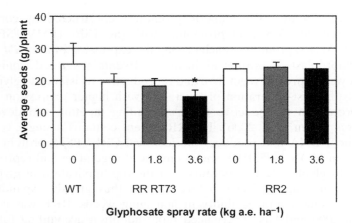

Figure 3.2. Average seed (g) per plant as a function of glyphosate rate (kg a.e. ha⁻¹) in WT (WT Ebony), Roundup Ready® (RR) RT73, and RR2 canola plants ($N = 5$). One-way analysis of variance (ANOVA) using the Student's test ($*p < 0.05$) (JMP software, SAS Institute, Cary, NC).

3.5.5   Glyphosate Hybridization System for Corn

Our understanding of glyphosate translocation and its effect on male reproductive tissues has led to new areas of research in hybrid seed production where a male-sterile parent is desired. Hybrid plants often exhibit heterosis resulting in enhanced growth uniformity, stress tolerance, and yield. The three key components for a typical hybrid seed production system include a male-sterile female parent to prevent self-pollination, a mechanism to propagate and maintain the male-sterile female parent, and a means to restore the male fertility in the F1 hybrid seeds for commercial planting. A number of methods have been tested for commercial production of hybrid seeds including hybrid corn (Williams 1995). Although mechanical and manual detasseling continues to be used successfully for hybrid corn seed production, it is extremely labor and equipment intensive. Cytoplasmic male sterility (*cms*) has also been used for hybrid corn seed production; however, its performance can be unpredictable in different germplasms and/or under different environments (Kaul 1988). Furthermore, the Texas cytoplasmic male sterility or *cms-T*, once widely adopted, was abandoned due to its linkage with the southern corn leaf blight disease (Ullstrup 1972).

We have been developing a technology for hybrid seed production by using glyphosate as a male gametocide, a technology we term Roundup Hybridization System (RHS). The technology is based on observations that male gametes in many crops are particularly sensitive to glyphosate resulting in male sterility from application of glyphosate (Dhingra et al. 1988). By precisely manipulating the expression profile of *CP4 EPSPS*, we have been able to achieve the

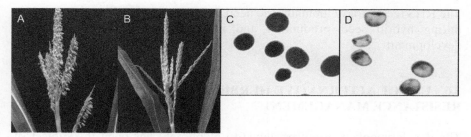

Figure 3.3. Glyphosate-induced male sterility in RHS transgenic corn. (A) A fully fertile tassel of an RHS plant sprayed with Roundup WeatherMax® at 0.84 kg a.e.ha⁻¹ at V3. (B) A sterile tassel of an RHS plant sprayed with Roundup WeatherMax at 0.84 kg a.e.ha⁻¹ at V3 and V10. Viable (C) and nonviable (D) pollen grains from the fertile (A) and sterile (B) tassels, respectively, stained with the Alexander stain.

desired glyphosate sensitivity in male gametes and insensitivity in the rest of the plant.

We have constructed an RHS transgene cassette in which the *CP4 EPSPS* gene expression is controlled by the enhanced cauliflower mosaic virus 35S promoter (Kay et al. 1987). It has been shown that the 35S promoter expresses poorly in the tapetum cells and microspores, both of which are critical for male gametocyte development (Plegt and Bino 1989). This RHS cassette was inserted into corn immature embryos via *Agrobacterium*-mediated transformation and the resulting plants showed vegetative resistance but were male sterile when sprayed with glyphosate at the V10 growth stage. The plants were female fertile producing normal seed set when cross-pollinated with WT pollen. Based on molecular analysis, events with a single copy, single insert of the RHS transgene were advanced for further efficacy analysis. Homozygous R2 plants demonstrated excellent vegetative resistance and complete male sterility when sprayed with glyphosate (0.84 kg a.e.ha⁻¹) at V3 and V10. Figure 3.3A shows one representative R2 event with a morphologically normal tassel after the V3 spray, whereas a double spray at V3 and V10 showed the desired sterile tassel with no anther extrusion or pollen shed (Fig. 3.3B). Pollen grains collected from the V3-sprayed plant were viable with normal phenotype (Fig. 3.3C), whereas those from the V3/V10-sprayed plant were nonviable with irregular shape and empty cytoplasm (Fig. 3.3D).

The RHS technology has several distinctive advantages over the current production systems. Glyphosate application at the appropriate growth stage induces male sterility without injury to the plant and provides added benefit of weed control. When sprayed at the early growth stage, the RHS plants are fully fertile and capable of self-pollination, which greatly simplifies the maintenance of the male-sterile female parent. And finally, hybrid seeds can be easily produced by application of glyphosate over RHS females and RR males as pollinators, and the resulting hybrid seeds are fully resistant to glyphosate.

The RHS technology eliminates the need for mechanical and hand detasseling during hybrid seed production and is currently in advanced product development.

3.6   USE OF ALTERNATIVE HERBICIDES FOR WEED RESISTANCE MANAGEMENT

The development of weed resistance to glyphosate has been the subject of many articles and reviews (Powles 2008). The first case of weed resistance was observed in ryegrass in 1996 (Powles et al. 1998), and since then, populations of 18 weed species have been added to the list (Heap 2010). Factors such as selection pressure, usage, and herbicide mode of action (MOA) can all contribute to the development of resistant weeds (Sammons et al. 2007), and these topics are discussed in detail in other chapters.

Because of its many attributes, glyphosate is expected to remain as the dominant herbicide in agriculture for the foreseeable future (Duke and Powles 2008). Relative to most other herbicides and taking into consideration the volume of usage, the rate of weed resistance development for glyphosate has been slow. Figure 3.4 shows selected herbicides representing various MOAs, the total number of weed species controlled according to the label and percentage of resistant weed species (I. Heap, pers. comm.). It can be seen that glyphosate controls the greatest number of weed species at 382 with 14 species or 3.7% as having developed resistance. Dicamba and glufosinate control fewer yet still a significant number of weeds and with low to no incidence of weed resistance. In contrast, atrazine, fluazifop, and chlorsulfuron control relatively fewer weeds and have high incidences of resistance. Analysis in Figure 3.4 indicates that glyphosate continues to be a valuable tool providing weed control for the vast majority of growers. In general, the highest incidences of weed resistance have been associated with herbicides belonging to three MOAs, which include inhibitors of the acetolactate synthase (ALS), acetyl coenzyme A carboxylase (ACCase), and photosystem II (PSII) (Heap, www.weedscience.org).

Weed resistance management strategies have been developed to control GR weeds and to delay the onset of resistance in other weed populations. One aspect of this strategy involves the use of herbicides with alternative MOA and with residual activity. Some selective herbicides are effective; however, choices can be limited depending on the crop, and many have weed resistance issues of their own. Another strategy is to engineer crop resistance to other broad-spectrum herbicides that can be used in conjunction with glyphosate. We identified dicamba and glufosinate as herbicides of interest based on numerous factors including efficacy, weed spectrum, safety, cost, and compatibility with glyphosate. Dicamba and glufosinate control a significant number of weed species with little to no weed resistance issues; however, both require genetic engineering to improve crop safety.

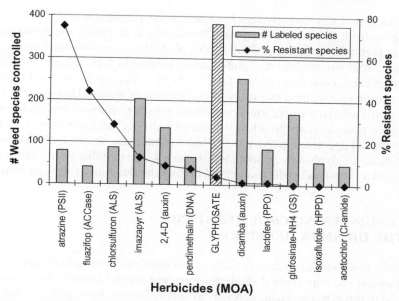

Figure 3.4. Herbicides representing various modes of action (MOAs), number of weed species controlled according to the label, and the percentages of weed-resistant species. ACCase, acetyl coenzyme A carboxylase; ALS, acetolactase synthase; Cl-amide, chloroacetamide; DNA, dinitroaniline; GS, glutamine synthetase; HPPD, hydroxyphenyl pyruvate dioxygenase; PSII, photosystem II; PPO, protoporphyrinogen oxidase.

3.7 ENGINEERING CROP RESISTANCE TO THE GLUFOSINATE HERBICIDE

Glufosinate is a broad-spectrum herbicide that inhibits glutamine synthetase (GS), which catalyzes assimilation of ammonia with glutamate to form glutamine. Glufosinate is a racemic DL-mixture of phosphinothricin, which was identified as a component of the tripeptide bialophos from *Streptomyces viridochromogenes*. Bialophos is hydrolyzed to L-phosphinothricin, which is a potent, irreversible inhibitor of GS. The herbicidal activity of glufosinate is attributed to build up of ammonia, which is measurable 1 h after treatment in the light (Donn 2007).

Glufosinate resistance was first commercialized in canola in 1995 (LibertyLink®, Bayer CropScience, Research Triangle Park, NC) and has since been commercialized in many other crops including corn and cotton, and is under development in soybeans (CaJacob et al. 2007). The development of glufosinate-resistant crops has been reviewed previously and is achieved via the deactivation mechanism (Donn 2007; Gressel 2002). Glufosinate selection has been commonly used in plant transformation experiments utilizing either the *BAR* gene from *Streptomyces hygroscopicus* (Thompson et al. 1987) or the *PAT* gene from *S. viridochromogenes* (Wohlleben et al. 1988). These two genes encode highly homologous (87%) L-phosphinothricin acetyltransferases that catalyze rapid *N*-acetylation of glufosinate. Because of high enzyme efficiency, a protein expression level of less than 0.1% of total protein was sufficient to confer resistance to field use rates of glufosinate in plants (Donn 2007). One strategy for weed resistance management is to stack glyphosate and glufosinate resistance traits in crops to allow the use of both herbicides; however, glufosinate's broad-spectrum activity is limited to small-size weeds and has shown antagonism to glyphosate in some weed species. The inability to use these herbicides in a tank-mix application could impact the flexibility for weed control.

3.8   ENGINEERING CROP RESISTANCE TO THE DICAMBA HERBICIDE

Dicamba is a member of the synthetic auxin family with broad efficacy against dicot weeds and with limited safety for monocot crops. Dicamba was commercialized in the 1960s and commercially available formulations include Clarity® and Banvel® (BASF, Florham Park, NJ), which are labeled for use in soybean, cotton, and corn. In soy and cotton, Clarity can be applied preemergence (PRE) at 0.27–0.55 kg a.e. ha^{-1} with a 15- to 30-day delayed planting to avoid crop injury. In corn, similar rates can be applied between PRE to early postemergence (POE). Corn is naturally tolerant to dicamba, although injury is possible depending on growth stage, germplasm, use rates, and environmental conditions. Other members of the synthetic auxin family include 2,4-D and 4-chloro-2-methylphenoxy acetic acid (MCPA), which are phenoxy acetic acids that differ structurally from dicamba, which is in the benzoic acid class.

3.8.1   Identification of a Dicamba Deactivation Enzyme

Krueger et al. (1989) reported the isolation of microorganisms capable of degrading dicamba from soil and water samples obtained from a dicamba manufacturing plant. They identified eight species of soil bacteria from five genera that were capable of utilizing dicamba as the sole carbon source. A *Pseudomonas maltophilia* strain was able to mineralize dicamba with 3,6-dichlorosalicylic acid (DCSA) as an intermediate. Subsequent work by Subramanian et al. (1997) reported three dicamba-degrading enzymes including an O-demethylase from *Clostridum thermoaceticum*, a monooxygenase from *P. maltophilia* DI-6, and a P-450 hydroxylase from corn endosperm culture.

    Further studies revealed that the activity in *P. maltophilia* DI-6 is derived from a multicomponent system comprised of an oxygenase, ferredoxin, and a reductase (Wang et al. 1997). Biochemical studies have shown that electrons are sequentially shuttled from NADH to the reductase, the ferredoxin, and ultimately to the oxygenase. All three proteins were purified and were required

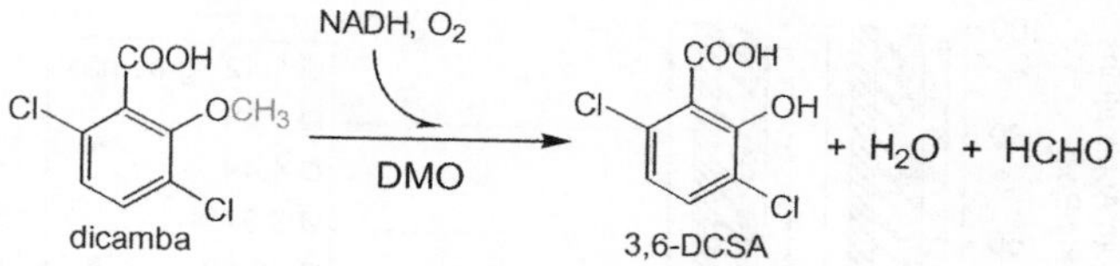

Figure 3.5. Metabolic deactivation of dicamba to 3,6-dichlorosalicylic acid (DCSA) by dicamba monooxygenase (DMO).

for activity *in vitro*. The resulting reaction (Fig. 3.5) converted dicamba to DCSA with water and formaldehyde as by-products, thus demonstrating that the O-demethylase activity was derived from a dicamba monooxygenase (DMO) (Chakraborty et al. 2005). The *DMO* gene encodes a 37.3-kDa protein composed of 339 amino acids and belongs to the family of Rieske nonheme iron oxygenases (Herman et al. 2005). The DMO protein showed conserved sequences for the Rieske (2Fe-2S) and nonheme Fe domains, and limited homology (36% and 34%, respectively) to other Rieske nonheme iron oxygenases such as toluene sulfonate methyl-monooxygenase and vanillate demethylase (Herman et al. 2005; Weeks et al. 2006). The DMO crystal structure revealed a homo-trimer in a head to tail configuration. The catalytic site is formed at the interface of two monomers from the nonheme iron and the Rieske domains of each monomoer. The dicamba-bound structure of DMO has led to the identification of amino acids in the active site that are responsible for substrate stabilization and catalysis (D'Ordine et al. 2009).

3.8.2 Soybean Transformation of DMO

Although *in vitro* activity required the presence of additional bacterial reductase and ferredoxin, DMO alone was sufficient for activity in plants. Soybean transformation of DMO was mediated by *Agrobacterium* and utilized a cassette containing a constitutive viral promoter from peanut chlorotic streak virus (Maiti and Shepperd 1998) with a chloroplast targeting sequence. Molecular analysis identified R0 transformants with single copy inserts. At the R1 generation, transformants were challenged with a POE spray of dicamba at $0.55\,kg\,a.e.ha^{-1}$. Further molecular analysis identified homozygous events with a clean single copy, single insert of the DMO cassette. Flanking sequences were determined for bioinformatics analysis of the insertion site to insure no disruption of endogenous genes, and analysis of *in planta* DMO sequence insured the integrity of the transgene. At subsequent generations, transformants were challenged with applications of dicamba PRE and/or POE for event selection and advancement.

Figure 3.6 shows a typical greenhouse titration of WT (or conventional) and dicamba-resistant (DR) soybeans. Injury was assessed at 28 days after treatment following two applications of dicamba ($0.55–5.0\,kg\,a.e.ha^{-1}$) at PRE and early POE (V3 or 3-trifoliate stage). The results clearly showed that WT soy

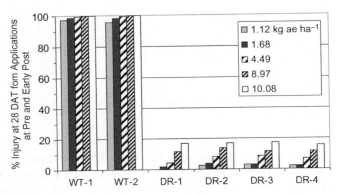

Figure 3.6. Percentages of injury from dicamba (0.55–5.0 kg a.e. ha^{-1}) applications pre-emergence at planting and early postemergence at V3 in wild-type (WT) and dicamba-resistant (DR) soybean events in the greenhouse. The labeled use rate for dicamba in soybeans is between 0.27 and 0.55 kg a.e. ha^{-1}. DAT, days after treatment.

is completely killed by the lowest rate, while DR plants showed little to no injury even at 5.0 kg a.e. ha^{-1}, which is far above the expected field use rate of dicamba (0.27–0.55 kg a.e. ha^{-1}). The greenhouse results have translated well to the field, and we have observed excellent resistance of DR soybeans to PRE and POE applications of dicamba.

3.9   STACKING OF HERBICIDE-RESISTANT TRAITS

The stacking of multiple herbicide-resistant traits can occur in a transformation vector or in breeding. In vector stacking, multiple herbicide-resistant traits are linked and introduced into the plant genome as a single insert. In breed stacking, independent events with different herbicide-resistant traits are brought together via cross-pollination. The selection of stacking method is a complicated business decision affected by commercialization timing and availability of events with the desired traits. Stacking is further complicated by the need to bring together other traits for insect control, disease control, drought tolerance, and yield.

For soybeans, work is underway to breed-stack dicamba and glyphosate resistance traits (DR × GR). Dicamba and glyphosate show excellent compatibility, and plants with double-stacked traits have demonstrated excellent resistance to dicamba and/or glyphosate in field trials; DR × GR soybean is currently in advanced development. In cotton, the vector stack of dicamba and glufosinate resistance traits is in early stage of testing, and has demonstrated good resistance to dicamba and/or glufosinate in greenhouse and field trials. In corn, the breed stack of glyphosate and glufosinate resistant traits along with insect control traits is in late stage development. The resulting SmartStax®

(Monsanto) product, with a total of eight genes, will serve as a platform for stacking other traits. We believe the option to use glyphosate and glufosinate along with selective preemergent herbicides such as acetochlor and atrazine will expand weed control options in corn and reduce the risk of weed resistance development.

Commercialization of first-generation GR crops revolutionized agriculture by enabling the use of glyphosate in-crop without the risk of crop injury. The development of second-generation products was aided by the identification of at-risk tissues to enhance the expression of CP4 EPSPS resulting in improved performance with enhanced glyphosate resistance and/or expanded window of application. Research is underway to develop third-generation products, which will stack multiple herbicide-resistant traits and provide the flexibility of using one or more herbicides for the purpose of expanding weed control options, maximizing yield potential, and managing resistant weeds.

ACKNOWLEDGMENTS

We are grateful to numerous colleagues at Monsanto who have contributed to this work. The development of a commercial product can easily involve hundreds of people across many organizations, and we regret that we were not able to list the names of all the contributors.

REFERENCES

Alexander, P. P. 1969. Differential staining of aborted and non-aborted pollen. *Stain Technology* 44:117–122.

Alibhai, M. F., C. CaJacob, P. C. C. Feng, G. Heck, Y. Qi, S. Flasinski, and W. C. Stallings, inventors; Monsanto Company, assignee. 2006. Glyphosate resistant class I 5-enolpyruvylshikimate-3-phosphate synthase (EPSPS). U.S. Patent Publication 20,060,143,727.

Anderson, J. A. and J. A. Kolmer. 2005. Rust control in glyphosate tolerant wheat following application of the herbicide glyphosate. *Plant Disease* 89:1136–1142.

Axelos, M., C. Bardet, T. Liboz, A. Le Van Thai, C. Curie, and B. Lescure. 1989. The gene family encoding the *Arabidopsis thaliana* translation elongation factor EF-1 alpha: molecular cloning, characterization and expression. *Molecular & General Genetics* 219(1–2):106–112.

Barry, G. F., G. M. Kishore, S. R. Padgette, M. Taylor, K. Kolacz, M. Weldon, D. Re, D. Eicholtz, K. Fincher, and L. Hallas. 1992. Inhibitors of amino acid biosynthesis: strategies for imparting glyphosate tolerance to crop plants. In B. K. Singh, H. E. Flores, and J. C. Shannon, eds. *Biosynthesis and Molecular Regulation of Amino Acids in Plants*. Madison, WI: American Society of Plant Physiologists, pp. 139–145.

Berg, B. J. V., P. E. Hammer, B. L. Chun, L. C. Schouten, B. Carr, R. Guo, C. Peters, T. K. Hinson, V. Beilinson, A. Shkita, R. Deter, Z. Chen, V. Samoylov, C. T. Bryant, M. E. Stauffer, T. Eberle, D. J. Moellenbeck, N. B. Carozzi, M. G. Koziel, and N. B. Duck. 2008. Characterization and plant expression of a glyphosate-tolerant enolpyruvyl-shikimate phosphate synthase. *Pest Management Science* 64:340–345.

CaJacob, C. A., P. C. C. Feng, G. R. Heck, M. F. Alibhai, R. D. Sammons, and S. R. Padgette. 2004. Engineering resistance to herbicides. In P. Christou and H. Klee, eds. *Handbook of Plant Biotechnology*. Chichester: John Wiley & Sons, pp. 353–372.

CaJacob, C. A., P. C. C. Feng, S. E. Reiser, and S. R. Padgette. 2007. Genetically modified herbicide resistant crops. In W. Krämer and U. Schirmer, eds. *Modern Crop Protection Compounds*. Weinheim, Germany: Wiley-VCH, pp. 283–316.

Castle, L. A., D. L. Siehl, R. Gorton, P. A. Patten, Y. H. Chen, S. Bertain, H. Cho, N. Duck, J. Wong, and D. Liu. 2004. Discovery and directed evolution of a glyphosate tolerance gene. *Science* 304:1151–1154.

Cerny, R. E., J. T. Bookout, C. A. CaJacob, J. R. Groat, J. L. Hart, G. R. Heck, S. A. Huber, J. Listello, A. B. Martens, M. E. Oppenhuizen, B. Sammons, N. K. Scanlon, Z. W. Shappley, J. X. Yang, and J. Xiao. Development and characterization of a cotton (*Gossypium hirsutum*) event with enhanced reproductive resistance to glyphosate *Crop Science* (in press).

Chakraborty, S., M. Behrens, P. I. L. Herman, A. F. Arendsen, W. R. Hagen, D. L. Carlson, X. Z. Wang, and D. P. Weeks. 2005. A three-component dicamba O-demethylase from *Pseudomonsas maltophilia*, strain DI-6: purification and characterization. *Archives of Biochemistry and Biophysics* 437:20–28.

Chen, Y. S., C. Hubmeier, M. Tran, A. Martens, R. E. Cerny, R. D. Sammons, and C. CaJacob. 2006. Expression of CP4 EPSPS in microspores and tapetum cells of cotton (*Gossypium hirsutum*) is critical for male reproductive development in response to late-stage glyphosate applications. *Plant Biotechnology Journal* 4:477–487.

Dhingra, O. P., J. E. Franz, G. Keyes, D. F. Loussaert, and C. S. Mamer, inventors; Monsanto Company, assignee. 1988. Gametocides. U.S. Patent 4,735,649, filed April 5, 1988.

Dill, G. M., C. A. CaJacob, and S. R. Padgette. 2008. Glyphosate-resistant crops: adoption, use and future considerations. *Pest Management Science* 64:326–331.

Donn, G. 2007. Glutamine synthetase inhibitors. In W. Krämer and U. Schirmer, eds. *Modern Crop Protection Compounds*. Weinheim, Germany: Wiley-VCH, pp. 302–316.

D'Ordine, R. L., T. J. Rydel, M. Storek, E. Sturman, F. Moshir, R. K. Bartlett, G. R. Brown, R. Eilers, C. Dart, Y. Qi, S. Flasinski, and S. J. Franklin. 2009. Dicamba mono-oxygenase: structural insights into a dynamic Rieske oxygenase that catalyzes an exocyclic monooxygenation. *Journal of Molecular Biology* 392:481-497.

Duke, S. O. and S. B. Powles. 2008. Glyphosate: a once-in-a-century herbicide. *Pest Management Science* 64:319–325.

Feng, P. C. C., G. J. Baley, W. P. Clinton, G. J. Bunkers, M. F. Alibhai, T. C. Paulitz, and K. K. Kidwell. 2005. Glyphosate controls rust diseases in glyphosate-resistant wheat and soybean. *Proceedings of the National Academy of Sciences of the United States of America* 48:17290–17295.

Feng, P. C. C. and T. Chiu. 2005. Distribution of ^{14}C-glyphosate in mature glyphosate-resistant cotton from application to a single leaf or over-the-top spray. *Pesticide Biochemistry and Physiology* 82:36–45.

Feng, P. C. C., T. Chiu, R. D. Sammons, and J. S. Ryerse. 2003. Droplet size affects glyphosate retention, absorption and translocation in corn. *Weed Science* 51: 443–448.

Feng, P. C. C., C. Clark, G. C. Andrade, M.C. Balbi, and P. Caldwell. 2008. The control of Asian rust by glyphosate in glyphosate-resistant soybeans. *Pest Management Science* 64:353–359.

Feng, P. C. C., J. S. Ryerse, and R. D. Sammons. 1998. Correlation of leaf damage with uptake and translocation of glyphosate in velvetleaf (*Abutilon theophrasti*). *Weed Technology* 12:300–307.

Feng, P. C. C., M. Tran, T. Chiu, R. D. Sammons, G. R. Heck, and C. A. CaJacob. 2004. Investigations into glyphosate resist horseweed (*Conyza canadensis*): retention, uptake and translocation. *Weed Science* 52:498–505.

Gianessi, L. P. 2008. Economic impacts of glyphosate-resistant crops. *Pest Management Science* 64:346–352.

Gressel, J. 2002. *Molecular Biology of Weed Control*. New York: Taylor & Francis, pp. 219–277.

Heap, I. 2010. The International Survey of Herbicide Resistant Weeds. http://www. weedscience.com (accessed April 8, 2010).

Heck, G. R., C. L. Armstrong, J. D. Astwood, C. F. Behr, J. T. Bookout, S. M. Brown, T. A. Cavato, D. L. DeBoer, M. Y. Deng, C. George, J. R. Hillyard, C. M. Hironaka, A. R. Howe, E. H. Jakse, B. E. Ledesma, T. C. Lee, R. P. Lirette, M. L. Mangano, J. N. Mutz, Y. Qi, R. E. Rodriguez, S. R. Sidhu, A. Silvanovich, M. A. Stoecker, R. A. Yingling, and J. You. 2005. Development and characterization of a CP4 EPSPS-based, glyphosate-tolerant corn event. *Crop Science* 45:329–339.

Herman, P. L., M. Behrens, S. Chakraborty, B. M. Chrastil, J. Barycki, and D. P. Weeks. 2005. A three component dicamba O-demethylase from *Pseudomonas maltophilia*, strain DI-6. *Journal of Biological Chemistry* 280:24759–24767.

Jones, M. A. and C. E. Snipes. 1999. Tolerance of transgenic cotton to topical applications of glyphosate. *Journal of Cotton Science* 3:19–26.

Kaul, M. L. H. 1988. *Male Sterility in Higher Plants*. Berlin: Springer-Verlag.

Kay, R., A. Chan, M. Daly, and J. McPherson. 1987. Duplication of CaMV 35S promoter sequences creates a strong enhancer for plant genes. *Science* 236:1299–1302.

Klee, H. J. and S. G. Rogers. 1987. Cloning of an *Arabidopsis* gene encoding 5-enolpyruvylshikimate-3-phosphate synthase: sequence analysis and manipulation to obtain glyphosate-tolerant plants. *Molecular & General Genetics* 210:437–442.

Krueger, J. P., R. G. Butz, Y. H. Atallah, and D. J. Cork. 1989. Isolation and identification of microorganisms for the degradation of dicamba. *Journal of Agricultural and Food Chemistry* 37:534–538.

Lundry, D. R., W. P. Ridley, J. J. Meyer, S. G. Riordan, M. A. Nemeth, W. A. Trujillo, M. L. Breeze, and R. Sorbet. 2008. Composition of grain, forage, and processed fractions from second-generation glyphosate-tolerant soybean, MON 89788, is equivalent to that of conventional soybean (*Glycine max* L.). *Journal of Agricultural and Food Chemistry* 56(12):4611–4622.

Maiti, I. B. and R. J. Shepperd, inventors; University of Kentucky, assignee. 1998. Promoter (FLT) for the full-length transcript of peanut chlorotic streak caulimovirus (PCLSV) and expression of chimeric genes in plants. U.S. Patent 5,850,019, filed December 15, 1998.

McElroy, D., W. Zhang, J. Cao, and R. Wu. 1990. Isolation of an efficient actin promoter for use in rice transformation. *The Plant Cell* 2:163–171.

Nida, D. L., K. H. Kolacz, R. E. Beuhler, W. R. Deaton, W. R. Schuler, T. A. Armstrong, M. L. Taylor, C. C. Ebert, G. J. Rogan, S. R. Padgette, and R. L. Fuchs. 1996. Glyphosate-tolerant cotton: genetic characterization and protein expression. *Journal of Agricultural and Food Chemistry* 44(7):1960–1966.

Padgette, S. R., K. H. Kolacz, X. Delannay, D. B. Re, B. J. Lavallee, C. N. Tinius, W. K. Rhodes, Y. I. Otero, G. F. Barry, D. A. Eichholtz, G. M. Peschke, D. L. Nida, N. B. Taylor, and G. M. Kishore. 1995. Development, identification, and characterization of a glyphosate-tolerant soybean line. *Crop Science* 35:1451–1461.

Padgette, S. R., D. B. Re, G. F. Barry, D. E. Eichhlotz, X. Delannay, R. L. Fuchs, G. M. Kishore, and R. T. Fraley. 1996. New weed control opportunities: development of soybeans with a Roundup Ready™ gene. In S. O. Duke, ed. *Herbicide-Resistant Crops: Agricultural, Environmental, Economic, Regulatory, and Technical Aspects.* New York: CRC Press, pp. 53–84.

Plegt, L. and R. J. Bino. 1989. β-Glucuronidase activity during development of the male gametophyte from transgenic and non-transgenic plants. *Molecular & General Genetics* 216:321–327.

Pline, W. A., J. W. Wilcut, K. L. Edminsten, J. Thomas, and R. Wells. 2002. Reproductive abnormalities in glyphosate-resistant cotton caused by lower CP4-EPSPS levels in the male reproductive tissues. *Weed Science* 50:438–437.

Powles, S. B. 2008. Evolved glyphosate-resistant weeds around the world: lessons to be learnt. *Pest Management Science* 64:360–365.

Powles, S. B., D. F. Lorraine-Colwill, J. J. Dellow, and C. Preston. 1998. Evolved resistance to glyphosate in rigid ryegrass (*Lolium rigidum*) in Australia. *Weed Science* 46:604–607.

Preston, C. and A. M. Wakelin. 2008. Resistance to glyphosate from altered herbicide translocation patterns. *Pest Management Science* 64:372–376.

Reddy, K. N., A. M. Rimando, and S. O. Duke. 2004. Aminomethylphosphonic acid, a metabolite of glyphosate, causes injury in glyphosate-treated, glyphosate-resistant soybean. *Journal of Agricultural and Food Chemistry* 52:5139–5143.

Richins, R., H. Scholthof, and R. Shepard. 1987. Sequence of figwort mosaic virus DNA (caulimovirus group). *Nucleic Acids Research* 15:8451–8466.

Sammons, R. D., D. C. Heering, N. Dinicola, H. Glick, and G. A. Elmore. 2007. Sustainability and stewardship of glyphosate and glyphosate-resistant crops. *Weed Technology* 21:347–354.

Sidhu R. S., B. G. Hammond, R. L. Fuchs, J. N. Mutz, L. R. Holden, B. George, and T. Olson. 2000. Glyphosate-tolerant corn: the composition and feeding value of grain from glyphosate-tolerant corn is equivalent to that of conventional corn (*Zea mays* L.). *Journal of Agricultural and Food Chemistry* 48:2305–2312.

Siehl D. L., L. A. Castle, R. Gorton, Y. H. Chen, S. Bertain, H. J. Cho, R. Keena, D. Liu, and M. W. Lassner. 2005. Evolution of a microbial acetyltransferase for modification of glyphosate: a novel tolerance strategy. *Pest Management Science* 61:235–240.

Subramanian, M. V., J. Tuckey, B. Patel, and P. J. Jensen. 1997. Engineering dicamba selectivity in crops: a search for appropriate degradative enzymes. *Journal of Industrial Microbiology and Biotechnology* 19:344–349.

Thompson, C. J., N. R. Movva, R. Tizard, R. Crameri, J. E. Davies, M. Lauwereys, and J. Botterman. 1987. Characerization of the herbicide resistance gene bar from *Streptomyces hygroscopicus. European Molecular Biology Organization Journal* 6: 2519–2523.

Ullstrup, A. J. 1972. The impacts of southern corn leaf blight epidemics of 1970–1971. *Annual Reviews of Phytopathology* 10:37–50.

Wang, X. Z., B. Li, P. L. Herman, and D. P. Weeks. 1997. A three-component enzyme system catalyzes the O-methylation of the herbicide dicamba in *Pseudomonas maltophilia* DI-6. *Applied and Environmental Microbiology* 63:1623–1626.

Weeks, D. P., X. Z. Wang, and P. L. Herman, inventors; University of Nebraska, assignee. 2006. Methods and materials for making and using transgenic dicamba-degrading organisms. U.S. Patent 7,105,724, filed September 12, 2006.

Williams, M. E. 1995. Genetic engineering for pollination control. *Trends in Biotechnology* 13:344–349.

Wohlleben, W., W. Arnold, J. Broer, D. Hillmann, E. Struch, and A. Puhler. 1988. Nucleotide sequence of the phosphinothricin *N*-acetyltransferase gene from *Streptomyces viridochromogenes* Tu494 and its expression in *Nicotiana tabacum. Gene* 70:25–37.

4

TRANSITIONING FROM SINGLE TO MULTIPLE HERBICIDE-RESISTANT CROPS

Jerry M. Green and Linda A. Castle

4.1 INTRODUCTION

Before the advent of glyphosate-resistant (GR) crops, many experts thought that the utility of herbicide-resistant (HR) crops, whether derived from conventional breeding practices or genetic engineering, would be limited to being just another tool to complement selective herbicides (Burnside 1996; Duvick 1996; Hess 1996). Most did not appreciate the full impact of HR crops until sales of GR soybean (*Glycine max* (L.) Merr) started in 1996. Since that time, adoption of GR crops has been rapid, increasing dramatically year after year such that in 2008 more than 90% of the soybeans in the United States and over 70% globally were HR (James 2008). Similar rates of adoption are occurring in corn (*Zea mays* L.), cotton (*Gossypium hirsutum* L.), and canola, both Argentine (*Brassica napus* L.) and Polish (*Brassica rapa* L.).

Success came despite an unpopular "grower contract" and strong objections by biotechnology opponents to the unknown impacts on the environment and human health, the ethics of interfering with nature, the patenting of life forms, and the lack of labeling on food products. This debate has been ongoing since the genesis of genetically modified (GM) technology and has been extensively reported in the press and in a number of books (Etine 2005; Fedoroff and Brown 2004; Lambrecht 2001; Ruse and Castle 2002). The GM debate often

Glyphosate Resistance in Crops and Weeds: History, Development, and Management
Edited by Vijay K. Nandula
Copyright © 2010 John Wiley & Sons, Inc.

centers on social and moral issues that are difficult for scientists and business leaders to address. Anti-biotech forces generally want actions delayed because of the lack of long-term data, while pro-biotech forces want fast action because of the societal and environmental benefits of GM crops and their role in solving world hunger. In Europe, the debate has resulted in delays that have translated into a de facto moratorium on growing GM crops. Fortunately, transgenes have not created the problems that critics predicted, and future advances promise direct benefits for both growers and consumers.

Good weed management is essential for growers to increase productivity and meet growing demands for food, feed, fiber, and fuel. However, no new commercial herbicide with a new mode of action has been discovered for 25 years (Stuebler et al. 2008). The introduction of HR crops broadens the number of weed management options with currently available herbicides. To date, five transgenic herbicide resistance traits to three herbicides have been commercialized: glyphosate resistance conferred by CP4 5-enolpyruvylshikimate-3-phosphate synthase (CP4 EPSPS; EC 2.5.1.19), glyphosate resistance conferred by mutated EPSPS from corn, glyphosate resistance conferred by CP4 EPSPS in combination with glyphosate oxidoreductase (GOX) metabolic mechanism, a nitrilase for metabolic resistance to bromoxynil (EC 3.5.5.6), and glufosinate resistance conferred by phosphinothricin acetyltransferase (PAT; EC 2.3.1.X). GR crops and glyphosate proved to be the most effective and flexible weed management system, enabling reductions in costs, energy use, and tillage that benefited both the grower and the environment.

Glyphosate was the ideal herbicide for crop resistance, controlling over 300 annual and perennial weeds at a wide range of application timings without any rotational crop restrictions (Franz et al. 1997). GR crops allowed growers to use glyphosate and replace more expensive, selective herbicides that controlled a narrower weed spectrum. Growers applied glyphosate alone over large areas with highly variable and prolific weeds, inevitably leading to the evolution of resistant weeds, just as the overuse of antibiotics led to resistant bacteria (Heap 2009). Glyphosate alone worked well for a decade, but now weeds are adapting and growers must face the consequences of their actions (Nichols et al. 2008; Powles 2008b). To solve weed problems, chemical and seed companies are developing a new generation of HR crops that will be resistant to multiple herbicide modes of action.

Multiple HR crops can be developed either through breeding or molecular stacks. Breeding stacks combine traits by crossing two existing lines, whereas molecular stacks combine multiple genes at a single locus. For example, herbicide resistance is commonly molecularly stacked with insect resistance and used as a selectable marker. Government agencies view such new transgene combinations the same as any new insertion of foreign DNA and require full regulatory approvals. Breeding stacks avoid some regulatory requirements, but managing multiple HR genes at different locations in the genome makes breeding more challenging. In this chapter, we review currently available herbicide resistance traits and discuss the potential to combine these traits with each other and new traits to create new weed management options and help growers transition from single to multiple HR crops (Castle et al. 2006; Green 2009).

4.2   THE FIRST HR CROPS

Growers have been using selective herbicides since 2,4-dichloroacetic acid (2,4-D) was discovered in the 1940s. Scientists discovered selective herbicides by screening large numbers of chemicals on a range of crops and weeds to find herbicides that were safe to crops while still controlling key weeds. The method was successful for decades and the level of tolerance was often very high, sometimes as high as HR crops (Green and Ulrich 1993). However, the number of chemicals that had to be tested to find a new selective herbicide increased dramatically. Fifty years ago, the number was only a few thousand while today it is more than 100,000 (Stuebler et al. 2008). This has made the discovery process very expensive and forced scientists to find new ways to make herbicides safe to crops. The outcome was to modify the biology instead of the chemistry to create crop selectivity with currently available nonselective herbicide technology.

Several traditional genetic methods to create HR crops have been successful: seed and pollen mutagenesis, tissue and cell culture selection, and even whole plant screening (Table 4.1). Crops resistant to herbicides that inhibit acetolactate synthase (ALS; EC 2.2.1.6), also known as acetohydroxyacid synthase, have been relatively easy to develop and the most successful in the marketplace. ALS-resistant weeds are widespread (Heap 2009) but before weeds evolve resistance, ALS herbicides can provide low-cost, broad-spectrum weed control with soil residual activity. Growers currently can buy at least seven different ALS-resistant crops (Shaner et al. 2007; Tan et al. 2005). The first was ALS-resistant corn generated from tissue culture selections (Anderson and Georgeson 1989). Later, another ALS-resistant corn was developed from pollen mutagenesis; microspore selection was used for canola; and seed mutagenesis was used for soybean, lentil (*Lens culinaris* Medik.), wheat (*Triticum aestivum* L.), and rice (*Oryza sativa* L.). A resistant *als* gene was transferred with traditional breeding from a weedy relative to create resistant sunflower (*Helianthus annuus* L.). In all cases, resistance is due to a mutation of the native ALS protein to reduce binding, and thus inhibition by ALS herbicides. The ALS amino acid changes (in reference to the *Arabidopsis thaliana* sequence) include Pro197Ser in soybean, Ala205Val in sunflower, Trp574Val in corn and canola, Ser653Asn in corn, canola, lentil, wheat, and rice, and Gly654Glu in rice.

The efficacy of nontransgenic HR traits was usually high, but the success of HR crops in the marketplace was modest until the introduction of transgenic glyphosate resistance. Transgenic GR soybean and canola were introduced in the mid-1990s and other GR crops followed rapidly. Bromoxynil- and

TABLE 4.1. Examples of Nontransgenic Herbicide-Resistant Crops

Selection Method	Herbicide Type	Crop(s)	Reference
Whole plant	Triazine	Canola	Vaughn (1986)
Seed mutagenesis	Terbutryne	Wheat	Pinthus (2006)
	Sulfonylurea	Soybean	Sebastian and Chaleff (1987)
	Imidazolinone	Wheat	Newhouse et al. (1991)
		Rice	Croughan (1998)
Tissue culture	Sulfonylurea	Canola	Charne et al. (2002)
	Atrazine	Soybean	Alfonso et al. (1996)
	Imidazolinone	Corn	Newhouse et al. (1991)
Cell selection	Imidazolinone	Sugarbeet	Wright and Penner (1998)
Pollen mutagenesis	Imidazolinone	Corn	Bright (1992)
Microspore selection	Imidazolinone	Canola	Swanson et al. (1989)
Transfer from weedy relative	ALS inhibitor	Sunflower	Al-Khatib and Miller (2000)
		Sorghum	Tuinstra and Al-Khatib (2008a)
	ACCase inhibitor	Sorghum	Tuinstra and Al-Khatib (2008b)

ACCase, acetyl CoA carboxylase; ALS, acetolactate synthase.

glufosinate-resistant transgenic crops were also introduced in the 1990s, but with the exception of glufosinate-resistant canola, have not yet enjoyed the level of adoption by growers as GR crops.

4.3   GR CROPS

Glyphosate is a strong competitive inhibitor of EPSPS, the sixth enzyme in the shikimate biosynthetic pathway that produces essential aromatic amino acids (tryptophan, tyrosine, and phenylalanine) (Senseman 2007). Within a decade after glyphosate became commercially available, the search began to find crop resistance. Nontransgenic approaches were not successful and transgenic approaches were difficult and not initially successful (Franz et al. 1997). Attempts to find any natural enzymes in crops that could metabolically inactivate or were insensitive to glyphosate failed. Finally, a new class of insensitive EPSPS enzyme with enzymatic characteristics similar to plant EPSPS was isolated from a soil bacterium that was surviving in a glyphosate manufacturing waste stream in Luling, Louisiana (Barry et al. 1992). The *epsps* gene from this bacterium, *Agrobacterium tumefaciens* strain CP4, was used to develop GR soybean, cotton, corn, Argentine canola, Polish canola, alfalfa (*Medicago sativa* L.), and sugarbeet (*Beta vulgaris* L.) (Table 4.2). These GR crops profoundly changed how many growers managed weeds.

TABLE 4.2. Summary of Current Transgenic Herbicide-Resistant Crop Technology

Crop	Resistance Trait	Trait Gene(s)	Trait Designation	Promoter(s)	Transit Peptide	First Sales
Alfalfa	Glyphosate	cp4 epsps	J101, J163	FMV 35S	CTP2	2005
Canola	Glyphosate	cp4 epsps and goxv247	GT73	FMV 35S and FMV 35S	CTP2 and CTP1	1996
Cotton	Glufosinate	pat	HCN92	CaMV 35S		1995
	Bromoxynil	bxn	ACS-BN011-5	CaMV 35S		2000[a]
	Bromoxynil	bxn	BXN			1995[a]
	Glyphosate	cp4 epsps	MON1445	FMV 35S	CTP2	1996
	Glyphosate	Two cp4 epsps	MON88913	FMV e35S/TSF1 and 35S/ACT8	CTP2 and CTP2	2006
	Glyphosate	zm-2mepsps	GHB614	Ph4a748At	TPotpC	2009
	Glufosinate	bar	LLCotton25	CaMV 35S	OTP	1998
Corn	Glyphosate	Three modified zm-2mepsps	GA21	Os-Act1		2001
	Glyphosate	Two cp4 epsps	NK603	Os-Act1 and CaMV e35S	CTP2 and CTP2	
	Glyphosate and ALS	gat4621 and zm-hra	98140	ubiZM1 and zmALS	ZM-ALS CTP	TBD
	Glufosinate	pat	T14, T25	CaMV 35S		1996
Soybean	Glyphosate	cp4 epsps	GTS 40-3-2	CaMV e35S	CTP4	1996
	Glyphosate	cp4 epsps	MON89788	FMV e35S/TSF1	CTP2	2009
	Glyphosate and ALS	gat4601 and gm-hra	356043	SCP1 and SAMS	GM-ALS CTP	TBD
	Glufosinate	pat	A2704-12	CaMV 35S		2009
Sugarbeet	Glyphosate	cp4 epsps	H7-1	FMV e35S	CTP2	2007

[a]Discontinued.

ALS, acetolactate synthase; TBD, to be determined.

The first GR crops were not perfect. The timing, rate, and number of glyphosate applications had to be restricted to ensure crop tolerance (Dill et al. 2008). Despite early reports that GR soybean varieties performed poorly under heat stress (Gertz et al. 1999) and had a "yield drag" (Elmore et al. 2001), grower acceptance was astonishing and GR crops became the most rapidly adopted farm technology in history. A new generation of GR crops currently in development will likely extend this trend. The new crop traits incorporate one or more of the following for improved performance:

1. different promoter(s), insertion site, number of copies, and other genetic elements to improve expression;
2. new *epsps* genes with increased resistance and/or improved enzymatic characteristics;
3. new metabolic glyphosate-inactivating genes;
4. combinations of target-site and metabolic inactivation genes; or
5. combinations with other herbicide resistance genes to expand weed management options.

4.3.1   GR Soybean

The first commercial GR soybean event, GTS 40-3-2 (Roundup Ready®, Monsanto Company, St. Louis, MO), was first field tested in 1991 and was sold to growers 5 years later (APHIS 2009). The marketing strategy to distribute GTS 40-3-2 soybean as broadly as possible was successful. GTS 40-3-2 soybean was produced by using particle acceleration transformation to insert the *cp4 epsps* gene into the cultivar "A5403" (Padgette et al. 1995). The *gus* gene for production of β-glucuronidase was the screenable marker, but was not integrated into the host genome. The enhanced cauliflower mosaic virus (CaMV) e35S promoter and duplicated enhancer regions constitutively drive gene expression. A chloroplast transit peptide (CTP4) from *Petunia hybrida* directs CP4 EPSPS to the chloroplast, the location of the shikimate pathway. The termination sequence is a nopaline synthase (*nos* 3′) transcriptional element from *A. tumefaciens*.

Some growers observed problems with the initial 40-3-2 varieties. In some situations, chlorosis appeared 4–10 days after glyphosate application. This "yellow flash" was in the upper canopy and usually associated with high rates of glyphosate, such as in a spray overlap. The degradation of glyphosate to aminomethylphosphonate (AMPA) reportedly causes this chlorosis (Reddy et al. 2004; Sammons and Tran 2008). Glyphosate, like many other phosphonic acids, can also act as a chelating agent and form stable complexes with divalent and trivalent micronutrients such as zinc (Zn) and manganese (Mn) (Bernards et al. 2005; Subramaniam and Hoggard 1988). Complexing of Mn with glyphosate inside the plant reduces its bioavailability and can cause similar symptoms (Bott et al. 2008).

Other growers observed lower yields. One explanation for the yield reduction is the preferential accumulation of glyphosate in meristems and male reproductive tissues such as the tapetum and developing pollen. These tissues are strong sinks and particularly sensitive to any residual glyphosate (CaJacob et al. 2007). Other possible explanations include over- or under-expression of EPSPS, genes associated with the insertion site, or the initial varieties retained too much of the lower yielding genetics of the "A5403" line used to insert the trait (Elmore et al. 2001). Independent research has never determined the basis for the "yield drag."

In 2009, MON89788 (Roundup Ready 2 Yield®, Monsanto Company) was introduced as the first new transgenic trait in soybean in more than a decade. MON89788 contains the same CP4 EPSPS as GTS 40-3-2, but with the gene inserted at a different site in an elite variety "A3244" with a different promoter and regulatory elements to enhance expression in the sensitive tissues. MON89788 was developed with *Agrobacterium*-mediated transformation of soybean meristematic tissue using the transformation vector, pV-GMGOX20 (AGBIOS 2009). The *cp4 epsps* gene is under the control of a chimeric promoter, FMW e35S/TSF1, which combines the enhancer sequences from the 35S promoter of the figwort mosaic virus (FMV) and the promoter from the *tsf1* gene from *A. thaliana*, which codes for the elongation factor, EF-1 alpha. A chloroplast transit peptide (CTP2) coding sequence from the *ShkG* gene of *A. thaliana* directs the protein to the chloroplast. The transcriptional termination sequence is derived from the T-E9 DNA sequence of pea (*Pisum sativum* L.), containing the 3' untranslated region of the ribulose-1,5-bisphosphate carboxylase/oxygenase (Rubisco) small subunit E9 gene. Initial promotions claim MON89788 varieties yield more because of the transformation technique and event selection process, but there is currently no scientific literature that formally compares negative and positive isolines or explains any yield effect.

In 2008, the United States Department of Agriculture (USDA) deregulated the first molecular HR stack in a soybean event called 356043 (Optimum® GAT®, Pioneer Hi-Bred, Johnston, IA). This stack combines a metabolic mechanism to inactivate glyphosate and a modified ALS for resistance to ALS-inhibiting herbicides. The metabolic mechanism is based on glyphosate acetyltransferase (GAT4601) that catalyzes the acetylation of glyphosate to the inactive N-acetyl glyphosate (NAG). Commercial release is projected for 2013 pending global regulatory approvals and continued field evaluation.

The *gat4601* gene is derived from the sequences of three weakly active N-acetyltransferase isozymes from the soil bacterium *Bacillus licheniformis* (Weigmann) Chester (Castle et al. 2004). To increase the glyphosate acetylation activity, a collection of recombined *gat* genes was expressed in *Escherichia coli* and screened. The vast majority of recombinants were discarded, but recombinants with improved glyphosate acetylation activity were selected to go through iterative rounds of genetic recombinations to further improve activity. This technique for molecular recombination between genes and directional screening to improve enzymatic properties is called gene shuffling (Crameri et al. 1998; Stemmer 1994). Shuffling of the *gat* genes provided an enzyme that conferred high resistance to glyphosate (Siehl et al. 2007).

The 356043 event was produced in "Jack" soybean using particle acceleration transformation (AGBIOS 2009). Expression of *gat4601* is under the control of the SCP1 promoter and the omega 5′ untranslated leader of the tobacco mosaic virus. The SCP1 promoter is a synthetic constitutive promoter containing portions of the CaMV 35S promoter and the Rsyn7-Syn II core synthetic consensus promoter. The termination sequence is from *pin*II, the proteinase inhibitor II gene of *Solanum tuberosum*. Field observations with 356043 have not noted the "yellow flash" symptom that is associated with GTS 40-3-2.

The 356043 stack also expresses a modified soybean *als* gene (*gm-hra*) that gives broad-spectrum resistance to ALS-inhibiting herbicides at the whole-plant level. During tissue culture selection, the GM-HRA trait was used as the selectable marker. The GM-HRA enzyme has two amino acid mutations, tryptophan to leucine at position 574 and proline to alanine at position 197 (in reference to the *A. thaliana* sequence) (Bedbrook et al. 1995). Expression of *gm-hra* with its native CTP (GM-ALS CTP) is under the control of the *S*-adenosyl-L-methionine synthetase (SAMS) promoter from soybean and an intron that interrupts the SAMS 5′ untranslated region (5′-UTR). Termination for *gm-hra* is provided by the native soybean ALS sequence.

4.3.2   GR Corn

The first GR corn, GA21 (Agrisure® GT, Syngenta Seeds, Golden Valley, MN, but initially sold in 1998 as Roundup Ready by Monsanto), used a modified maize *epsps* gene (*zm-2mepsps*) that was insensitive to glyphosate (Spencer et al. 2000). GA21 has a single insertion of a cassette with three complete copies of *zm-2mepsps* in tandem and three incomplete copies. GA21 corn does not express the three incomplete copies. The rice actin I promoter (Os-Act1) controls expression and the termination sequence is a *nos* 3′ transcriptional element from *A. tumefaciens*. The optimized chloroplast transit peptide (OTP) is from DNA sequences isolated from maize and sunflower *Rubisco* genes. Particle acceleration transformation inserted a fragment from the pDPG434 transformation vector into a cell culture of the inbred "AT" with glyphosate used to select the transformed cells. This technology and legal disputes as to its ownership led to a number of mergers and acquisitions (Charles 2001). GA21 is still commercial and is being stacked with glufosinate and insect resistance traits.

A new event, NK603 (Roundup Ready® 2, Monsanto Company), was developed to avoid disputes over ownership of GA21 as well as to improve crop tolerance. NK603 has gene expression copies to enhance expression, particularly in vulnerable meristematic tissues. Particle acceleration was used to transform the inbred line "AW × CW" with two copies of a slightly modified *epsps cp4* gene, one driven by the Os-Act1 promoter and the other by an enhanced CaMV e35S promoter with duplicated enhancer sequences (CaJacob et al. 2007). Both cassettes contain the CTP2 chloroplast transit peptide from the *A. thaliana* EPSPS. The termination sequence is a *nos* 3′ transcriptional element from *A. tumefaciens*. Commercial release of NK603 in a breeding stack with glufosinate and four insect resistance traits occurred in 2010.

GR corn technology continues to evolve with a resistance mechanism based on metabolic inactivation. The event, 98140 (Optimum GAT), was produced by *Agrobacterium*-mediated transformation of a proprietary Pioneer inbred "PHWVZ." The 98140 line expresses the *gat4621* gene that codes for an enhanced glyphosate acetyltransferase that catalyzes the inactivation of glyphosate to NAG (Castle et al. 2004). The *gat4621* gene is derived from the same shuffling process used to produce *gat4601*. The 98140 corn has a molecular stack of *gat4621* with a highly resistant *als* gene from corn, *zm-hra*. The ZM-HRA enzyme has two amino acid mutations, tryptophan to leucine at position 574 and proline to alanine at position 197 (in reference to the *A. thaliana* sequence) (Bedbrook et al. 1995). A corn ubiquitin (ubiZM1) promoter drives expression of *gat4621* with a 5′-UTR and an intron, and with transcription terminated by the proteinase inhibitor II (*pin*II) from *S. tuberosum*. A corn ALS promoter drives expression of *zm-hra* and its native CTP (ZM-ALS CTP) with transcription terminated by the *pin*II from *S. tuberosum*. The vector, PHP24279, also has three copies of the CaMV 35S enhancer region that contribute to the constitutive expression of *gat4621* and *zm-hra*. This molecular stack of glyphosate and ALS herbicide resistance does not change any natural tolerance mechanisms to selective corn herbicides, and thus creates more options to combine herbicide modes of action to delay the evolution of HR weeds (Green et al. 2009).

4.3.3 GR Cotton

Cotton growers needed an effective postemergence herbicide when GR cotton MON1445 (Roundup Ready) was commercialized in 1997, enabling glyphosate as a new over-the-top option. The MON1445 trait was inserted into "Coker 312" with *Agrobacterium*-mediated transformation (AGBIOS 2009). The transformation vector contained two glyphosate resistance genes, *epsps cp4* and *gox*, a gene that encodes GOX to catalyze the degradation of glyphosate to AMPA and glyoxylate. However, the *gox* gene did not integrate into the cotton genome. The *epsps cp4* sequence from *A. tumefaciens* strain CP4 was modified slightly to improve expression, but the modification did not change the amino acid sequence. Expression is controlled by a caulimovirus (CMoVb) promoter from a modified FMV, and the transcriptional termination sequence was derived from the T-E9 DNA sequence of pea containing the 3′ nontranslated region of the Rubisco small subunit E9 gene. The chloroplast transit peptide is CTP2 from *A. thaliana*. Antibiotic resistance from the neomycin phosphotransferase II gene was the selectable marker and the gene is present and expressed in the commercial event. A second antibiotic marker gene, *aad*, used for bacterial selection is also present but not expressed. Tolerance to glyphosate was marginally acceptable. Over-the-top glyphosate applications after the four-leaf stage occasionally caused fruit abortion and yield loss because of insufficient expression of the *epsps cp4* gene in developing pollen and tapetum (Dill et al. 2008). However, growers valued the early weed control and rapidly adopted the technology.

The second GR cotton, MON88913 (Roundup Ready® Flex, Monsanto Company), was commercialized in 2006 to improve crop tolerance and allow growers the flexibility to apply glyphosate at later growth stages. MON88913 contains two *cp4 epsps* genes with chimeric promoters that more strongly express the trait in the 4- to 12-leaf vegetative stages and in the sensitive reproductive tissue (CaJacob et al. 2007). The genes from the transformation vector pV-GHGT35 with two tandem *cp4 epsps* gene cassettes were inserted into the variety "Coker 312" using *Agrobacterium*-mediated transformation (AGBIOS 2009). In the first cassette, the FMV e35S/TSF1 chimeric promoter regulates gene expression. In the second cassette, CaMV 35S/ACT8, a chimeric promoter from the actin *act8* gene of *A. thaliana*, and the CaMV 35S promoter, regulate the expression of the *cp4 epsps* gene. The transcriptional termination sequence was derived from the T-E9 DNA sequence of pea, containing the 3′ nontranslated region of the Rubisco small subunit E9 gene. MON88913 made managing weeds easier and cotton growers rapidly transitioned to it (Dill et al. 2008).

A more recent GR cotton event, GHB614 (Glytol®, Bayer CropScience, Research Triangle Park, NC), uses a modified maize *epsps* gene, *zm-2mepsps* (CFIA 2008; Trolinder et al. 2008). The ZM-2mEPSPS enzyme differs from the naturally occurring maize EPSPS by two amino acids and is the same EPSPS as in GA21 corn. The gene was inserted into "Coker 312" using *Agrobacterium*-mediated transformation. The event has a single copy of the *zm-2mepsps* gene and a constitutive promoter of the *histone H4* gene from *A. thaliana* (Ph4a748At) (Chaboute et al. 1987) and an optimized chloroplast transit peptide derived from genes of corn and sunflower (TPotpC) (Lebrun et al. 1996). Limited commercial sales in the Southwestern region of the United States began in 2009.

4.3.4   GR Canola

In some ways, canola was an ideal crop for herbicide resistance. Weed management options were limited and expensive and growers needed to reduce tillage and conserve water. Young canola plants are not very competitive and weeds substantially reduce yields. In addition, grain buyers penalize growers for weed contamination that reduces grain and oil quality by increasing erucic acid and glucosinolate levels (Devine and Buth 2001). Canadian and U.S. growers rapidly accepted GR canola, GT73 (Roundup Ready), when it was launched in 1996.

GT73 canola has two resistance mechanisms to protect against glyphosate, the *epsps cp4* gene from *A. tumefaciens* and a *gox* gene, *goxv247*, from strain LBAA of the bacterium *Ochrobactrum anthropi* (AGBIOS 2009). Putting a target-site system to withstand glyphosate before it is inactivated with a metabolic system to eliminate any residual glyphosate within the plant conceptually is logical. Indeed, the resistance mechanisms together confer commercially acceptable crop safety, whereas either alone does not (Gressel 2008). The *goxv247* gene produces a modified GOX enzyme that catalyzes the degradation of glyphosate to AMPA and glyoxylate. A 35S promoter from a modified virus (FMV) drives constitutive expression of both genes. Both use the transcriptional termination sequence derived from the T-E9 DNA sequence of pea, containing the 3′ nontranslated region of the Rubisco small subunit E9 gene. Transit peptides, CTP2 and CTP1, from the small subunit of the Rubisco gene of *A. thaliana* and the *epsps* gene of *A. thaliana*, target both genes to chloroplasts. *Agrobacterium*-mediated transformation inserted the genes into the cultivar "Westar" (AGBIOS 2009).

An important issue to consider for canola is its ability to transfer transgenes via pollen to conventional canola and related weed species such as wild mustard (*Sinapis arvensis* L.), India mustard (*Brassica juncea* (L.) Czern.), Ethiopian mustard (*Brassica carinata* A. Braun), black mustard (*Brassica nigra* (L.) W.D.J. Kock), annual wallrocket (*Diplotaxis muralis* (L.) DC.), wild radish (*Raphanus raphanistrum* L.), and common dogmustard (*Erucastrum gallicum* (Willd.) O.E. Schulz). The frequency of outcrossing to weed species is low and any hybrids are usually less fit, but at least one canola transgene has become established in a stable wild population (Warwick et al. 2008).

4.3.5 New EPSPS Resistance Mechanisms

Research continues to discover novel insensitive EPSPS enzymes. A new class of EPSPS has high resistance to glyphosate (high K_i) and maintains affinity (K_m) to the natural substrate phosphoenolpyruvate (PEP) (Vande Berg et al. 2008). The *epsps* gene, *aroA$_{1398}$*, was cloned from a bacterial strain designated ATX1398. The primary sequence for the AroA$_{1398}$ protein has only 22% homology with the CP4 EPSPS protein while exhibiting 800-fold more resistance to glyphosate at the enzyme level than corn EPSPS. Scientists at Athenix Corporation (Research Park Triangle, NC) used *Agrobacterium*-mediated transformation to create resistant corn plants. The expression cassette contains a novel constitutive promoter from eastern gamagrass (*Tripsacum dactyloides* (L.) L.) and a chloroplast transit peptide.

Another example of an insensitive EPSPS employed directed evolution to improve an EPSPS isolated from *Enterobacter aerogenes* (Heinrichs et al. 2007). The mutant EPSPS was expressed in *E. coli* and screened. The enzymatic activity and resistance to glyphosate of purified EPSPS variants were determined *in vitro*. After two iterations, some enzymes had a 32-fold higher K_i value for glyphosate and a 105-fold higher K_i/K_m ratio. Further mutagenesis improved enzyme solubility and stability. Corn expressing this gene is reported to be resistant to glyphosate.

The most common EPSPS mutation in GR weeds is proline to leucine at position 106 (P106L) (Perez-Jones et al. 2007). Zhou et al. (2006) created this mutation in a rice *epsps* gene by directed mutagenesis. In *E. coli*, the P106L mutation decreased affinity to glyphosate about 70-fold and affinity to PEP 4.6-fold. Resistance to glyphosate was observed when this gene was expressed in tobacco (*Nicotiana tabacum* L.). It is not clear that any of these new *epsps* genes will be able to displace *cp4 epsps* as the market leader, but they do provide a means for other companies to enter the glyphosate resistance business.

4.4   OTHER MECHANISMS THAT CAN BE STACKED WITH GLYPHOSATE RESISTANCE

GR crops have been very successful for more than a decade, but the evolution of GR weeds was faster than many expected and the lack of any new selective herbicides with a new mode of action are forcing HR crops to step forward again. Sequentially using single herbicide technologies until they are no longer effective is the fastest way to evolve multiple HR weeds (Nichols et al. 2008; Powles 2008b). Growers must periodically change their weed management practices and use combinations of cultural, mechanical, biological, and chemical methods to sustain the utility of existing technologies (Powles 2008a). The next wave of technology will combine glyphosate resistance with resistance to other herbicides in crops to give growers more flexibility to choose herbicide options that combine different modes of action and improve foliar and soil residual activity (Green et al. 2008). Multiple HR crops will extend the life span of glyphosate use. Fortunately, a number of herbicide resistance traits have already been characterized and can be developed as needed for resistant weed management (Table 4.3).

4.4.1   Glufosinate Resistance

Glufosinate, also known as phosphinothricin, is a broad-spectrum herbicide that controls important weeds such as morningglories (*Ipomoea* spp.), hemp sesbania (*Sesbania herbacea* (P. Mill.) McVaugh), Pennsylvania smartweed (*Polygonum pensylvanicum* L.), and yellow nutsedge (*Cyperus esculentus* L.) better than glyphosate. There are currently no weeds known to be glufosinate resistant. Glufosinate inhibits glutamine synthetase (GS; EC 6.3.1.2), an enzyme that catalyzes the conversion of glutamate plus ammonium to glutamine as part of nitrogen metabolism (Senseman 2007). Glufosinate behaves like a contact herbicide; it needs to be applied to younger plants than glyphosate and is not as effective on perennials that require extensive translocation for complete control.

TABLE 4.3. Nonglyphosate Herbicide Resistance Traits That Have Not Been Commercialized

Herbicide/Herbicide Class	Characteristics	References
2,4-Dichloroacetic acid	Microbial degradation enzyme	Streber and Willmitzer (1989)
Aryloxyphenoxypropionate ACCase inhibitor and phenoxy acid (auxin)	Microbial, aryloxyalkanoate dioxygenase	Wright et al. (2005)
Asulam	Microbial dihydropteroate synthase	Surov et al. (1998)
Dalapon	Microbial degradation enzyme	Buchanan-Wollaston et al. (1992)
Dicamba	*Pseudomonas maltophilia*, O-demethylase	Herman et al. (2005)
4-Hydroxyphenylpyruvate dioxidase (HPPD) inhibitors	Over-expression, alternate pathway, and increasing flux of pathway	Matringe et al. (2005)
Phenylurea	P450, *Glycine max*, and *Helianthus tuberosus*	Siminszky et al. (1999); Didierjean et al. 2002
Paraquat	Chloroplast superoxide dismutase	Sen Gupta et al. (1993)
Phenmedipham	Microbial degradation enzyme	Streber et al. (1994)
Phytoene desaturase (PDS) inhibitors	Resistant microbial and *Hydrilla verticillata* PDS	Arias et al. (2005)
Protoporphyrinogen oxidase (PPO) inhibitors	Resistant microbial and *Arabidopsis thaliana* PPO	Li and Nicholl (2005)
Multiple herbicides	Glutathione *S*-transferase, *Escherichia coli*	Skipsey et al. (2005)
	P450, *Zea mays*	Dam et al. (2007)

Glufosinate-resistant crops (LibertyLink®, Bayer CropScience) were developed in parallel with GR crops. Besides allowing glufosinate applications for weed control, glufosinate resistance is also useful as a selectable marker when molecularly stacked with other transgenes. Consequently, glufosinate resistance is fortuitously present in many insect-resistant crops. Glufosinate resistance is due to metabolic inactivation by an acetyltransferase enzyme that catalyzes the acetylation of glufosinate. Two glufosinate resistance genes, *bar* and *pat*, encode homologous enzymes (Herouet et al. 2005). Both genes were isolated from soil microorganisms, *pat* from *Streptomyces viridochromogenes* and *bar* from *Streptomyces hygroscopicus*. In 1995, canola became the first commercial crop resistant to glufosinate. Glufosinate resistance in canola was launched before glyphosate resistance and has been widely adopted.

Sales of glufosinate-resistant cotton with the *bar* gene and the CaMV 35S promoter started in 2004. The *bar* gene was integrated into "Coker 312" with *Agrobacterium*-mediated transformation (AGBIOS 2009). Sales of glufosinate-resistant corn started in 1996 in a stack with a Bt insect resistance trait and in 1997 as a stand-alone trait. Commercial corn lines use both the *bar* and *pat* genes in the several different events that have been deregulated. Certainly, the use of glufosinate resistance as the selectable marker for insect resistance traits has helped it become widely available in cotton and corn. Grower adoption and stacking with glyphosate resistance will likely increase as GR weeds become more problematic.

Sales of glufosinate-resistant soybeans with the *pat* gene and the CaMV 35S promoter started in 2009. This technology will enable the use in soybean of an urgently needed new mode of action with no known resistant weeds. When glufosinate is applied at the proper time, it will help growers control tough weed species such as common waterhemp (*Amaranthus rudis* Sauer) and Palmer amaranth (*Amaranthus palmeri* S. Watson) that are rapidly evolving resistance to all other currently available herbicide options. The value of glufosinate and glufosinate-resistant crops will grow as the number of multiple HR weeds increases (Legleiter and Bradley 2008; Powles 2009).

4.4.2   ALS-Inhibiting Herbicide Resistance

ALS, also known as acetohydroxyacid synthase, has two substrates, 2-ketobutyrate and pyruvate, and is required for the production of the branched-chain amino acids valine, leucine, and isoleucine (Senseman 2007). ALS is a nuclear-encoded enzyme that moves to the chloroplast with the help of a chloroplast transit peptide. Because the chemistry of ALS-inhibiting herbicides is very diverse, many crops are naturally tolerant to ALS herbicides. In addition, a number of permutations in the amino acid sequence of ALS give resistance while still allowing the enzyme to retain its functionality. These mutations can be broadly grouped into three phenotypes: broad cross-resistance to sulfonylureas, imidazolinones, triazolopyrimidines, pyrimidinylthiobenzoates, and sulfonylamino-carbonyl-triazolinones; resistance only to imidazolinones and pyrimidinylthiobenzoates; or resistance only to sulfonylureas and triazolopyrimidines (Duggleby et al. 2008; Tranel and Wright 2002). A highly resistant *als* gene (*hra*) with two mutations, tryptophan to leucine at position 574 and proline to alanine at position 197, gives resistance to all commercial classes of ALS-inhibiting herbicides (Lee et al. 1988). As previously discussed, two *hra* transgenes are under development in 98140 corn and 356043 soybean.

Other ALS-resistant traits are also under development. Scientists at BASF Ag Products (Research Triangle Park, NC) and the Brazilian research institute Embrapa (Brasilia, DF, Brazil) are codeveloping a transgenic ALS-resistant trait in soybean and sugarcane (*Saccharum officinarum* L.). Commercialization in soybean is projected for 2011. BASF is also collaborating with Cibus LLC (San Diego, CA), a plant science company, to produce other ALS-resistant crops with directed mutagenesis, a technique that does not insert any foreign genes, and thus may avoid significant political and regulatory hurdles. Commercialization is projected in canola by 2013.

4.4.3 4-hydroxyphenylpyruvate dioxygenase (HPPD)-Inhibiting Herbicide Resistance

HPPD (EC 1.13.11.27) converts 4-hydroxyphenyl pyruvate to homogentisate, a key step in plastoquinone biosynthesis, and its inhibition causes bleaching symptoms on new growth (Matringe et al. 2005). These symptoms result from an indirect inhibition of carotenoid synthesis due to the involvement of plastoquinone as cofactor of phytoene desaturase (PDS; EC 1.14.99.-). Tissue damage is slower to appear on older tissue as it depends on carotenoid turnover (Senseman 2007). In many ways, HPPD-inhibiting herbicides are ideal to complement glyphosate. They have soil residual activity, control key weeds, and no weeds have evolved resistance yet. Crops resistant to HPPD herbicides have been in field tests since 1999 (APHIS 2009). Bayer CropScience recently announced a collaboration with Mertec LLC (Adel, IA) and M.S. Technologies LLC (West Point, IA) to develop soybeans that are resistant to three herbicides, glyphosate, glufosinate, and HPPD inhibitors, particularly isoxaflutole (Stuebler et al. 2008). Syngenta (Basel, Switzerland) has also announced a native trait HPPD-resistant soybean product projected to be available in 2012 followed by a transgenic trait. Metabolic deactivation systems that give resistance to HPPD herbicides are discussed later.

4.4.4 Auxin Resistance

Auxin herbicides mimic the natural plant growth hormone, indole-3-acetic acid, and disrupt growth and development processes that can eventually cause plant death, particularly in broadleaf species (Senseman 2007). Auxin herbicides have been used for over 60 years (Marth and Mitchell 1944), but their mode of action has not been understood until recently (Hayashi et al. 2008; Kelley and Riechers 2007). Imparting resistance to auxin herbicides would be useful as auxin herbicides control a wide spectrum broadleaf weeds, including most known GR broadleaf weeds. Because auxin herbicides act at multiple receptors, making crops resistant by modifying the site of auxin action may be difficult. In addition, these sites of action respond differently to different auxin herbicide classes, for example, phenoxyacetates (e.g., 2,4-D), pyridinyloxyacetates (e.g., fluroxypyr), benzoates (e.g., dicamba), picolinates (e.g., picloram), and quinolinecarboxylates (e.g., quinclorac) (Walsh 2008). The diversity of auxin receptors makes metabolic inactivation a more attractive approach to develop resistant crops. At present, researchers have identified three different metabolic systems to inactivate auxin herbicides. Each is specific, only inactivating certain herbicide classes.

4.4.4.1 Dicamba Resistance Dicamba is a widely used auxin herbicide that controls important broadleaf weeds in corn (Senseman 2007). Corn is naturally tolerant to dicamba, but soybeans and cotton are sensitive and scientists have long sought a resistance transgene for these crops to expand weed control options (Subramanian et al. 1997). In 2003, a gene encoding a deactivation enzyme dicamba monooxygenase (DMO) cloned from a soil bacterium, *Stenotrophomonas maltophilla*, was used to make resistant plants (Behrens et al. 2007). The DMO enzyme encodes a Rieske nonheme monooxygenase that deactivates dicamba to 3,6-dichlorosalicylic acid (DCSA). The complete bacterial dicamba O-demethylase complex consists of the monooxygenase, a reductase, and a ferredoxin. Electrons are shuttled from reduced nicotinamide adenine dinucleotide (NADH) through the reductase to the ferredoxin, and finally to the terminal component DMO. The ferredoxin component is similar to that found in chloroplasts. DMO is effective when expressed from the nucleus with a CTP. Expression is better with a CTP and best when present in the chloroplast genome where the monooxygenase would have a steady stream of electrons from reduced ferredoxin produced by photosynthesis and where transgenic proteins can often be expressed at higher levels.

Monsanto is pursuing the commercialization of dicamba resistance stacked with glyphosate resistance in soybean and cotton. Dicamba-resistant plants may also have potential to increase tolerance to 2,4-D drift (Feng and Brinker 2007) and other plant stresses (Bhatti et al. 2008).

4.4.4.2 2,4-D Resistance The family of *tfdA* genes from many bacterial species is well known to produce proteins that degrade 2,4-D and provide some auxin herbicide resistance in transgenic plants (Lyon et al. 1993; Streber and Willmitzer 1989). More recently, Laurent et al. (2000) reported isolating a *tfdA* gene from the bacterium *Alcaligenes eutrophus* that coded for an oxygenase that catalyzed the degradation of 2,4-D to 2,4-dichlorophenol (2,4-DCP). This transgene conferred resistance to 2,4-D when expressed in cotton. Skirvin et al. (2007) independently discovered this gene in bacteria isolated from soil that had been exposed to 2,4-D. Grapes (*Vitis vinifera* L.) expressing this gene were reported to withstand an application of 2,4-D at $0.5\,kg\,ha^{-1}$, 100 times the rate that normally kills grape. These resistance traits could have utility stacked with glyphosate resistance both as a mechanism to enable 2,4-D application to control key GR weeds and as a protection mechanism from its nontarget spray drift or volatility.

A new family of genes, aryloxyalkanoate dioxygenase (*aad*), with low homology to the *tfdA* genes, provides resistance to certain auxin and acetyl coenzyme A carboxylase (ACCase; EC 6.4.1.2) inhibiting herbicide chemical classes (Müller et al. 2006; Schleinitz et al., 2004; Wright et al. 2005, 2007). ACCase is the first step of fatty acid synthesis and catalyzes the adenosine triphosphate (ATP)-dependent carboxylation to form acetyl coenzyme A from malonyl coenzyme A in the cytoplasm, chloroplasts, mitochondria, and peroxisomes (Vila-Aiub et al. 2007). The *aad-1* gene isolated from a gram-negative soil bacteria, *Sphingobium herbicidovorans*, codes for a Fe(II) and 2-ketoglutarate-dependent dioxygenase that degrades the alkanoate side chains of both 2,4-D and members of the aryloxyphenoxypropionate (AOPP) class of ACCase inhibitors, such as diclofop, to a hydroxyl (Wright et al. 2005). Another gene sequence called *aad-12*, isolated from *Delftia acidovorans*, codes for a 2-ketoglutarate-dependent dioxygenase that inactivates phenoxyacetate auxins such as 2,4-D and pyridinyloxyacetate auxins such as triclopyr or fluroxypyr, but not commercial AOPPs (Wright et al. 2007). The 2,4-D resistance traits under development by Dow AgroSciences (Indianapolis, IN) are coded DHT1 for corn and DHT2 for soybean and are being stacked with glyphosate and glufosinate resistance traits (Simpson et al. 2008).

4.4.5 P450 Metabolic Resistance

Cytochrome P450 monooxygenases (P450) inactivate a wide range of herbicides, and thus have potential to stack with glyphosate resistance. Plants have a diverse array of P450 enzymes, with functions that include the biosynthesis of lignins, ultraviolet protectants, pigments, defense compounds, fatty acids, hormones, and signaling molecules, and the degradation of internally and externally produced toxic compounds by catalyzing ring hydroxylation, epoxidation, sulfoxidation, dealkylation, or alkyl oxidation reactions. Herbicides that native P450 enzymes metabolically inactivate include acetanilides, bentazon, dicamba, some ALS-inhibiting herbicides, isoxazoles, and urea herbicides (Barrett et al. 1997). P450 transgenes increase resistance to a similar range of herbicides (Dam et al. 2007; Didierjean et al. 2002). The chemical specificity of the P450s and other native metabolic systems, which do not inactivate all herbicides within a mode of action, offers the potential to allow growers to use different herbicides in the same mode of action class to control weeds in one season and any feral crop in the next season.

4.4.6 Nontransgenic Multiple Herbicide Resistance

Nontransgenic HR crops are still being developed. For example, sorghum (*Sorghum bicolor* (L.) Moench) growers currently have few weed management options and need help to control the worst weed problem, shattercane. Shattercane is the same species as sorghum and impossible to control with selective herbicides. However, the sorghum market is not large enough to support the high costs of transgene development. Since shattercane and sorghum can readily interbreed, researchers used weedy populations as the source of resistance to ALS-inhibiting and ACCase-inhibiting herbicides (Tuinstra and Al-Khatib 2008a, 2008b) to create a HR crop. Since the traits are not transgenic, most countries will not require regulatory approval and the technology can be commercialized rapidly. Ironically, transgenes may still be useful with nontransgenic traits by helping prevent gene flow back to weedy populations. Linking standard crop traits such as nonshattering, short stature, uniform flowering, and germination to the herbicide resistance traits would significantly reduce the fitness of any feral escapes or weedy hybrids formed (Gressel 2008).

4.4.7   Using Traits Sequentially

If HR traits cannot be stacked together because of any biological, regulatory, or business reasons, growers can still obtain most of the benefit if they use the traits singly in a controlled rotation. For example, the technology exists to make wheat resistant to herbicides with three different modes of action. Each resistance mechanism gives growers an option to control their weeds with a different mode of action. In a 3-year trait rotation, even continuous wheat growers would be able to diversify their weed management practices and obtain environmental benefits such as less tillage and energy use, reduced soil erosion, improved water infiltration and soil structure, and increased habitat diversity (Friedman 2008). However, only imidazolinone-resistant wheat is commercially available and using imidazolinone herbicides alone is not sustainable because of the prevalence of ALS-resistant weeds. The traits for resistance to glyphosate and glufosinate remain undeveloped because of opposition to GM wheat. Growers need all three herbicide resistance systems now to get the full economic and environmental benefits (Cook 2000; Jacquemin et al. 2009).

4.5   SUMMARY

The development of transgenic GR crops required a large amount of resources to navigate through the complex scientific, regulatory, legal, and business issues. Growers did not like the "grower contract" associated with GR crops and losing the freedom to replant the seed that they harvested, but they still signed the contract and rapidly adopted the technology (Charles 2001). The initial GR crops transformed how many growers managed weeds. However, many growers overused the technology by planting only GR crops and using only glyphosate to control highly variable and prolific weeds over wide areas. Such overuse of glyphosate made the evolution of resistant weeds inevitable and now glyphosate alone is not effective in many areas. These growers urgently need new weed management technology. To answer this need, industry is developing a new generation of HR crops with resistance to glyphosate and herbicides with other modes of action. Transitioning from single to multiple HR crops will give growers new weed management options with existing herbicide technology and help extend the utility of glyphosate and GR crops.

REFERENCES

AGBIOS. 2009. Agbios GM Database. http://www.agbios.com/dbase.php (accessed January 8, 2009).

Alfonso, M., J. J. Pueyo, K. Gaddour, A. L. Etienne, D. Kirilovsky, and R. Picorel. 1996. Induced new mutation of D1 serine-268 in soybean photosynthetic cell cultures produced atrazine resistance, increase stability of S2QB- and S3QB-state, and increased sensitivity to light stress. *Plant Physiology* 112:1499–1508.

Al-Khatib, K. and J. F. Miller. 2000. Registration of four genetic stocks of sunflower resistant to imidazolinone herbicides. *Crop Science* 40:869–870.

Anderson, P. C. and M. Georgeson. 1989. Herbicide-tolerant mutants of corn. *Genome* 34:994–999.

Animal and Plant Health Inspection Service (APHIS). 2009. Status of Permits, Notifications, and Petitions. http://www.aphis.usda.gov/biotechnology/status.shtml (accessed March 5, 2009).

Arias, R. S., M. D. Netherland, P. Atul, and F. D. Dayan. 2005. Biology and molecular evolution of resistance to phytoene desaturase inhibitors in *Hydrilla verticillata* and its potential use to produce herbicide-resistant crops. *Pest Management Science* 61:258–268.

Barrett, M., N. Polge, R. Baerg, L. D. Bradshaw, and C. Poneleit. 1997. Role of cyto-chrome P450s in herbicide metabolism and selectivity and multiple herbicide metabolizing cytochrome P450 activities in maize. In K. K. Hatzios, ed. *Regulation of Enzymatic Systems Detoxifying Xenobiotics in Plants*. Dordrecht, the Netherlands: Kluwer Academic, pp. 35–50.

Barry G., G. Kishore, S. Padgette, M. Taylor, K. Kolacz, M. Weldon, D. Re, D. Eichholtz, D. Fincher, and L. Hallas. 1992. Inhibitors of amino acid biosynthesis: strategies for imparting glyphosate tolerance to crop plants. In B. K. Singh, H. E. Flores, and J. C. Shannon, eds. *Biosynthesis and Molecular Regulation of Amino Acids in Plants*. Rockville, MD: American Society of Plant Physiologists, pp. 139–145.

Bedbrook, J. R., R. S. Chaleff, S. C. Falco, B. J. Mazur, C. R. Somerville, and N. S. Yadav, inventors; E. I. Du Pont de Nemours and Company, assignee. 1995. Nucleic acid fragment encoding herbicide resistant plant acetolactate synthase. U.S. Patent 5,378,824, filed January 3, 1995.

Behrens, M. R., N. Mutlu, S. Cjairabprtu, W. Jiang, B. J. LaVallee, P. L. Herman, T. E. Clemente, and D. P. Weeks. 2007. Dicamba resistance: enlarging and preserving biotechnology-based weed management strategies. *Science* 316:1185–1188.

Bernards, M. L., K. D. Thelen, D. Penner, R. B. Muthukumaran, and J. L. McCraken. 2005. Glyphosate interaction with manganese in tank mixtures and its effect on glyphosate absorption and translocation. *Weed Science* 53:787–794.

Bhatti, M., P. C. C. Feng, J. Pitkin, and S.-W. Hoi, inventors; Monsanto Technology LLC, assignee. 2008. Methods for improving dicamba monooxygenase transgene encoding plant resistance to stress and disease by application of dicamba herbicides and its metabolites. World Intellectual Property Organization Patent Application WO/2008/048964 A2, 1-50, filed April 24, 2008.

Bott, S., T. Tesfamariam, H. Candan, I. Cakmak, V. Römheld, and G. Neumann. 2008. Glyphosate-induced impairment of plant growth and micronutrient status in glyphosate-resistant soybean (*Glycine max* L.). *Plant and Soil* 312:185–194.

Bright, S. W. J. 1992. Herbicide-resistant crops. *Current Topics in Plant Physiology* 7:184–194.

Buchanan-Wollaston, V., A. Naser, and F. C. Cannon. 1992. A plant selectable marker gene based on the detoxification of the herbicide Dalapon. *Plant Cell Reports* 11:627–631.

Burnside, O. C. 1996. An agriculturalist's viewpoint of risks and benefits of herbicide-resistant cultivars. In S. O. Duke, ed. *Herbicide-Resistant Crops Agricultural, Environmental, Economic, Regulatory, and Technical Aspects*. Boca Raton, FL: CRC Press, pp. 391–406.

CaJacob, C. A., P. C. C. Feng, S. E. Reiser, and S. R. Padgette. 2007. Genetically modified herbicide-resistant crops. In W. Krämer and U. Schirmer, eds. *Modern Crop Protection Chemicals*, Vol. 1. Weinheim, Germany: Wiley-VCH, pp. 283–302.

Canadian Food Inspection Agency (CFIA). 2008. Canadian Food Inspection Agency Notices of Submission. Notices of Submission for Approval of Novel Food and Livestock Use That Includes an Environmental Safety Assessment of Cotton Genetically Modified for Herbicide Tolerance from Bayer CropScience Inc. http://www.inspection.gc.ca/english/plaveg/bio/dd/dd0872e.shtml (accessed March 17, 2008).

Castle, L. A., D. L. Siehl, R. Groton, P. A. Patten, Y. H. Chen, S. Bertain, H. J. Cho, N. Duck, J. Wong, D. Liu, and M. W. Lassner. 2004. Discovery and directed evolution of a glyphosate tolerance gene. *Science* 304:1151–1154.

Castle, L. A., G. Wu, and D. McElroy. 2006. Agricultural input traits: past, present and future. *Current Opinion in Biotechnology* 17:105–112.

Chaboute, M., N. Chaubet, G. Philipps, M. Ehling, and C. Gigot. 1987. Genomic organization and nucleotide sequences of two histone H3 and two histone H4 genes from *Arabidopsis thaliana*. *Plant Molecular Biology* 8:179–191.

Charles, D. 2001. *Lords of the Harvest: Biotech, Big Money, and the Future of Food*. Cambridge, MA: Perseus Publishing.

Charne, D. G., J. D. Patel, and A. Grombacher, inventors; Pioneer Hi-Bred International, assignee. 2002. *Brassica napus* with early maturity (early napus) and resistance to an AHAS-inhibitor herbicide. U.S. Patent Application 2002/012962 A1, 1–8, filed August 29, 2002.

Cook, R. J. 2000. Science-based risk assessment for the approval and use of plants in agricultural and other environments. In G. J. Persley and M. M. Lantin, eds. *Agricultural Biotechnology and the Poor: Proceedings of an International Conference*. October 21–22, 1999. Washington, DC: Consultative Group on International Agricultural Research (CIGAR), pp. 123–130.

Crameri, A., E. Bermudez, S. Raillard, and W. P. Stemmer. 1998. DNA shuffling of a family of genes from diverse species accelerates directed evolution. *Nature* 391:284–290.

Croughan, T. P., inventor; Board of Supervisors of Louisiana State University and Agricultural and Mechanical College, assignee. 1998. Herbicide resistant rice. U.S. Patent 5,773,704, filed June 30, 1998.

Dam, T., A. D. Guida, C. B. Hazel, B. Li, M. E. Williams, inventors; E. I. Du Pont de Nemours and Company, assignee. 2007. A maize gene for cytochrome P450 conferring resistance to a wide range of herbicide types and its use. U.S. Patent Application 7214515 A1, 1-64, filed March 9, 2007.

Devine, M. D. and J. L. Buth. 2001. Advantages of genetically modified canola: a Canadian perspective. In *Proceedings of the Brighton Crop Protection Conference— Weeds*. Surrey, UK: British Crop Protection Council, pp. 367–372.

Didierjean, L., L. Gondet, R. Perkins, S. M. C. Lau, H. Schaller, D. P. O'Keefe, and D. Werck-Reichart. 2002. Engineering herbicide metabolism in tobacco and *Arabidopsis* with CYP76B1, a cytochrome P450 enzyme from Jerusalem artichoke. *Plant Physiology* 130:179–189.

Dill, G. M., C. A. CaJacob, and S. R. Padgette. 2008. Glyphosate-resistant crops: adoption, use and future considerations. *Pest Management Science* 64:326–331.

Duggleby, R. G., J. A. McCourt, and L. W. Guddat. 2008. Structure and mechanism of inhibition of plant acetohydroxyacid synthase. *Plant Physiology and Biochemistry* 46:309–324.

Duvick, D. N. 1996. Seed company perspectives. In S. O. Duke, ed. *Herbicide-Resistant Crops Agricultural, Environmental, Economic, Regulatory, and Technical Aspects.* Boca Raton, FL: CRC Press, pp. 253–262.

Elmore, G. A., F. W. Roeth, L. A. Nelson, C. A. Shapiro, R. N. Klein, S. Z. Knezevic, and A. R. Martin. 2001. Glyphosate-resistant soybean cultivar yields compared with sister lines. *Agronomy Journal* 93:408–412.

Etine, J., ed. 2005. *Let Them Eat Precaution*. Washington, DC: The AEI Press.

Fedoroff, N. V. and N. M. Brown. 2004. *Mendel in the Kitchen: A Scientist's View of Genetically Modified Foods*. Washington, DC: John Henry Press.

Feng, P. C. C. and R. J. Brinker, inventors; Monsanto Technology LLC, assignee. 2007. Methods for weed control. World Intellectual Property Organization Patent Application WO/2007/1433690 A2, 1-61, filed December 13, 2007.

Franz, J. E., M. K. Mao, and J. A. Sikorski. 1997. *Glyphosate: A Unique Global Pesticide*. Washington, DC: American Chemical Society.

Friedman, T. L. 2008. *Hot, Flat, and Crowded: Why We Need a Green Revolution—And How It Can Renew America*. New York: Farrar, Strauss, and Giroux.

Gertz, J. M., W. K. Vencill, and N. S. Hill. 1999. Tolerance of transgenic soybean (*Glycine max*) to heat stress. In *Proceedings of the Brighton Crop Protection Conference—Weeds*. Surrey, UK: British Crop Protection Council, pp. 835–840.

Green, J. M. 2009. Evolution of glyphosate-resistant crop technology. *Weed Science* 57:108–117.

Green, J. M., T. Hale, M. A. Pagano, J. L. Andreassi II, and S. A. Gutteridge. 2009. Response of 98140 corn with *gat4621* and *hra* transgenes to glyphosate and ALS-inhibiting herbicides. *Weed Science* 57:142–148.

Green, J. M., C. B. Hazel, D. R. Forney, and L. M. Pugh. 2008. New multiple-herbicide crop resistance and formulation technology to augment the utility of glyphosate. *Pest Management Science* 64:332–339.

Green, J. M. and J. F. Ulrich. 1993. Response of maize (*Zea mays* L.) inbreds and hybrids to sulfonylurea herbicides. *Weed Science* 41:508–516.

Gressel, J. 2008. *Genetic Glass Ceilings: Transgenics for Crop Biodiversity*. Baltimore, MD: Johns Hopkins University Press.

Hayashi, K., X. Tan, N. Zheng, T. Hatate, Y. Kimura, S. Kepinski, and H. Nozaki. 2008, Small-molecule agonist and antagonists of F-box protein-substrate interactions in auxin perception and signaling. *Proceedings of the National Academy of Sciences of the United States of America* 105:5632–5637.

Heap, I. M. 2009. The International Survey of Herbicide Resistant Weeds. http://www.weedscience.com (accessed March 5, 2009).

Heinrichs, V., C. T. Bryant, T. K. Hinson, V. Beilinson, V. Samoylov, and Z. Chen. 2007. Improved resistance of evolved 5-enolpyrovoylshikimate 3-phosate synthases (EPSPS) to the herbicide glyphosate. In R. Kazlauskas, S. Lutz, and D. Estell, eds. *Enzyme Engineering XIX*. Brooklyn, NY: Engineering Conferences International, p. 47.

Herman. P. L., M. Behrens, S. Chakraborty, B. M. Chrastil, J. Barycki, and D. P. Weeks. 2005. A three-component dicamba O-demethylase from Pseudomonas maltophilia, Strain DI6: gene isolation, characterization, and heterozygous expression. *Journal of Biological Chemistry* 280:24759–24767.

Herouet, C., D. J. Esdaile, B. A. Mallyon, E. Debruyne, A. Schulz, T. Currier, K. Hendicks, R. J. van der Klis, and D. Rouan. 2005. Safety evaluation of the phosphinothricin acetyltransferase proteins encoded by the pat and bar sequences that confer tolerance to glufosinate-ammonium herbicide in transgenic plants. *Regulatory Toxicology and Pharmacology* 41:134–149.

Hess, F. D. 1996. Herbicide-resistant crops: perspective from a herbicide manufacturer. In S. O. Duke, ed. *Herbicide-Resistant Crops Agricultural, Environmental, Economic, Regulatory, and Technical Aspects*. Boca Raton, FL: CRC Press, pp. 263–270.

Jacquemin, B., J. Gasquez, and X. Reboud. 2009. Modelling binary mixtures of herbicides in populations resistant to one of the components: evaluation of resistance management. *Pest Management Science* 65:113–121.

James, C. 2008. *Global Status of Commercialized Biotech/GM Crops: 2008*. ISAAA Brief No. 39. Ithaca, NY: ISAAA.

Kelley, K. B. and D. E. Riechers. 2007. Recent developments in auxin biology and new opportunities for auxinic herbicide research. *Pesticide Biochemistry and Physiology* 89:1–11.

Lambrecht, B. 2001. *Dinner at the New Gene Café*. New York: St. Martin's Griffin.

Laurent, F., L. Debrauwer, E. Rathahao, and R. Scalla. 2000. 2,4-Dichlorophenoxyacetic acid metabolism in transgenic tolerant cotton (*Gossypium hirsutum*). *Journal of Agricultural and Food Chemistry* 48:5307–5311.

Lebrun, M., B. Leroux, and A. Sailland, inventors; Rhone-Poulenc Agrochimie, assignee. 1996. Chimeric gene for the transformation of plants. U.S. Patent 5,510,471, filed April 23, 1996.

Lee, K. Y., J. Townsend, J. Tepperman, M. Black, C. F. Chui, B. Mazur, P. Dunsmuir, and J. Bedbrook. 1988. The molecular-basis of sulfonylurea resistance in tobacco. *European Molecular Biology Organization Journal* 7:1241–1248.

Legleiter, T. R. and K. W. Bradley. 2008. Glyphosate and multiple herbicide resistance in common waterhemp (*Amaranthus rudis*) populations from Missouri. *Weed Science* 56:582–587.

Li, X. and D. Nicholl. 2005. Development of PPO inhibitor-resistant cultures and crops. *Pest Management Science* 61:277–285.

Lyon, B. R., Y. L. Cousins, D. J. Llewellyn, and E. S. Dennis. 1993. Cotton plants transformed with a bacterial degradation gene are protected from accidental spray drift damage by the herbicide 2,4-dichlorophenoxyacetic acid. *Transgenic Research* 3:162–169.

Marth, P. C. and J. W. Mitchell. 1944. 2,4-Dichlorophenoxyacetic acid as a differential herbicide. *Botanical Gazette* 106:224–232.

Matringe, M., A. Sailland, B. Pelissier, A. Roland, and O. Zind. 2005. p-Hydroxyphenylpyruvate dioxygenase inhibitor-resistant plants. *Pest Management Science* 61:269–276.

Müller, T. A., T. Fleischmann, J. R. van der Meer, and H.-P. E. Kohler. 2006. Purification and characterization of two enantioselective α-ketoglutarate-dependent dioxygenases, RdpA and SdpA, from *Sphingomonas herbicideovarans* MH. *Applied Environmental Microbiology* 72:4853–4861.

Newhouse, K. E., B. K. Singh, D. L. Shaner, and M. A. Stidham. 1991. Mutations in corn (*Zea mays* L.) conferring resistance to imidazolinone herbicides. *Theoretical and Applied Genetics* 83:65–70.

Nichols, R. N., S. Culpepper, C. Main, M. Marshall, T. Mueller, J. Norsworthy, R. Scotts, K. Smith, L. Steckel, A. York. 2008. Glyphosate-resistant populations of *Amaranthus palmeri* prove difficult to control in the southern United States. *Proceedings of the 5th International Weed Science Congress* 1:227.

Padgette, S. R., K. H. Kolac, X. Delannay, D. B. Re, B. J. LaVallee, C. N. Tinius, W. K. Rhodes, Y. I. Otero, G. F. Barry, and D. A. Eicholtz. 1995. Development, identification, and characterization of glyphosate-tolerant soybean line. *Crop Science* 35:1451–1461.

Perez-Jones, A., K. Park, N. Polge, J. Colquhoun, and C. A. Mallory-Smith. 2007. Investigating the mechanisms of glyphosate resistance in *Lolium multiflorum*. *Planta* 226:395–404.

Pinthus, M. J. 2006. The effect of chlormequat seed-parent treatment on the resistance of wheat seedlings to terbutryne and simazine. *Weed Research* 12:241–247.

Powles, S. B. 2008a. Evolved glyphosate-resistant weeds around the world: lessons to be learnt. *Pest Management Science* 64:360–365.

Powles, S. B. 2008b. Evolution in action: glyphosate-resistant weeds threaten world crops. *Outlooks on Pest Management* 19:256–259.

Reddy, K. N., A. M. Rimando, and S. O. Duke. 2004. Aminomethylphosphonic acid, a metabolite of glyphosate, causes injury in glyphosate-treated, glyphosate-resistant soybean. *Journal of Agricultural and Food Chemistry* 52:5139–5143.

Ruse, M. and D. Castle. 2002. *Genetically Modified Foods: Debating Biotechnology*. New York: Prometheus Books.

Sammons, R. D. and M. Tran. 2008. Examining yellow flash in Roundup Ready soybean. *Proceedings of the North Central Weed Science Society* 63:120.

Schleinitz, K. M., S. Kleinsteuber, T. Vallaeys, and W. Babel. 2004. Localization and characterization of two novel genes encoding for stereospecific dioxygenases catalyzing 2(2,4-dichlorophenoxy)propionate cleavage in *Delftia acidovorans* MC1. *Applied Environmental Microbiology* 70:5351–5365.

Sebastian, S. A. and R. S. Chaleff. 1987. Soybean mutants with increased tolerance for sulfonylurea herbicides. *Crop Science* 27:948–952.

Sen Gupta, A., J. L. Heinen, A. S. Holaday, J. J. Burke, and R. D. Allen. 1993. Increased resistance to oxidative stress in transgenic plants that overexpress chloroplast Cu/

Zn superoxide dismutase. *Proceedings of the National Academy of Sciences of the United States of America* 90:1629–1633.

Senseman, S. A., ed. 2007. *Herbicide Handbook*, 9th ed. Lawrence, KS: Weed Science Society of America.

Shaner, D. L., M. Stidham, and B. Singh. 2007. Imidazolinone herbicides. In W. Krämer and U. Schirmer, eds. *Modern Crop Protection Compounds*, Vol. 1. Weinheim, Germany: Wiley-VCH, pp. 82–92.

Siehl, D. L., L. A. Castle, R. Groton, and R. J. Keenan. 2007. The molecular basis of glyphosate resistance by an optimized microbial acetyltransferase. *Journal of Biological Chemistry* 282:1146–1155.

Siminszky, B., F. T. Corbin, E. R. Ward, T. J. Fleischmann, and R. E. Dewey. 1999. Expression of a soybean cytochrome P450 monooxygenase cDNA in yeast and tobacco enhances metabolism of phenylurea herbicides. *Proceedings of the National Academy of Sciences of the United States of America* 96:1750–1755.

Simpson, D. M., T. R. Wright, R. S. Chambers, M. A. Peterson, C. Cui, A. E. Robinson, J. S. Richburg, D. C. Ruen, S. Ferguson, B. E. Maddy. 2008. Introduction to Dow AgroSciences herbicide tolerance traits. *Abstracts of Weed Science Society of America* 48:115.

Skipsey, M., I. Cummins, C. J. Andrews, I. Jepson, and R. Edwards. 2005. Manipulation of plant tolerance to herbicides through coordinated metabolic engineering of a detoxifying glutathione transferase and thiol cosubstrate. *Plant Biotechnology Journal* 3:409–420.

Skirvin, R. M., M. A. Norton, S. K. Farrand, and R. M. S. Mulwa, inventors; University of Illinois, assignee. 2007. Grape plant named "Improved Chancellor." U.S. Patent Application 20070044185.

Spencer, M., R. Mumm, J. Gwyn, inventors; DeKalb Genetic Corporation, assignee. 2000. Glyphosate resistant maize lines. U.S. Patent 6040497, filed March 21, 2000.

Streber, W. R., U. Kutschka, F. Thomas, and H.-D. Pohlenz. 1994. Expression of a bacterial gene in transgenic plants confers resistance to the herbicide phenmedipham. *Plant Molecular Biology* 25:977–987.

Streber, W. R. and L. Willmitzer. 1989. Transgenic tobacco expressing a bacterial detoxifying enzyme are resistant to 2,4-D. *Bio/Technology* 8:811–816.

Stemmer, W. P. 1994. DNA shuffling by random fragmentation and reassembly: in vitro recombination for molecular evolution. *Proceedings of the National Academy of Sciences of the United States of America* 91:10747–10751.

Stuebler, H., H. Kraehmer, M. Hess, A. Schulz, and C. Rosinger. 2008. Global changes in crop production and impact trends in weed management—an industry view. *Proceedings of the 5th International Weed Science Congress* 1:309–319.

Subramaniam, V. and P. E. Hoggard. 1988. Metal complexes of glyphosate. *Journal of Agricultural and Food Chemistry* 36:1326–1329.

Subramanian, M. V., J. Tuckey, B. Patel, and P. J. Jensen. 1997. Engineering dicamba selectivity in crops: a search for appropriate degradative enzyme(s). *Journal of Industrial and Microbiology and Biotechnology* 19:344–349.

Surov, T., D. Aviv, R. Aly, D. M. Joel, T. Goldman-Guez, and J. Gressel. 1998. Generation of transgenic asulam-resistant potatoes to facilitate eradications of parasitic broomrapes (*Orobanche* spp.) with the su gene as the selectable marker. *Theoretical and Applied Genetics* 96:132–137.

Swanson, E. B., M. J. Herrgesell, M. Arnold, D. W. Sippell, and R. S. C. Wong. 1989. Microspore mutagenesis and selection: canola plants with field tolerance to the imidazolinones. *Theoretical and Applied Genetics* 78:535–530.

Tan, S. Y., R. R. Evans, M. L. Dahmer, B. K. Singh, and D. L. Shaner. 2005. Imidazolinone-tolerant crops: history, current status and future. *Pest Management Science* 61:246–257.

Tranel, P. J. and T. R. Wright. 2002. Resistance of weeds to ALS-inhibiting herbicides: what have we learned? *Weed Science* 50:700–712.

Trolinder, L., J. M. Ellis, J. Holloway, and S. Baker. 2008. Glytol cotton—new herbicide tolerant cotton from Bayer CropScience. In *Proceedings of the Beltwide Cotton Production Research Conference*. Nashville, TN: National Cotton Council, pp. 1719–1723.

Tuinstra, M. R. and K. Al-Khatib, inventors; Kansas State University Research Foundation, assignee. 2008a. Acetolactate synthase herbicide resistant sorghum. U.S. Patent Application 2008/0216187 A1, filed June 19, 2008.

Tuinstra, M. R. and K. Al-Khatib, inventors; Kansas State University Research Foundation, assignee. 2008b. Acetyl-CoA carboxylase herbicide resistant sorghum. WO Patent Application 2008/089061 A1, filed July 24, 2008.

Vande Berg, B. J., P. E. Hammer, B. L. Chun, L. C. Schouten, B. Carr, R. Guo, C. Peters, T. K. Hinson, V. Beilinson, A. Shekita, R. Deter, Z. Chen, V. Samoylov, C. T. Bryant, M. E. Stauffer, T. Eberle, D. J. Moellenbeck, N. B. Carozzi, M. G. Koziel, and N. B. Duck. 2008. Characterization and plant expression of a glyphosate-tolerant enolpyruvylshikimate phosphate synthase. *Pest Management Science* 64:340–345.

Vaughn, K. C. 1986. Characterization of triazine-resistant and -susceptible isolines of canola (*Brassica napus*). *Plant Physiology* 82:859–863.

Vila-Aiub, M. M., P. B. Neve, and S. B. Powles. 2007. Resistance cost of a cytochrome P450 herbicide-metabolism but not an ACCase target site mutation in multiple resistant *Lolium rigidum* populations. *New Phytologist* 167:787–796.

Walsh, T. 2008. The molecular mode of action of picolinate auxin herbicides. *Proceedings of the 5th International Weed Science Congress* 1:330.

Warwick, S. I., A. Légère, M.-J. Simard, and T. James. 2008. Do escaped genes persist in nature? The case of an herbicide resistance transgene in a weedy *Brassica rapa* population. *Molecular Ecology* 17:1387–1395.

Wright, T. R., J. M. Lira, D. J. Merlo, N. Hopkins, inventors; Dow AgroSciences, assignee. 2005. Novel herbicide resistance genes. World Intellectual Property Organization Patent WO/2005/107437, filed November 17, 2005.

Wright, T. R., J. M. Lira, T. A. Walsh, D. J. Merlo, P. S. Jayakumar, G. Lin, inventors; Dow AgroSciences, assignee. 2007. Novel herbicide resistance genes. World Intellectual Property Organization Patent WO/2007/053482, filed October 5, 2007.

Wright, T. R. and D. Penner. 1998. Cell selection and inheritance of imidazolinone resistance in sugarbeet (*Beta vulgaris*). *Theoretical and Applied Genetics* 96:612–620.

Zhou, M., H. Xu, X. Wei, Z. Ye, L. Wei, W. Gong, Y. Wang, and Z. Zhu. 2006. Identification of a glyphosate-resistant mutant of rice 5-enolpyruvylshikimate 3-phosate synthase using a directed evolution strategy. *Plant Physiology* 140:184–195.

5

TESTING METHODS FOR GLYPHOSATE RESISTANCE

Dale L. Shaner

5.1 INTRODUCTION

Determining whether or not a weed population is resistant to glyphosate is an important aspect for managing this problem. Testing methods for genetically engineered glyphosate-resistant (GR) crops are relatively straightforward because the mutations are known and the level of resistance is extremely high. However, the methods to test for glyphosate resistance in weed biotypes are not as clear-cut. One of the first considerations for determining glyphosate resistance in weed biotypes is to clearly define the meaning of resistance and to differentiate resistance from tolerance. The level of glyphosate resistance in many weed populations is relatively low. In many cases, there is only a 3- to 15-fold difference between a GR and a glyphosate-susceptible (GS) population (Heap 2008) compared with more than a thousandfold difference with acetolactate synthase (ALS) inhibitors (Heap 2008).

Many factors affect glyphosate efficacy and make it difficult to separate low levels of resistance from limited activity due to other factors. Natural variability within weed populations can have a profound effect on the response to glyphosate. For example, tall waterhemp (*Amaranthus tuberculatus* (Moq.) Ex DC JD Sauer) populations show inherent variability in response to glyphosate (Zelaya and Owen 2005). A study comparing the response of contemporary and historical tall waterhemp accessions to glyphosate showed that approximately 5% of the populations were insensitive to $200\,g\,a.e.\,ha^{-1}$ of

Glyphosate Resistance in Crops and Weeds: History, Development, and Management
Edited by Vijay K. Nandula
Copyright © 2010 John Wiley & Sons, Inc.

glyphosate, although all of the populations were controlled at 870g ha^{-1} (Volenberg et al. 2007). Lambsquarters (*Chenopodium album* L.) (Hite et al. 2008), giant ragweed (*Ambrosia trifida* L.) (Westhoven et al. 2008), field bindweed (*Convolvulus arvensis* L.) (DeGennaro and Weller 1984; Westwood et al. 1997), and pitted morningglory (*Ipomoea lacunosa* L.) (Stephenson et al. 2007) populations have also shown variable response to glyphosate.

Glyphosate's efficacy is also affected by application variables such as formulation, adjuvants, spray volume, and time of application as well as the growth stage, nutrient, and water status of the plants (Ahmadi et al. 1980; Chase and Appleby 1979; De Ruiter and Meinen 1998; McWhorter and Azlin 1978; Mithila et al. 2008). Glyphosate is most effective when applied at low spray volumes with a surfactant (Ramsdale et al. 2003). Divalent cations such as calcium can interact with glyphosate preventing its penetration into the leaf (Hall et al. 2000; Thelen et al. 1995). The addition of ammonium sulfate can overcome this antagonism and is often added to the spray solutions (Thelen et al. 1995). Plants that are nitrogen stressed or drought stressed are more tolerant to glyphosate compared with well-watered and well-fertilized plants (Mithila et al. 2008; Zhou et al. 2008).

For this chapter, glyphosate resistance will be defined as "the evolved capacity of a previously herbicide-susceptible weed population to withstand an herbicide and complete its life cycle when the herbicide is used at its normal rate in an agricultural situation" (Heap 2008). Since there are many factors that can affect glyphosate efficacy, it is difficult to develop simple methods to screen for glyphosate resistance. This chapter will focus on procedures for screening for glyphosate resistance in weed populations by describing the different methods that have been used and will summarize the strengths and weaknesses of each approach.

5.2   TESTING METHODS

5.2.1   Greenhouse and Field Assays

If a weed population is suspected to be GR, the initial characterization requires a detailed dose-response. This test confirms resistance by subjecting the plants to normal field application conditions as closely as possible. The first consideration for a dose–response curve is how to grow the plants and the growth stage at which to treat the plants. The rates used for the dose response will depend on the level of resistance and the inherent sensitivity of the species to glyphosate. Table 5.1 summarizes the rates used in many of the studies on GR weed populations/biotypes. The amount of resistance in the alleged resistant populations/biotypes ranges from a low of threefold to a high of 15-fold (Table 5.1).

Most of these studies were done in the greenhouse. In two of the studies, the populations were tested under field conditions. The LD_{50} (the dose that

TABLE 5.1. Methods and Rate Ranges Used to Screen Whole Plants for Glyphosate Resistance

Species	Location	Test Site	Rate Range (kg ha⁻¹)	Spray Volume (L ha⁻¹)	Growth Stage	Assessment		Resistance Level	Reference
						Method	Time after Treatment (days)		
Lambsquarters	United States	GH	0.21–6.72	230	5-cm tall	FW	30	3-fold	Hite et al. (2008)
Horseweed	United States	GH	0.28–13.4	234	1.5- to 4-cm rosette	Visual	28	8- to 13-fold	VanGessel (2001)
Goosegrass	United States	GH	0.1–13.4	190	10- to 12-leaf	Shoot DW	21	10-fold	Koger et al. (2004)
	Malaysia	Field	0.5–4.25	450	25-cm tall	Visual	28	7- to 11-fold	Lee and Ngim (2000)
Rigid ryegrass	Australia	Outdoor pot	0.05–7.5	139	2- to 3-leaf	Shoot DW	21	7- to 11-fold	Powles et al. (1998)
	Australia	GH	0.38–2.25	225	2- to 3-leaf	Visual	21	9.5-fold	Pratley et al. (1999)
	Australia	GH	0.25–3.95	100	2- to 3-leaf	Shoot DW	21	4-fold	Neve et al. (2004)
	Chile	GH	0.36–4.32	179	Tillering	Shoot FW	21	3- to 6-fold	Perez and Kogan (2003)
Palmer amaranth	United States	Field	1.25–10.00	140	5–13 cm	Visual	28	10-fold	Culpepper et al. (2006)
	United States	GH	0.04–7.2	140	5–13 cm	Shoot FW	20	5-fold	Culpepper et al. (2006)
	United States	GH	0.02–2.24	94	5- to 7-leaf	Visual	28	7-fold	Norsworthy et al. (2008)
Italian ryegrass	United States	GH	0.01–3.37	187	3-leaf	Shoot DW	21	5-fold	Perez-Jones et al. (2005)
Hairy fleabane	Spain	GH	0.04–9	200	12- to 15-leaf	Biomass	23–28	3- to 8-fold	Urbano et al. (2007)
Waterhemp	United States	GH	0.21–2.5	187	NA	Shoot DW	14	2- to 4-fold	Zelaya and Owen (2005)
Wild Poinsettia	Brazil	GH	0.05–1.2	220	6- to 8-leaf	Visual	14–21	3-fold	Vidal et al. (2007)

DW, dry weight; FW, fresh weight; GH, greenhouse; NA, not applicable.

kills 50% of a population) for GR goosegrass (*Eleusine indica* L.) and Palmer amaranth (*Amaranthus palmeri* L.) was 7- to 10-fold higher than their respective susceptible biotypes. In both of these cases, herbicide injury was visually assessed (Culpepper et al. 2006; Lee and Ngim 2000). Interestingly, Culpepper et al. (2006) also evaluated the resistant biotypes in the greenhouse and found a similar level of glyphosate resistance, but there was a big difference in the LD_{50} between the greenhouse and field. In the field, the LD_{50} for the susceptible and resistant biotypes were 0.15 and 1.2 kg ha^{-1}, respectively, whereas in the greenhouse the LD_{50} was 0.09 and 0.56 kg ha^{-1}, respectively. The plants were approximately twofold more tolerant to glyphosate in the field compared with greenhouse-grown plants. While this difference is not unexpected, the results do indicate that care must be taken in extrapolating greenhouse results to the field.

It is important to have controlled and repeatable applications in the greenhouse under conditions as close to the field as possible. The time after treatment and the methods of assessing the activity of glyphosate reported in the literature are quite variable. Glyphosate is a relatively slow-acting herbicide, and phytotoxicity may not be apparent for several weeks after application. Visual assessments are usually taken between 14 and 28 days after treatment (DAT). While effects on shoot fresh or dry weight have also been measured between 14 and 28 DAT, most measurements have been taken at 21 DAT (Table 5.1). If plant biomass is measured too soon after application, the differences between susceptible and resistant biotypes may be masked by the amount of plant material that is present at the time of application. On the other hand, if measured many weeks after application, then the resources needed to assess plant populations may become burdensome and fewer populations can be assessed.

The growth stage of the plants at the time of application can also have a profound effect on the plant's response to the herbicide. In general, plants are treated at a relatively young age (e.g., 2- to 5-leaf stage). However, this early timing may not truly reflect the level of resistance to glyphosate. Shrestha et al. (2007) found that the level of resistance of horseweed (*Conyza canadensis* (L.) Cronq.) is highly influenced by the growth stage of the plant at the time of glyphosate application. They examined two GR biotypes compared with a susceptible biotype. Plants sprayed at the 18- to 21-leaf (rosette) stage exhibited approximately a sixfold difference in the LD_{50} between susceptible and resistant biotypes. Plants sprayed at the 5- to 8-leaf stage were the most susceptible to glyphosate, whereas both susceptible and resistant plants that were bolting required higher rates of the herbicide to be killed. For the susceptible biotypes, 1.12 kg ha^{-1} gave 100% mortality when they were sprayed at the 5- to 8-leaf stage, but this same rate provided less than 40% mortality when the plants were sprayed at late bolting. The resistant biotypes required 4.48 kg ha^{-1} for complete control at the 5- to 8-leaf stage, but there was no mortality at this same rate when the plants were in the late bolting stage. Similar observations have been made for lambsquarters (Schuster et al. 2007), hairy fleabane

(*Conyza bonariensis* (L.) Cronq) (Dinelli et al. 2008), and johnsongrass (*Sorghum halepense* (L.) Pers.) (Vila-Aiub et al. 2007). These results illustrate the need to standardize the growth stages and to use multiple rates in order to clearly determine if a plant population is resistant to glyphosate as well as the necessity of having a susceptible standard. Choosing the wrong growth stage could lead one either to missing a resistant population if the plants sprayed are too young or to wrongly conclude that a plant population is resistant if they are sprayed too late.

5.2.1.1 *Discriminating Dose Screens*

One method to screen large numbers of populations for glyphosate resistance is to use one rate, called a discriminating dose, that will qualitatively separate resistant from susceptible biotypes based on visual assessments. The discriminating dose has to be high enough to kill susceptible biotypes but low enough to allow resistant biotypes to survive, including heterozygous resistant plants. In many cases, the level of resistance in GR biotypes is less than 10-fold (Table 5.1). This makes it difficult to determine the optimal single discriminating dose. The level of expected glyphosate resistance and the inherent sensitivity of the weed species to glyphosate dictate the discriminating dose (Table 5.2). The discriminating dose was 0.24 kg ha^{-1} for hairy fleabane (Urbano et al. 2007), 1.72 kg ha^{-1} for horseweed (Davis et al. 2008), and 3.36 kg ha^{-1} for lambsquarters (Westhoven et al. 2008). Such screens should also include a known resistant and susceptible biotype to ensure that the dose is correct.

The problem with the use of a single rate to identify resistant from the susceptible biotypes is the inherent variability within a weed population to glyphosate. This difference can be as much as twofold (Volenberg et al. 2007), and it is questionable if a twofold difference in sensitivity is really great enough to consider one biotype resistant and another susceptible, particularly if the discriminating dose is less than the normal field rate. The results from multiple studies show that every weed species will require its own dose rate and that there are no rules of thumb that can be applied. In addition, experiments using discriminating dose response should be followed by a more intense procedure with multiple doses on several of the most and least resistant biotypes to clearly show that the discriminating dose truly separated resistant from susceptible biotypes.

5.2.2 Other Whole Plant Screens

Some drawbacks with the use of greenhouse or field screens include (1) access to adequate greenhouse facilities and spraying equipment, (2) space, (3) time, (4) personnel, and (5) weed biology. Most greenhouse screens, particularly for multiple populations, require access to greenhouse facilities and spraying equipment, which are needed to grow the plants and treat them. If many populations are to be treated, then one either has to have plenty of greenhouse space or prolonged access to greenhouse facilities. Because glyphosate

TABLE 5.2. Discriminating Doses Used for Screening for Glyphosate-Resistant Weed Biotypes

Species	Test Site	Rates (kg ha[-1])	Spray Volume (L ha[-1])	Growth Stage	Analysis	Reference
Horseweed	GH	1.72	187	5- to 10-cm rosette	Visual, 21 DAT	Davis et al. (2008)
Lambsquarters	GH	1.68	190	7- to 8-node	Visual, 21 DAT	Westhoven et al.
		4.2			Dead Moderately tolerant Tolerant	(2008)
Giant ragweed	GH	2.5	95	3- to 4-node	Visual, 21 DAT Dead Moderately tolerant Tolerant	Westhoven et al. (2008)
Horseweed	GH	0.8 3.2	190	4- to 8-cm rosette	Shoot FW, 21 DAT	Trainer et al. (2005)
Palmer amaranth	GH	0.87	94	5- to 7-leaf	Visual survival, 28 DAT	Norsworthy et al. (2008)
Hairy fleabane	GH	0.238	200	8- to 9-cm rosette	Shoot FW, DW, 21 DAT	Urbano et al. (2007)

DAT, days after treatment; DW, dry weight; FW, fresh weight; GH, greenhouse.

is relatively slow acting, plants need to grow for a minimum of 14 days (preferably 21–28 days) before they can be adequately evaluated. Finally, the plants to be tested are usually from seed that have to germinate, and this can be a major time consumer if the seed have strong dormancy that has to be broken. All of this requires significant amount of space and time, as well as the personnel to take care of the plants.

Other methods used to screen weed biotypes for glyphosate resistance include using plant cuttings, germinating and growing seedling in Petri plates or 24-cell culture cluster plate, floating excised leaves on glyphosate solutions, and applying glyphosate through the transpiration stream of excised seedlings (Table 5.3). These assays will be discussed in more detail below.

5.2.2.1 Greenhouse Test with Plant Cuttings A "quick test" was developed by Boutsalis (2001) to screen for herbicide resistance in plant material that

TABLE 5.3. Whole Plant Bioassays for Detecting Glyphosate Resistance

Species	Procedure	Rate	Analysis	Reference
Horseweed	Leaf dip	600–4800 mg L⁻¹	Visual injury: 48 HAT	Koger et al. (2005b)
Barley	Petri plate	0.01–4 mM	Coleoptile length: 4 DAT	Escorial et al. (2001)
Soybean	4-h seed soak	0.01–0.05 mM	Growth in pots: 14 DAT	Mann et al. (2004)
Italian ryegrass	Cuttings	225–675 kg ha⁻¹	Visual rating: 21 DAT	Wakelin and Preston (2006)
	Petri plate	10–160 ppm	Shoot length: 8 DAT	Perez and Kogan (2003)
	Petri plate	40–1280 mg L⁻¹	Shoot length: 8 DAT	Michitte et al. (2005)
	Petri plate	12.5–400 mg L⁻¹	Germination: 7 DAT	Perez-Jones et al. (2007)
Cotton	Pollen viability	1.12 kg ha⁻¹	Treated at 4- and 8-leaf and pollen collected three times weekly Image analysis of pollen	Pline et al. (2002)
Chinese foldwing	Detached leaf float	0.2–8 mM	Visual rating: 2.5–24.5 HAT	Yuan et al. (2002)

DAT, days after treatment; HAT, hours after treatment.

was received from fields where there was an herbicide failure. In this test, cuttings are made from plants that are initially grown in the field or the greenhouse and the cuttings are repotted and allowed to reestablish before testing. Wakelin and Preston (2006) used this assay to confirm the resistance of plants that contained a mutation in the 5-enolpyruvylshikimate-3-phosphate synthase (EPSPS) gene. The assay has the advantage of not requiring seed, and by making multiple cuttings from the same mother plant, multiple herbicides or multiple rates of the same herbicide can be applied. In addition, the results from the assay can be obtained within 4 weeks after receiving the material (Boutsalis 2001), which may be soon enough to allow alternative control methods to be used within the same season to manage the resistant population.

5.2.3 Seedling-Petri Plate Assays

Petri plate assays have been widely used to screen large numbers of seeds for glyphosate resistance. In most procedures, seeds are germinated on filter paper that has been wetted with different concentrations of glyphosate and then measuring germination, shoot length, and/or root length 6–8 DAT. Escorial et al. (2001) describe a method to screen cereals for glyphosate tolerance by measuring the effects of glyphosate on coleoptile elongation 4 DAT. Their seedling assay on wheat (*Triticum aestivum* L.) and barley (*Hordeum vulgare* L.) correlated well with plant-sprayed assays. Perez and Kogan (2003) and Michitte et al. (2005) exposed Italian ryegrass (*Lolium perenne* L. ssp. *multiflorum* (Lam.) Husnot) seed populations from different orchards in Chile to a range of glyphosate concentrations (10–160 mg a.e. L^{-1} and 40–1280 mg L^{-1}, respectively) and measured shoot length 8 DAT. Both labs could easily differentiate between the GS and GR biotypes. Additionally, Perez and Kogan (2003) found good agreement in the estimated levels of resistance in the Petri plate assay compared with a greenhouse pot assay and noted that root growth was not as sensitive as shoot growth to glyphosate in their assay. Perez-Jones et al. (2007) used Petri plate assays to screen populations of Italian ryegrass in Oregon for glyphosate resistance by recording the percent germination at 7 DAT. They found good agreement between the Petri plate assay and a whole plant response based on shoot biomass.

Neve et al. (2004) screened rigid ryegrass (*Lolium rigidum* Gaudin) populations from Australia with a seed germination-based assay. They compared a range of volumes and concentrations of aqueous solutions of glyphosate applied to either filter paper or sand and measured germination and shoot and root length at 6 DAT. Sand was a better medium than filter paper and root length of germinated seeds provided the best indication of glyphosate resistance in these experiments, which also had close agreement between the Petri plate assay and whole plant assays.

Zelaya and Owen (2005) used a similar type of assay to screen waterhemp seed for glyphosate resistance. Seeds were placed on paper disks moistened with seven different concentrations of glyphosate within 24-well cell culture cluster plates. Germination and seedling hypocotyl and root length were recorded 14 DAT. The seedling assays were also good estimates of the visual variability in glyphosate resistance assessed in the field.

A variant on the above assays is to imbibe seed in glyphosate solutions, plant the seed, and then assess emergence and growth of the seedlings (Mann et al. 2004). This procedure was used to screen soybean (*Glycine max* Merr.) seed for the *CP4 EPSPS* gene and worked very well in differentiating resistant seed from susceptible seed in mixed populations. However, this procedure has not been reported for screening weed populations.

There are several advantages of the Petri plate assays compared with the field and greenhouse assays. The Petri plate assays, in general, are 2- to 10-fold more rapid than the field or greenhouse assays, and require less maintenance and cost. It may be possible to select a single rate of glyphosate to discriminate between resistant and susceptible populations, although determining this will take time and experimentation. Potential problems with a Petri plate assay are getting consistent seed germination and coping with the inherent variability within a weedy population. Zelaya and Owen (2005) found that there was up to an 18-fold variance in the response of waterhemp populations to glyphosate. This type of variance makes it difficult to assess the level of resistance of a weed population.

5.2.3.1 Excised Leaves Excised leaf assays have been used to screen for glyphosate resistance. Yuan et al. (2002) compared the glyphosate tolerance of two species, Chinese foldwing (*Dicliptera chinensis* (L.) Juss.) and floss flower (*Ageratum houstonianum* P. Mill), by floating excised leaves on 2 mM glyphosate solutions and measuring leaf necrosis at 12.5 and 24.5 hours after treatment (HAT). The sensitive species, floss flower, showed necrosis at 12.5 HAT, whereas the tolerant species, Chinese foldwing, showed no injury at 12.5 HAT and only minor injury at 24.5 HAT.

Koger et al. (2005b) exposed excised leaf tissue from corn (*Zea mays* L.), cotton (*Gossypium hirsutum* L.), soybean, and horseweed to a range of glyphosate concentrations (300–1200 mg a.e. L^{-1}) by immersing the bottom 2.5 cm of corn leaf segments and the petiole along with the bottom one-fourth of the cotton, soybean, and horseweed leaves in 6.8 mL of glyphosate solution contained in a 7-mL vial. The plants were exposed for 72 h and percent leaf injury was assessed. There was a significant difference in the amount of injury due to glyphosate on susceptible cultivars compared with GR cultivars of corn, cotton, and soybeans at all rates, but susceptible corn was much less injured than susceptible cotton or soybeans. The same was true for susceptible versus GR horseweed, although the differences were much greater for field-grown plants compared with greenhouse-grown plants. Greenhouse-grown GR horseweed was more injured by all the rates of glyphosate compared with field-grown plants.

Excised leaf assays have potential to be used as screens for glyphosate resistance, but more studies need to be done to determine their effectiveness. The two assays described above were very rapid (1–3 days) and appeared to be able to differentiate tolerance to glyphosate, but not enough species have been tested to determine the broad utility of this type of assay.

5.2.4 Metabolism-Based Assays

There are other types of assays for screening for glyphosate efficacy that are based on measuring different metabolic processes such as amino acid biosynthesis, photosynthesis, transpiration, and others that could be used to test for glyphosate resistance.

5.2.4.1 Transpiration and Photosynthesis Among the earliest physiological responses to glyphosate treatment are a reduction of transpiration and subsequent inhibition of photosynthesis. A decrease in transpiration and photosynthesis can be measured within hours after treatment in pinto beans (*Phaseolus vulgaris* L.), pea (*Pisum sativum* L.), cocklebur (*Xanthium* *pensylvanicum* L.), sugarbeet (*Beta vulgaris* L.), wheat (*T. aestivum* L.), john-songrass (*S. halepense* L.), and velvetleaf (*Abutilon theorphrasti* medikus) (Brecke and Duke 1980; Fernandez et al. 1994; Ferrell et al. 2003; Fuchs et al. 2002; Gougler and Geiger 1984; Shaner 1978). In velvetleaf, both of these processes continue to decline until they reach zero by 5 DAT (Fuchs et al. 2002). This response is caused by the phytotoxicity of glyphosate because the effects on transpiration in pinto beans can be prevented by supplying plants with tyrosine and phenylalanine along with glyphosate (Shaner and Lyon 1980). Although neither of these parameters has been measured in GR weeds, the rapid effect of glyphosate on transpiration and photosynthesis might be a quick screen for glyphosate resistance.

5.2.4.2 Chlorophyll Biosynthesis Glyphosate treatment causes bleaching and chlorosis in metabolically active sink tissue, such as immature leaves (Franz et al. 1997). This effect has been used to assess the phytotoxicity of glyphosate in soybeans, johnsongrass, yellow and purple nutsedge (*Cyperus esculentus* L. and *Cyperus rotundus* L.), lambsquarters, Florida beggarweed (*Desmodium tortuosum* (Sw.) DC.), barley, and velvetleaf (Abu-Irmaileh and Jordan 1978; Ferrell et al. 2003; Fuchs et al. 2002; Ketel 1996; Kitchen et al. 1981; Sharma and Singh 2001; Villanueva et al. 1985). Chlorosis can be detected within 2 DAT in soybeans, velvetleaf, and sugarbeet (Fuchs et al. 2002; Kitchen et al. 1981). Changes in chlorophyll levels were used by Donahue et al. (1994) to monitor the resistance of genetically modified hybrid poplar (*Populus* spp.). Leaf disks from each line were floated on different concentrations of glyphosate, and chlorophyll was subsequently quantified in these tissues. The chlorophyll content in the resistant lines was much less affected than in the untransformed, sensitive lines. These results suggest that monitoring chloro-phyll levels in plants after glyphosate treatment might be a way to screen for resistance.

5.2.4.3 Shikimate Accumulation Glyphosate kills plants by inhibiting EPSPS (EC 2.5.1.19), which leads to rapid accumulation of shikimate-3-phos-phate (S3P) and shikimate. The endogenous shikimate level in most plants is very low, ranging from 0.04 to $0.06 \, \text{mg g}^{-1}$ fresh weight (Yoshida et al. 1975). In susceptible plants, shikimate accumulates primarily in the actively growing parts of the plant such as the meristematic zones, young expanding leaves, reproductive tissue, and roots (Marchiosi et al. 2008; Pline et al. 2002; Schultz et al. 1990). On the other hand, shikimate levels can reach up to 16% of the dry weight of the tissue in susceptible plants treated with lethal rates of glyphosate (Schulz et al. 1990). Sublethal doses of glyphosate cause a transient rise in shikimate levels, reaching a peak between 4 and 7 DAT and decreasing thereafter (Anderson et al. 2001; Henry et al. 2005, 2007).

The effect of glyphosate on shikimate levels has been used to measure glyphosate efficacy and drift. Harring et al. (1998) measured shikimate accu-mulation in oilseed rape (*Brassica napus* L. cv. Iris) to distinguish between formulations of glyphosate containing different surfactants. The ED_{50} (the dose that caused a 50% increase in shikimate accumulation) estimates were a good measure of the relative strength of the different formulations.

Koger et al. (2005a) examined the relationship between shikimate accumulation in different rice (*Oryza sativa* L.) cultivars and the effect of sublethal rates on glyphosate on injury and yield. Shikimate levels in the glyphosate-treated plants were strongly correlated to yield reduction in two varieties of rice and were a better predictor of yield reduction than visual injury.

The effect of glyphosate on shikimate accumulation in resistant plants has been well studied. Genetically engineered crop varieties that contain *CP4 EPSPS* do not accumulate shikimate at normal glyphosate use rates (Pline et al. 2002; Singh and Shaner 1998). Shikimate does accumulate in resistant weed populations, but the level and the duration of accumulation differs from susceptible populations. Mueller et al. (2003) compared the response of two GR horseweed populations to glyphosate with that of a GS population. The GR populations were approximately fourfold more tolerant to glyphosate than the GS population. The GR populations also showed transitory injury to glyphosate, from which they recovered. Shikimate accumulated in both GS and GR horseweed populations from Tennessee by 2 DAT, but declined 40% in the GR populations between 2 and 4 DAT, while it increased 35% in the GS populations (Mueller et al. 2003). Similar responses have been reported in GR biotypes of hairy fleabane, rigid ryegrass, and Italian ryegrass (Dinelli et al. 2008; Michitte et al. 2007; Simarmata et al. 2003; Wakelin and Preston 2006). Shikimate accumulated to similar levels in both GS and GR biotypes after treatment with glyphosate at 2 to 4 DAT, but the shikimate level decreased or remained the same in the GR biotypes; whereas, it continued to increase in the GS biotypes. In most of these cases, the mechanism of resistance was due to an alteration in the translocation of glyphosate to the meristematic zones of the plants. However, even in the cases where the mechanism of resistance is due to an alteration in EPSPS, glyphosate treatment results in an increase in shikimate levels in GR biotypes, but the levels are much less than in GS plants (Baerson et al. 2002a; Perez-Jones et al. 2007).

The pattern of transient accumulation of shikimate in GR weed biotypes appears to be similar to that of GS plants exposed to sublethal rates of glyphosate. There is an initial increase, followed by a decrease. This phenomenon needs to be taken into consideration if one wants to use shikimate accumulation as a screen for glyphosate resistance. It is critical to determine the glyphosate use rate and to clearly define the time period after application when shikimate levels will be measured. One also should have both a resistant and susceptible standard population to be sure the assay is working properly.

Another method to screen for glyphosate resistance is to measure shikimate accumulation in excised leaf disks. In this assay (Shaner et al. 2005), disks are excised from young, rapidly expanding leaves, floated on solutions containing different glyphosate concentrations, and incubated for 16–48 h under light. Incubation is stopped by acidification of the solution and then extracting the cellular contents by freeze thawing. The level of shikimate released from the tissue is then measured either spectrophotometrically (Cromartie and Polge 2000) or by high pressure liquid chromatography (HPLC) analysis (Singh and Shaner 1998).

Shikimate accumulation in leaf disks depends on a number of factors, including light or an external carbon source, such as sucrose, the concentration of glyphosate, age of tissue, and species (Shaner et al. 2005). All of these factors need to be determined for each species. However, in all species tested thus far, the highest shikimate accumulation occurred in disks taken from the youngest tissue (Shaner et al. 2005), which agrees with the observation that the shikimate pathway is predominantly located in the meristematic tissue (Schmid and Amrhein 1995; Weaver and Herrmann 1997).

This assay has been successfully used to identify GR biotypes of Palmer amaranth, horseweed, and Italian ryegrass (Table 5.4) (Culpepper et al. 2006; Koger et al. 2005b; Perez-Jones et al. 2005). In Palmer amaranth, shikimate accumulated in leaf disks from the GS biotype at all of the glyphosate concentrations tested (0.05–5 mM), whereas there was no shikimate accumulation in the tissues from the GR biotypes at any glyphosate concentration (Culpepper et al. 2006). The responses of horseweed and Italian ryegrass were somewhat different. In horseweed, the GS biotype accumulated shikimate at the lowest concentration tested (0.002 mM), but the GR biotype did not accumulate shikimate until the concentration reached 0.063 mM, a 30-fold difference (Koger et al. 2005b). A similar observation was made with Italian ryegrass.

TABLE 5.4. Relationship between Shikimate Accumulation and Mechanism of Glyphosate Resistance

	Shikimate Concentration $(\text{mg mL}^{-1}$ or $\text{mg g}^{-1})$					
	Glyphosate Concentration (mM)					
	10–25		250–350		Mechanism of	
Species	GS	GR	GS	GR	Resistance	Reference
Horseweed	20–35[a]	0–5	24–37	29–37	Reduced translocation	Koger et al. (2005b); Koger and Reddy (2005)
Italian ryegrass	1000[b]	0	1500	1500	Reduced translocation	Perez-Jones et al. (2005, 2007)
Palmer amaranth	39	0	48	0	Over-expression of EPSPS	Culpepper et al. (2006); Gaines et al. (2008)

[a] mg mL^{-1}.
[b] mg g^{-1}.

There was significant accumulation of shikimate in leaf disks from a GS biotype treated with 0.001 mM of glyphosate, but the GR biotype disks did not accumulate shikimate until 0.03 mM of glyphosate was applied, also a 30-fold difference between GR and GS. In both of these species, the mechanism of resistance is due to decreased mobility of glyphosate in the whole plant. A survey of California horseweed populations for glyphosate resistance using the leaf disk assay resulted in similar results (Hanson et al. 2009). Disks from a known GS biotype accumulated the maximum level of shikimate at 0.016 mM, while disks from the GR biotype did not reach a maximum level until the concentration reached 0.125 mM. For the screen, excised leaf disks from the newest fully expanded leaves were incubated with 0.016 mM of glyphosate. The plants were scored as resistant if the disks accumulated less than 12.1 µg shikimate L^{-1} or as susceptible if it accumulated more than 20.5 mg L^{-1}. Plants that fell in between these two extremes were classified as intermediate. Approximately 60% of the 141 locations tested were GR.

5.2.4.4   Methods for Measuring Shikimate

Several methods can be used to measure changes in shikimate levels to screen for glyphosate resistance. While shikimate is easily extracted from plant tissue by grinding and extracting in either 0.25 N HCl or H_2SO_4, an even simpler way is to place tissues in dilute acid and extracting shikimate by freeze thawing. The freeze thawing disrupts the cell membranes, and the shikimate diffuses out of the tissue.

There are two methods for measuring shikimate in the extracts. One method is via HPLC and the other is via spectrophotometry. Each method has its advantages and disadvantages.

There are several different procedures published on detecting shikimate via HPLC analysis. NH_2 or C_{18} columns are commonly used with acidic isocratic mobile phases, and shikimate is detected at a wavelength between 210 and 215 nm (Anderson et al. 2001; Mueller et al. 2003; Pline et al. 2002; Singh and Shaner 1998).

There are also different methods for detecting shikimate spectrophotometrically. In the spectrophotometric assay, shikimic acid is converted to trans-aconitic acid by treating the extract with periodic acid, then stopping the reaction with sodium hydroxide, and measuring absorbance at 320 nm. The original assay was described by Gaitonde and Gordon (1958) and required the use of glycine to stabilize the transaconitic acid and the assay needed to be done in a timely fashion because of the instability of the transaconitic acid. The assay was improved by Cromartie and Polge (2000) by adding m-periodate during the initial incubation and sodium sulfite with the sodium hydroxide. This assay is much more stable than the original assay and the results are more reliable.

Mueller et al. (2003) and Pline et al. (2002) compared the results of shikimate levels from the HPLC versus the spectrophotometric assay and both concluded that the HPLC assay was more accurate than the spectrophotometric assay. In addition, the HPLC assay directly measures shikimate not only in treated tissue but also in untreated tissue. The spectrophotometric assay can only make a relative measurement between treated and untreated tissue. One has to use an extract from untreated tissue for the background absorption at 380 nm, and the absorption may be due to multiple cellular components besides transaconitic acid. However, the spectrophotometric assay has a much higher throughput than the HPLC assay because it only requires access to a spectrophotometer and can be done quickly on multiple samples, which can be contained in an array of containers, including 96-well microtiter plates. Hence, it depends on the objectives of the assay which method is better. For absolute measurements of shikimate levels, the HPLC method is the best, but for screening large numbers of individuals, the spectrophotometric assay is the screen of choice because it is takes much less time and is less expensive than the HPLC assay.

5.2.5 *In Vitro* Assays

5.2.5.1 **In Vitro *EPSPS Assay***   EPSPS is the target site for glyphosate. It is not easy to extract EPSPS from plant material nor is the assay straightforward. While the primary mechanism of resistance to ALS and acetyl coenzyme A carboxylase (ACCase) inhibitors is an alteration of the target site at multiple sites within the enzyme, this is not the case for glyphosate resistance. Only one of the several mutation sites reported with EPSPS has been associated with resistance in weed species. This is at Pro_{106}, which has been replaced by a Ser, Thr, or Ala (Baerson et al. 2002b; Jasieniuk et al. 2008; Ng et al. 2003, 2004; Perez-Jones et al. 2007; Wakelin and Preston 2006; Yu et al. 2007). EPSPS with the Pro106Ser mutation in goosegrass is approximately fivefold less sensitive to glyphosate compared with the wild-type EPSPS (Baerson et al. 2002b). Kaundun et al. (2008) found that goosegrass populations in the Philippines with this mutation were approximately twofold more resistant to glyphosate than the wild type. Wakelin and Preston (2006) reported a similar level of resistance in rigid ryegrass populations carrying a similar mutation in EPSPS.

There are several methods for extracting EPSPS from plant material. Because the enzyme is primarily located in the chloroplast (Mousdale and Coggins 1985), enriching the extraction with semi-intact chloroplasts may provide enriched samples. Because the mRNA for EPSPS is primarily located in the young, growing parts of the plant (Weaver and Herrmann 1997), these are the tissues that should be extracted. In general, tissue is extracted in a buffer, pH 7–7.5, containing 10 mM dithiothreitol, 0.25 mM ethylenediaminetetraacetic acid (EDTA), 1 mM phenylmethylsulfonylfluoride, and 1 mM benzamidine (Mousdale and Coggins 1985; Nafziger and Slife 1983; Ream et al. 1988; Smith et al. 1986). If the assay for EPSPS is measuring the release of phosphate, ammonium heptamolybdate is also added to inhibit general phosphatase activity. The extractions have been used directly for assaying EPSPS or further purified by ammonium sulfate precipitation and chromatography.

There are several ways to assay EPSPS *in vitro*. The simplest is to monitor the release of phosphate from phosphoenolpyruvate (PEP) and S3P using the malachite green method (Lanzetta et al. 1979). In this assay, phosphate release from PEP and S3P by general phosphatases is inhibited greater than 98% by the addition of ammonium molybdate. The assay mixture contains 50 mM Hepes, 1 mM S3P, 1 mM PEP, and 0.1 mM ammonium heptamolybdate, pH 7.0, in a final volume of 100 μL. After preincubation (5 min, 30°C), the reaction is started by addition of the enzyme. After an appropriate incubation period at 30°C, the reaction is terminated by the addition of phosphate reagent (9.2 mM malachite green and 8.5 mM ammonium heptamolybdate in 1 M HCl with 2 g L^{-1} 3-[(3-cholamidopropyl)dimethylammonio]-1-propanesulfonate followed by 34% sodium citrate solution after 1 min (Nafziger and Slife 1983; Padgette et al. 1988; Steinrucken et al. 1986). One of the difficulties with this assay is the availability and cost of substrates, particularly S3P. Most commonly, the substrate is isolated from bacterial cultures that overproduce S3P (Bondinell et al. 1971). There is a company (Toronto Research Chemicals, Inc., Toronto, Canada) that sells S3P, and there may be other commercial sources for this compound.

Other assays have also been used. Padgette et al. (1988) describe an assay that measures the formation of ^{14}C-EPSP from ^{14}C-PEP utilizing an HPLC analysis. Boocock and Coggins (1983) measured EPSPS in the reverse direction by coupling the release of PEP to the pyruvate kinase and lactate hydrogenase reactions and monitoring the oxidation of NADH. In the same paper, the authors describe measuring the forward reaction by coupling the formation of EPSP to the chorismate synthase reaction and monitoring chorismate formation at 275 nm. It is doubtful that the *in vitro* EPSPS assay will be used for wide-scale screening for glyphosate resistance due to the cost of the assay and the fact that in many cases the mechanism of resistance is not due to an altered EPSPS.

5.2.6 DNA-Based Methods

Many of the cases of glyphosate resistance described to date are not due to an alteration of the EPSPS, but to a reduction of the translocation of glyphosate to the meristematic regions of the plant and the EPSPS mutations that have been identified only reduce the sensitivity of the enzyme to glyphosate by two- to fivefold (Wakelin and Preston 2006). Running an *in vitro* EPSPS assay may not be a top priority in determining the mechanism of resistance to glyphosate, particularly if there are other, more rapid methods to determine if the mechanism of glyphosate resistance is due to an altered EPSPS.

There are many molecular methods available to extract either DNA or RNA from plant material and to analyze gene sequences. The ones that are most applicable to testing for glyphosate resistance are the quantitative or real-time polymerase chain reaction (PCR) and single nucleotide polymorphisms (SNP) analysis. The gene sequence for EPSPS is highly conserved across plant species, particularly at the sites that are critical for catalysis. As noted previously, the only mutation site for GR weed biotypes that has been identified is Pro_{106}. A simple nucleotide base pair mutation at this site can result in substituting Ser, Thr, or Ala for Pro. Pro_{106} is not actually involved in the binding of PEP or glyphosate. Instead, a substitution at this site changes the orientation of two other amino acids which are involved in the binding of glyphosate, reducing the ability of glyphosate to bind to the enzyme (Healy-Fried et al. 2007). However, these changes in orientation do not alter the S3P and PEP binding sites and so have a minimal effect on the catalytic efficiency of the enzyme (Healy-Fried et al. 2007).

The DNA sequence of EPSPS has been determined in more than 70 species (Sammons et al. 2007). Thus, it is relatively easy to design primers to allow the PCR amplification of the section of the EPSPS gene containing Pro_{106}, which can then be sequenced to determine if there have been any mutations at this site. Chong et al. (2008) used this method to determine the relationship between nucleotide variability in EPSPS in goosegrass populations in Malaysia and glyphosate resistance. They examined six GR populations and found that there was a mutation at Pro_{106} in four of the populations, which resulted in either Ser or Thr substitutions. However, two of the GR populations did not have a mutation at Pro_{106}, suggesting that either there are other mutations that have not been identified or the mechanism of resistance in these GR populations is not due to an alteration of EPSPS. Jasieniuk et al. (2008) did a similar survey of GR Italian ryegrass populations in California and found that all the GR populations exhibited mutations at Pro_{106}, resulting in either a Ser or Ala substitution for Pro.

Although mutations at Pro_{106} have not yet been identified in GR broad-leaved weeds, it is possible that SNP or pyrosequencing analysis, described in the next paragraph, could be used to screen for glyphosate resistance in certain species. Warwick et al. (2008) used a TaqMan genotyping assay using real-time PCR (RT-PCR) for an SNP analysis for a Trp to Leu point mutation at amino acid position 574 in the ALS gene in kochia (*Kochia scoparia* L.). A similar type of assay could be used to identify mutations at Pro_{106} in EPSPS.

Another relatively new procedure is pyrosequencing. Pyrosequencing is a procedure that combines four enzymes (Klenow fragment of DNA polymerase I, ATP sulfurylase, luciferase, and apyrase) (Ahmadian et al. 2006). Each base is added one at a time to a mixture containing these four enzymes plus a DNA template. When the right nucleotide is put into the mixture, pyrophosphate is released as the nucleotide is incorporated into the DNA chain. The pyrophosphate is used to convert adenosine phosphosulfate to adenosine triphosphate (ATP), which then drives the luciferin–luciferase reaction and the release of light (Ahmadian et al. 2006). The light that is released is detected. The apyrase removes the unincorporated nucleotides and ATP between the additions of different nucleotides. This procedure has been used to screen for mutations of ACCase in blackgrass (*Alopecurus myosuroides* L.) collected from fields with ACCase herbicide failures (Wagner et al. 2008). More than 100 sites were sampled, and a mutation that substituted an isoleucine for leucine was detected in 91% of the cases of target site-based resistance. A similar approach could be used to screen for target-site resistance in EPSPS.

Another technique that can be used to test for glyphosate resistance is RT-PCR (also called quantitative PCR). RT-PCR detects and quantifies a fluorescent signal after each amplification cycle in the PCR. It can be used to determine relative gene expression levels in different plants or plant parts (Chao 2008). This technique has been used to detect two- to threefold higher expression levels of EPSPS in GR versus GS biotypes of rigid ryegrass (Baerson et al. 2002b), horseweed (Dinelli et al. 2006), and hairy fleabane (Dinelli et al. 2008). Gaines et al. (2008) used quantitative PCR to show that a GR Palmer amaranth biotype had 64–128 more copies of genomic EPSPS than GS biotypes.

5.2.7 Absorption and Translocation of ^{14}C-Glyphosate

Glyphosate kills plants by inhibiting EPSPS, which is primarily located in the meristematic regions of the plant (Mollenhauer et al. 1987). One of the reasons that glyphosate is such an effective herbicide is its mobility within the plant. Glyphosate translocation within the plant is similar to the movement of the photoassimilates and it accumulates in the meristematic tissue (Arnaud et al. 1994; Bromilow et al. 1993). Changes in glyphosate translocation patterns affect its efficacy. The predominant mechanism of resistance to glyphosate across a number of species is differential translocation. The mechanism of resistance in GR biotypes of rigid ryegrass (Wakelin et al. 2004), Italian ryegrass (Michitte et al. 2005; Nandula et al. 2008; Perez-Jones et al. 2007), horseweed (Dinelli et al. 2006; Feng et al. 2004; Koger and Reddy 2005), and hairy fleabane (Dinelli et al. 2008) have all been shown to be related to less translocation of glyphosate in the GR biotype compared with the GS biotype (Table 5.5). In all of these cases, glyphosate movement out of the treated leaf to the meristematic zones was measured by treating a source leaf of the plant with ^{14}C-glyphosate and measuring the movement of radioactivity out of the treated leaf to the rest of the plant. In the GR biotypes, glyphosate moved in the transpiration stream to the tips of the leaves, but there was less movement out of the treated leaf to the rest of the plant. While there are clear differences in the amount of glyphosate translocated out of the treated leaf between the GR and GS biotypes, the differences can become less with time. Michitte et al. (2007) found inconsistent results in their studies on GR Italian ryegrass. In one experiment, there was a clear difference in the translocation pattern between GR and GS biotypes, but these differences were not as evident in a second experiment. The researchers suggested that the differences between their results and those of others who clearly showed that differences in translocation between GR and GS biotypes could have been due to the fact that they did not treat the whole plant with glyphosate, but only treated one leaf with ^{14}C-glyphosate. C. Preston (pers. comm.) also found that translocation

TABLE 5.5. Translocation of ^{14}C-Glyphosate in GS and GR Biotypes

Species	Biotype	Translocation (%) (24–48 HAT)	Reference
Rigid ryegrass	GS	90.5[a]	Wakelin et al.
	GR	69	(2004)
Italian ryegrass	GS	40.7	Perez-Jones et al.
	GR	31.2	(2007)
	GS	55[a]	Michitte et al.
	GR	36	(2007)
Horseweed	GS	30[a]	Koger and
	GR	18	Reddy (2005)
	GS	60[b]	Dinelli et al.
	GR	45	(2006)
	GS	40	Feng et al. (2004)
	GR	20	

[a]Average of GS and GR biotypes tested.
[b]Estimate from figure in paper.
GR, glyphosate-resistant; GS, glyphosate-susceptible.

patterns between the GR and GS biotypes were highly variable and difficult to interpret unless the whole plant was treated with unlabeled glyphosate except for the leaf that was treated with ^{14}C-glyphosate. These results suggest that it is important to treat all of the plant, except for the treated leaf, with unlabeled glyphosate in order to see clear differences between GR and GS biotypes.

5.2.8 Testing Methods—Strengths and Weaknesses

There are multiple tools available for testing for glyphosate resistance. Each tool has its strengths and weaknesses. Whole plant screens, whether in the field or greenhouse, should be used as an initial method to determine if a biotype is GR. Whole plant screens should detect any type of resistance, and the level of resistance may be an indication of the mechanism (Fig. 5.1). Greenhouse or field screens may be relatively expensive in terms of time, space, and manpower. The results are also dependent on the growth stage and physiological state of the plants as well as on the physical properties of the glyphosate that is applied in terms of formulation, adjuvants, and other factors.

Screening for resistance using seedling assays such as in Petri plates, sand culture, plant parts, and others are more rapid than whole plant screens, but they are limited in their applicability. They will only detect resistance if it is manifested at the seedling stage. However, for screening large numbers of biotypes, seedling-based screens may provide an initial way to rank the level of resistance in each biotype.

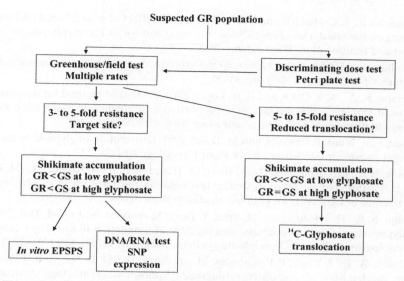

Figure 5.1. Sequence of methods for testing glyphosate resistance in weeds.

The use of physiologically based screens, such as shikimate accumulation or loss of chlorophyll, may be more rapid than either greenhouse or seedling based assays and they can be relatively high throughput, so large numbers of biotypes can be screened quickly. The shikimate assay has been shown to work on both site of action and translocation-based mechanisms of resistance, but it requires access to a laboratory and needs to be tailored to each species. The results of the shikimate assay do provide some information on the mechanism of resistance, although it is not definitive by itself (Fig. 5.1).

DNA/RNA-based assays could be a very robust way to screen large populations for glyphosate resistance if the mechanism of resistance is due to either differential expression of EPSPS or a mutation in EPSPS (Fig. 5.1). However, these types of assays also require access to a laboratory and the expense of such assays may prohibit their widespread use.

The use of ^{14}C-glyphosate is essential for determining if reduced translocation is the mechanism of resistance. This technique requires a specialized laboratory that can safely handle radioactive materials, and care needs to be taken to design the experiments correctly to identify differential rates of translocation. Therefore, it is unlikely that using ^{14}C-glyphosate will be used as a broad-scale screening method.

REFERENCES

Abu-Irmaileh, B. E. and L. S. Jordan. 1978. Some aspects of glyphosate action in purple nutsedge (*Cyperus rotundus*). *Weed Science* 26:700–703.

Ahmadi, M. S., L. C. Haderlie, and G. A. Wicks. 1980. Effect of growth stage and water stress on barnyardgrass (*Echinochloa crus-galli*) control and on glyphosate absorption and translocation. *Weed Science* 28:277–282.

Ahmadian, A., M. Ehn, and S. Hober. 2006. Pyrosequencing: history, biochemistry, and future. *Clinica Chimica Acta* 363:83–94.

Anderson, K. A., W. T. Cobb, and B. R. Loper. 2001. Analytical method for determination of shikimic acid: shikimic acid proportional to glyphosate application rates. *Communications in Soil Science and Plant Analysis* 32:2831–2840.

Arnaud, L., F. Nurit, P. Ravanel, and M. Tissut. 1994. Distribution of glyphosate and its target enzyme inside wheat plants. *Plant Physiology* 40:217–223.

Baerson, S. R., D. J. Rodriguez, N. A. Biest, M. Tran, J. You, R. W. Kreuger, G. M. Dill, J. E. Pratley, and K. J. Gruys. 2002a. Investigating the mechanism of glyphosate resistance in rigid ryegrass (*Lolium rigidum*). *Weed Science* 50:721–730.

Baerson, S. R., D. J. Rodriguez, M. Tran, Y. Feng, N. A. Biest, and G. M. Dill. 2002b. Glyphosate-resistant goosegrass. identification of a mutation in the target enzyme 5-enolpyruvylshikimate-3-phophsate synthase. *Plant Physiology* 129:1265–1275.

Bondinell, W. E., J. Vnek, P. F. Knowles, M. Sprecher, and D. B. Sprinson. 1971. On the mechanisms of 5-enolpyruvylshikimate-3-phosphate synthetase. *Journal of Biological Chemistry* 246:6191–6196.

Boocock, M. R. and J. R. Coggins. 1983. Kinetics of 5-enolpyruvylshikimate-3-phosphate synthase inhibition by glyphosate. *FEBS Letters* 154:127–133.

Boutsalis, P. 2001. Syngenta quick test: a rapid whole-plant test for herbicide resistance. *Weed Technology* 15:257–263.

Brecke, B. J. and W. B. Duke. 1980. Effect of glyphosate on intact bean plants (*Phaseolus vulgaris* L.) and isolated cells. *Plant Physiology* 66:656–659.

Bromilow, R. H., K. Chamberlain, A. J. Tench, and R. H. Williams. 1993. Phloem translocation of strong acids—glyphosate, substituted phosphonic and sulfonic acids—in *Ricinus communis*. *Pesticide Science* 37:39–47.

Chao, W. S. 2008. Real-time PCR as a tool to study weed biology. *Weed Science* 56:290–296.

Chase, R. L. and A. P. Appleby. 1979. Effect of humidity and moisture stress on glyphosate control of *Cyperus rotundus* L. *Weed Research* 19:241–246.

Chong, J. L., R. Wickneswari, B. S. Ismail, and S. Salmijah. 2008. Nucleotide variability in the 5-enoypyruvylshikimate-3-phosphate synthase gene from *Eleusine indica* (L.) Gaertn. *Pakistan Journal of Biological Science* 11:476–479.

Cromartie, T. H. and N. D. Polge. 2000. An improved assay for shikimic acid and its use as a monitor for the activity of sulfosate. *Abstracts of the Weed Science Society of America* 40:291.

Culpepper, A. S., T. L. Grey, W. K. Vencill, J. M. Kichler, T. M. Webster, S. M. Brown, A. C. York, J. W. Davis, and W. W. Hanna. 2006. Glyphosate-resistant Palmer amaranth (*Amaranthus palmeri*) confirmed in Georgia. *Weed Science* 54:620–626.

Davis, V. M., K. D. Gibson, and W. G. Johnson. 2008. A field survey to determine distribution and frequency of glyphosate-resistant horseweed (*Conyza canadensis*) in Indiana. *Weed Technology* 22:331–338.

De Ruiter, H. and E. Meinen. 1998. Influence of water stress and surfactant on the efficacy, absorption and translocation of glyphosate. *Weed Science* 46:289–296.

DeGennaro, F. P. and S. C. Weller. 1984. Differential susceptibility of field bindweed (*Convolvulus arvensis*) biotypes to glyphosate. *Weed Science* 32:472–476.

Dinelli, G., I. Marotti, A. Bonetti, P. Catizone, J. M. Urbano, and J. Barnes. 2008. Physiological and molecular basis of glyphosate resistance in *Conyza bonariensis* biotypes from Spain. *Weed Research* 48:257–265.

Dinelli, G., I. Marotti, A. Bonetti, M. Mineli, P. Catizone, and J. Barnes. 2006. Physiological and molecular insight on the mechanisms of resistance to glyphosate in *Conyza canadensis* (L.) Cronq. biotypes. *Pesticide Biochemistry and Physiology* 86:30–41.

Donahue, R. A., T. D. Davis, C. H. Michler, D. E. Riemenschneider, D. R. Carter, P. E. Marquardt, N. Sankhla, B. E. Haissig, and J. G. Isebrands. 1994. Growth, photosynthesis, and herbicide tolerance of genetically modified hybrid poplar. *Canadian Journal of Forest Research* 24:2377–2383.

Escorial, M. C., H. Sixto, J. M. Garcia-Baudin, and M. C. Chueca. 2001. A rapid method to determine cereal plant response to glyphosate. *Weed Technology* 15:697–702.

Feng, P. C. C., M. Tran, T. Chiu, R. D. Sammons, G. R. Heck, and C. A. CaJacob. 2004. Investigations into glyphosate-resistant horseweed (*Conyza canadensis*): retention, uptake, translocation, and metabolism. *Weed Science* 52:498–505.

Fernandez, C. J., K. J. McInnes, and J. T. Cothren. 1994. Carbon balance, transpiration, and biomass partitioning of glyphosate-treated wheat (*Triticum aestivum*) plants. *Weed Science* 42:333–339.

Ferrell, J. A., H. J. Earl, and W. K. Vencill. 2003. The effect of selected herbicides on CO_2 assimilation, chlorophyll fluorescence, and stomatal conductance in johnsongrass (*Sorghum halepense* L.). *Weed Science* 51:28–31.

Franz, J. E., M. K. Mao, and J. A. Sikorski. 1997. *Glyphosate: A Unique Global Herbicide.* ACS Monograph No. 189. Washington, DC: American Chemical Society, p. 653.

Fuchs, M. A., D. R. Geiger, T. L. Reynolds, and J. E. Bourque. 2002. Mechanisms of glyphosate toxicity in velvetleaf (*Abutilon theorphrasti* medikus). *Pesticide Biochemistry and Physiology* 74:27–39.

Gaines, T. A., P. Westra, J. E. Leach, S. M. Ward, B. Bukum, S. T. Chisholm, D. L. Shaner, C. Preston, A. S. Culpepper, T. L. Grey, T. M. Webster, W. K. Vencill, and P. J. Tranel. 2008. Molecular genetics of glyphosate resistance in Palmer amaranth. *Proceedings of the North Central Weed Science Society* 63:125.

Gaitonde, M. and M. Gordon. 1958. A microchemical method for the detection and determination of shikimic acid. *Journal of Biological Chemistry* 230:1043–1050.

Gougler, J. A. and D. R. Geiger. 1984. Carbon partitioning and herbicide transport in glyphosate treated sugarbeet (*Beta vulgaris*). *Weed Science* 32:546–551.

Hall, G. J., C. A. Hart, and C. A. Jones. 2000. Plants as sources of cations antagonistic to glyphosate efficacy. *Pest Management Science* 56:351–368.

Hanson, B. D., A. Shrestha, and D. L. Shaner. 2009. Distribution of glyphosate-resistant horseweed (*Conyza canadensis*) and relationship to cropping systems in the Central Valley of California. *Weed Science* 57:48–53.

Harring, T., J. C. Streibig, and S. Husted. 1998. Accumulation of shikimic acid: a technique for screening glyphosate efficacy. *Journal of Agricultural and Food Chemistry* 46:4406–4412.

Healy-Fried, M. L., T. Funke, M. A. Priestmann, H. Han, and E. Schonbrunn. 2007. Structural basis of glyphosate tolerance resulting from mutations of Pro101 in

Escherichia coli 5-enolpyruvylshikimate-3-phosphate synthase. *Journal of Biological Chemistry* 282:32949–32955.

Heap, I. 2008. *The International Survey of Herbicide Resistant Weeds.* http://www.weedscience.com (accessed December 15, 2008).

Henry, W. B., C. H. Koger, and D. L. Shaner. 2005. Accumulation of shikimate in corn and soybean exposed to various rates of glyphosate. *Crop Management* doi:10.1094/CM-2005-1123-01-RS.

Henry, W. B., D. L. Shaner, and M. W. West. 2007. Shikimate accumulation in sunflower, wheat and proso millet after glyphosate application. *Weed Science* 55:1–5.

Hite, G. A., S. R. King, E. S. Hagood, and G. I. Holtman. 2008. Differential response of a Virginia common lambsquarters (*Chenopodium album*) collection to glyphosate. *Weed Science* 56:203–209.

Jasieniuk, M., R. Ahmad, A. M. Sherwood, J. L. Firestone, A. Perez-Jones, W. T. Lanini, C. Mallory-Smith, and Z. Stednick. 2008. Glyphosate-resistant Italian ryegrass (*Lolium multiflorum*) in California: distribution, response to glyphosate, and molecular evidence for an altered target enzyme. *Weed Science* 56:496–502.

Kaundun, S. S., I. A. Zelaya, R. P. Dale, A. J. Lycett, P. Carter, K. R. Sharples, and E. McIndoe. 2008. Importance of the P106S target-site mutation in conferring resistance to glyphosate in a goosegrass (*Eleusine indica*) population from the Philippines. *Weed Science* 56:637–646.

Ketel, D. H. 1996. Effects of low doses of metamitron and glyphosate on growth and chlorophyll content of common lambsquarters (*Chenopodium album*). *Weed Science* 44:1–6.

Kitchen, L. M., W. W. Witt, and C. E. Rieck. 1981. Inhibition of chlorophyll accumulation by glyphosate. *Weed Science* 29:513–516.

Koger, C. H. and K. N. Reddy. 2005. Role of absorption and translocation in the mechanism of glyphosate resistance in horseweed (*Conyza canadensis*). *Weed Science* 53:84–89.

Koger, C.H., D. H. Poston, R. M. Hayes, and R. F. Montgomery. 2004. Glyphosate resistant horseweed (*Conyza canadensis*) in Mississippi. *Weed Science* 18:820–825.

Koger, C. H., D. L. Shaner, W. B. Henry, T. Nadler-Hassar, W. E. Thomas, and J. W. Wilcut. 2005b. Assessment of two nondestructive assays for detecting glyphosate resistance in horseweed (*Conyza canadensis*). *Weed Science* 53:559–566.

Koger, C. H., D. L. Shaner, L. J. Krutz, T. W. Walker, N. Buehring, W. B. Henry, W. E. Thomas, and J. W. Wilcut. 2005a. Rice (*Oryza sativa*) response to drift rates of glyphosate. *Pest Management Science* 61:1161–1167.

Lanzetta, P. A., L. J. Alvarez, P. S. Reinach, and O. A. Candia. 1979. An improved assay for nanomole amounts of inorganic phosphate. *Analytical Biochemistry* 100:95–97.

Lee, L. J. and J. Ngim. 2000. A first report of glyphosate-resistant goosegrass (*Eleusine indica* (L) Gaertn) in Malaysia. *Pest Management Science* 56:336–339.

Mann, C. L., V. R. Pantalone, and T. C. Mueller. 2004. A novel approach to determine the glyphosate tolerant trait in soybeans. *Journal of Agricultural and Food Chemistry* 52:1224–1227.

Marchiosi, R. M., D. L. Ferarese, E. A. Bonini, N. G. Fernandes, A. P. Ferro, and O. Ferarese-Fiolho. 2008. Glyphosate-induced metabolic changes in susceptible and glyphosate-resistant soybean (*Glycine max*) roots. *Pesticide Biochemistry and Physiology* doi:10.1016/j.pestbp.2008.09.03.

McWhorter, C. G. and W. R. Azlin. 1978. Effects of environment on the toxicity of glyphosate to johnsongrass (*Sorghum halepense*) and soybeans (*Glycine max*). *Weed Science* 26:605–608.

Michitte, P., R. De Prado, N. Espinosa, and C. Gauvrit. 2005. Glyphosate resistance in a Chilean *Lolium multiflorum*. *Communications in Agricultural and Applied Biological Sciences* 70:507–513.

Michitte, P., R. De Prado, N. Espinoza, J. P. Ruiz-Santaella, and C. Gauvrit. 2007. Mechanisms of resistance to glyphosate in a ryegrass (*Lolium multiflorum*) biotype from Chile. *Weed Science* 55:435–440.

Mithila, J., C. J. Swanton, R. E. Blackshaw, R. J. Cathcart, and J. C. Hall. 2008. Physiological basis for reduced glyphosate efficacy on weeds grown under low soil nitrogen. *Weed Science* 56:12–17.

Mollenhauer, C., C. C. Smart, and N. Amrhein. 1987. Glyphosate toxicity in the shoot apical region of the tomato plant. I. Plastid swelling is the initial ultrastructural feature following in vivo inhibition of 5-enolpyruvylshikimic acid 3-phosphate synthase. *Pesticide Biochemistry and Physiology* 29:55–65.

Mousdale, D. M. and J. R. Coggins. 1985. Subcellular localization of the common shikimate-pathway enzymes in *Pisum sativum* L. *Planta* 163:241–249.

Mueller, T. C., J. H. Massey, R. M. Hayes, C. L. Main, and C. N. Stewart, Jr., 2003. Shikimate accumulates in both glyphosate-sensitive and glyphosate-resistant horseweed (*Conyza canadensis* L. Cronq.). *Journal of Agricultural and Food Chemistry* 51:680–684.

Nafziger, E. D. and F. W. Slife. 1983. Physiological response of common cocklebur (*Xanthium pensylvanicum*) to glyphosate. *Weed Science* 31:874–878.

Nandula, V. K., K. N. Reddy, D. H. Poston, A. M. Rimando, and S. O. Duke. 2008. Glyphosate tolerance mechanism in Italian ryegrass (*Lolium multiflorum*) from Mississippi. *Weed Science* 56:344–349.

Neve, P., J. Sadler, and S. B. Powles. 2004. Multiple herbicide resistance in a glyphosate-resistant rigid ryegrass (*Lolium rigidum*) population. *Weed Science* 52:920–928.

Ng, C. H., W. Ratnam, S. Surif, and B. S. Ismail. 2004. Inheritance of glyphosate resistance in goosegrass (*Eleusine indica*). *Weed Science* 52:564–570.

Ng, C. H., R. Wickneswari, S. Salmijah, Y. T. Teng, and B. S. Ismail. 2003. Gene polymorphisms in glyphosate-resistant and -susceptible biotypes of *Eleusine indica* from Malaysia. *Weed Research* 43:108–115.

Norsworthy, J. K., R. C. Scott, K. L. Smith, and L. R. Oliver. 2008. Response of northeastern Arkansas Palmer amaranth (*Amaranthus palmeri*) accessions to glyphosate. *Weed Technology* 22:408–413.

Padgette, S. R., Q. K. Huynh, S. Aykent, R. D. Sammons, J. A. Sikorski, and G. M. Kishore. 1988. Identificaiton of the reactive cysteines of *Escherichia coli* 5-enolpyruvylshikimate-3-phosphate synthase and their nonessentiality for enzymatic catalysis. *Journal of Biological Chemistry* 263:1798–1802.

Perez, A. and M. Kogan. 2003. Glyphosate-resistant *Lolium multiflorum* in Chilean orchards. *Weed Research* 43:12–19.

Perez-Jones, A., K. W. Park, J. Colquhoun, C. Mallory-Smith, and D. Shaner. 2005. Identification of glyphosate-resistant Italian ryegrass (*Lolium multiflorum*) in Oregon. *Weed Science* 53:775–779.

Perez-Jones, A., K. Park, N. Polge, J. Colquhoun, and C. A. Mallory-Smith. 2007. Investigating the mechanisms of glyphosate resistance in *Lolium multiflorum*. *Planta* 226:395–404.

Pline, W. A., J. W. Wilcut, S. O. Duke, K. L. Edmisten, and R. W. Wells. 2002. Tolerance and accumulation of shikimic acid in response to glyphosate applications in glyphosate-resistant and conventional cotton (*Gossypium hirsutum* L.). *Journal of Agricultural and Food Chemistry* 50:506–512.

Powles, S. B., D. F. Lorraine-Colwill, J. J. Dellow, and C. Preston. 1998. Evolved resistance to glyphosate in rigid ryegrass (*Lolium rigidum*) in Australia. *Weed Science* 46:604–607.

Pratley, J., N. Urwin, R. Stanton, P. Baines, J. Broster, K. Cullis, D. Schafer, J. Bohn, and R. Krueger. 1999. Resistance to glyphosate in *Lolium rigidum*. I. Bioevaluation. *Weed Science* 47:405–411.

Ramsdale, B. K., C. G. Messersmith, and J. D. Nalewaja. 2003. Spray volume, formulation, ammonium sulfate and nozzle effects on glyphosate efficacy. *Weed Technology* 17:589–598.

Ream, J. E., H. C. Steinrucken, C. A. Porter, and J. A. Sikorski. 1988. Purification and properties of 5-enolpyruvylshikimate-3-phosphate synthase from dark-grown seedlings of *Sorghum bicolor*. *Plant Physiology* 87:232–238.

Sammons, R. D., D. C. Heering, N. Dinicola, H. Glick, and G. A. Elmore. 2007. Sustainability and stewardship of glyphosate and glyphosate-resistant crops. *Weed Technology* 21:347–354.

Schmid, J. and N. Amrhein. 1995. Molecular organization of the shikimate pathway in higher plants. *Phytochemistry* 39:737–749.

Schulz, A., T. Munder, H. Hollander-Czytko, and N. Amrhein. 1990. Glyphosate transport and early effects on shikimate metabolism and its compartmentation in sink leaves of tomato and spinach plants. *Zeitschrift für Naturforschung* 45c:529–534.

Schuster, C. L., D. E. Shoup, and K. Al-Khatib. 2007. Response of common lambsquarters (*Chenopodium album*) to glyphosate as affected by growth stage. *Weed Science* 55:147–151.

Shaner, D. L. 1978. Effects of glyphosate on transpiration. *Weed Science* 26:513–515.

Shaner, D. L. and J. L. Lyon. 1980. Interaction of glyphosate with aromatic amino acids on transpiration in *Phaseolus vulgaris*. *Weed Science* 28:31–35.

Shaner, D. L., T. Nadler-Hassar, W. B. Henry, and C. H. Koger. 2005. A rapid in vivo shikimate accumulation assay with excised leaf discs. *Weed Science* 53:769–774.

Sharma, S. D. and M. Singh. 2001. Environmental factors affecting absorption and bio-efficacy of glyphosate in Florida beggarweed (*Desmodium tortuosum*). *Crop Protection* 20:511–516.

Shrestha, A., K. J. Hembreee, and N. Va. 2007. Growth stage influences level of resistance in glyphosate-resistant horseweed. *California Agriculture* 61:67–70.

Simarmata, M., J. E. Kaufmann, and D. Penner. 2003. Potential basis of glyphosate resistance in California rigid ryegrass (*Lolium rigidum*). *Weed Science* 51:678–682.

Singh, B. K. and D. L. Shaner. 1998. Rapid determination of glyphosate injury to plants and identification of glyphosate resistant plants. *Weed Technology* 12:527–530.

Smith, C. M., D. Pratt, and G. A. Thompson. 1986. Increased 5-enolpyruvylshikimic acid-3-phosphate synthase activity in a glyphosate-tolerant variant strain of tomato cell. *Plant Cell Reports* 5:298–301.

Steinrucken, H. C., A. Schulz, N. Amrhein, C. A. Porter, and R. T. Fraley. 1986. Overproduction of 5-enolpyruvylshikimate-3-phosphate synthase in a glyphosate-tolerant *Petunia hybrida* cell line. *Archives of Biochemistry and Biophysics* 244:169–179.

Stephenson, D. O. IV, L. R. Oliver, and J. A. Bond. 2007. Response of pitted morning glory (*Ipomoea lacunosa*) accessions to chlorimuron, fomesafen, and glyphosate. *Weed Technology* 17:179–185.

Thelen, K. D., E. P. Jackson, and D. Penner. 1995. The basis for the hard-water antagonism of glyphosate activity. *Weed Science* 43:541–548.

Trainer, G. D., M. M. Loux, S. K. Harrison, and E. Regnier. 2005. Response of horseweed biotypes to foliar applications of cloransulam-methyl and glyphosate. *Weed Technology* 19:231–236.

Urbano, J. M., A. Borrego, V. Torres, J. M. Leon, C. Jimenez, G. Dinelli, and J. Barnes. 2007. Glyphosate-resistant hairy fleabane (*Conyza bonariensis*) in Spain. *Weed Technology* 21:396–401.

VanGessel, M. J. 2001. Glyphosate-resistant horseweed from Delaware. *Weed Science* 49:703–705.

Vidal, R. A., M. M. Trezzi, R. De Prado, J. P. Ruiz-Santaella, and M. Vila-Aiub. 2007. Glyphosate resistant biotypes of wild poinsettia (*Euphorbia heterophylla* L.) and its risk analysis on glyphosate-tolerant soybeans. *Journal of Food and Agricultural Chemistry* 5:265–269.

Vila-Auib, M. M., M. B. Balbi, P. E. Gundel, C. M. Ghersa, and S. B. Powles. 2007. Evolution of glyphosate-resistant Johnsongrass (*Sorghum halepense*) in glyphosate-resistant soybean. *Weed Science* 55:566–571.

Villanueva, M. J. C., B. F. Muniz, and R. S. Tames. 1985. Effects of glyphosate on growth, chlorophyll, and carotenoid levels of yellow nutsedge (*Cyperus esculentus*). *Weed Science* 33:751–754.

Volenberg, D. S., W. L. Patzoldt, A. G. Hager, and P. J. Tranel. 2007. Reponses of contemporary and historical waterhemp (*Amarnathus tuberculatus*) accessions to glyphosate. *Weed Science* 55:627–333.

Wagner, J., B. Laber, H. Menne, and H. Kraehmer. 2008. Rapid analysis of target-site resistance in blackgrass using pyrosequencing® technology. *Abstracts of the 5th International Weed Science Congress* 5:23.

Wakelin, A. M., D. F. Lorraine-Colwill, and C. Preston. 2004. Glyphosate resistance in four different populations of *Lolium rigidum* is associated with reduced translocation of glyphosate to meristematic zones. *Weed Research* 44:453–459.

Wakelin, A. M. and C. Preston. 2006. A target-site mutation is present in a glyphosate-resistant *Lolium rigidum* population. *Weed Research* 46:432–440.

Warwick, S. I., R. Xu, C. Sauder, and H. J. Beckie. 2008. Acetolactate synthase target-site mutations and single nucleotide polymorphism genotyping in ALS-resistant kochia (*Kochia scoparia*). *Weed Science* 56:797–806.

Weaver L. M. and K. M. Hermann. 1997. Dynamics of the shikimate pathway in plants. *Trends in Plant Science* 2:346–351.

Westhoven, A. M., V. M. Davis, K. D. Gibson, S. C. Weller, and W. G. Johnson. 2008. Field presence of glyphosate-resistant horseweed (*Conyza canadensis*), common lambsquarters (*Chenopodium album*) and giant ragweed (*Ambrosia trifida*) biotypes with elevated tolerance to glyphosate. *Weed Technology* 22:544–548.

Westwood, J. H., C. N. Yerkes, G. D. DeGennaro, and S. C. Weller. 1997. Absorption and translocation of glyphosate in tolerant and susceptible biotypes of field bindweed (*Convolvulus arvensis*). *Weed Science* 45:658–663.

Yoshida, S., K. Tazaki, and T. Minamikawa. 1975. Occurrence of shikimic and quinic acids in angiosperms. *Phytochemistry* 14:195–197.

Yu, Q., A. Cairns, and S. Powles. 2007. Glyphosate, paraquat, and ACCase mulitiple resistance evolved in a *Lolium rigidum* biotype. *Planta* 225:499–513.

Yuan, C. I., M. Y. Chaing, and Y. M. Chen. 2002. Triple mechanisms of glyphosate resistance in a naturally occurring glyphosate-resistant plant *Diclepta chinensis*. *Plant Science* 163:543–554.

Zelaya, I. A. and M. D. K. Owen. 2005. Differential response of *Amaranthus tuberculatus* (Moq ex DC) JD Sauer to glyphosate. *Pest Management Science* 61:936–950.

Zhou, J., B. Tao, C. G. Messersmith, and J. D. Nalewaja. 2008. Glyphosate efficacy on velvetleaf (*Abutilon theophrasti*) is affected by stress. *Weed Science* 55:240–244.

6

BIOCHEMICAL MECHANISMS AND MOLECULAR BASIS OF EVOLVED GLYPHOSATE RESISTANCE IN WEED SPECIES

ALEJANDRO PEREZ-JONES AND CAROL MALLORY-SMITH

6.1 INTRODUCTION

Glyphosate (N-(phosphonomethyl)glycine) was first introduced as a commercial herbicide in 1974 under the trade name of Roundup® (Monsanto Company, St. Louis, MO) (Franz et al. 1997). Glyphosate is a postemergent, systemic, nonselective, broad-spectrum herbicide that controls annual and perennial weeds and volunteer crops in a wide range of situations. Although glyphosate was initially used as a noncrop and plantation crop (e.g., orchards and vineyards) herbicide, it is now widely used in no-till crop production systems, and for selective weed control in transgenic glyphosate-resistant (GR) Roundup Ready® crops such as soybean (*Glycine max* (L.) Merr.), cotton (*Gossypium hirsutum* L.), canola (*Brassica napus* L.), maize (*Zea mays* L.), sugar beet (*Beta vulgaris* L.), and alfalfa (*Medicago sativa* L.) (Baylis 2000; Dill et al. 2008; Woodburn 2000).

With the introduction of GR crops in 1996, along with a rising adoption of no-till farming techniques, glyphosate use has increased dramatically worldwide. In 2006, GR crops comprising five species were grown on over 74 million ha in 13 countries, representing one of the more rapidly adopted weed management technologies in recent history (Dill et al. 2008). Reduction of

Glyphosate Resistance in Crops and Weeds: History, Development, and Management
Edited by Vijay K. Nandula
Copyright © 2010 John Wiley & Sons, Inc.

production cost, simple, flexible, efficient, and broad-spectrum weed control, and adoption of no-tillage have been described as reasons for this rapid and widespread adoption (Dill 2005; Dill et al. 2008; Gianessi 2005, 2008). In the United States, where 88% of soybean, 80% of cotton, and 31% of maize were GR in 2005 (Gianessi 2008), glyphosate use rose from 17 million lb in 1992 to 35 million lb in 1997 to 102 million lb in 2002 (Gianessi and Reigner 2006). The massive adoption of GR crops in the United States and other countries (e.g., Argentina and Brazil) has lead to a decrease in diversity of weed control methods and increase of glyphosate use, which has played a significant role in the selection pressure for the evolution of GR weeds (Powles 2008).

The biochemical and genetic basis of glyphosate resistance in plants have been reviewed previously (Powles and Preston 2006). The number of weed species evolving glyphosate resistance continues to increase, and now there is new evidence of a novel mechanism that can confer resistance to glyphosate. Here, we will examine the three mechanisms that are known to confer glyphosate resistance in weed species: target-site mutation, limited or reduced glyphosate translocation, and gene amplification. The genetics and inheritance of the mechanisms of glyphosate resistance have been reviewed in Chapter 7 and will not be covered here.

6.2   MODE OF ACTION OF GLYPHOSATE

Glyphosate has a unique mode of action. It inhibits the enzyme 5-enolpyruvylshikimate-3-phosphate (EPSP) synthase (EC 2.5.1.19) (Steinrücken and Amrhein 1980). EPSP synthase is the sixth enzyme of the shikimic acid pathway, in which phosphoenolpyruvate (PEP) and erythrose 4-phosphate are converted to chorismate, the precursor of the aromatic amino acids (phenylalanine, tyrosine, and tryptophan) and many aromatic secondary metabolites (e.g., auxins, phytoalexins, anthocyanins, and lignin) (Herrmann and Weaver 1999; Kishore and Shah 1988). EPSP synthase catalyzes the transfer of the enolpyruvyl moiety from PEP to shikimate-3-phosphate (S3P) to yield EPSP and inorganic phosphate (Pi) (Geiger and Fuchs 2002). Glyphosate is a transition state analog of PEP (Siehl 1997) and inhibits EPSP synthase through the formation of an EPSP synthase–S3P–glyphosate ternary complex, only binding to the enzyme after the formation of EPSP synthase–S3P binary complex (Alibhai and Stalling 2001). Thus, glyphosate acts as a competitive inhibitor with PEP as it occupies its binding site (Schönbrunn et al. 2001). EPSP synthase inhibition by glyphosate prevents the biosynthesis of aromatic amino acids. However, a more rapid and dramatic effect is the reduction of feedback inhibition and increase of carbon flow through the shikimic acid pathway, which results in a rapid increase and accumulation of shikimic acid and, to a lesser extent, shikimate-derived benzoic acids (Duke et al. 2003).

Glyphosate salts are highly polar, water-soluble molecules with low lipophilicity that probably penetrate the overall lipophilic cuticle via diffusion through a hydrophilic pathway (hydrated cutin and pectin strands) into the apoplast (Caseley and Coupland 1985; Franz et al. 1997; Hess 1985). There is generally a rapid initial phase of uptake, followed by a period of slower uptake. The absorption of glyphosate by plant cells through the plasma membrane into the symplast involves a passive diffusion mechanism, and also an active transport mechanism (phosphate carrier) (Caseley and Coupland 1985; Franz et al. 1997; Sterling 1994). Glyphosate is rapidly translocated in most plants. It readily enters the symplast and is extensively translocated throughout all parts of the plant via the phloem, following the same distribution pattern as photoassimilates (i.e., source to sink relationship). It is the ionizable functionality of glyphosate, which has three acid groups and a strong amine base, that confers good symplastic mobility, and loss of one or more of these ionizable functionalities would affect its zwitterionic properties and decrease movement in the phloem (Bromilow and Chamberlain 2000). Although most glyphosate transport appears to be symplastic, sufficient apoplastic movement occurs to consider the herbicide as an ambimobile compound (Franz et al. 1997).

The slow or lack of metabolism of glyphosate in plants has been described as one characteristic associated with its high activity in plants (Franz et al. 1997). Glyphosate is known to be exuded from the roots of treated plants as the intact molecule. In fact, several studies have found no metabolism, or no metabolites, of glyphosate in plants (Coupland 1985). Nevertheless, other reports indicated that some plant species, including maize and soybean, are able to slowly degrade glyphosate to carbon dioxide and aminomethylphosphonic acid (AMPA), following a similar pathway of glyphosate metabolism proposed for soil bacteria (Coupland 1985; Franz et al. 1997). AMPA has been shown to be produced and be moderately toxic in glyphosate-treated GR soybean (Reddy et al. 2004). However, no glyphosate oxidoreductase (GOX), the enzyme that degrades glyphosate to AMPA and glyoxylate in some soil microbes, has been identified in plants (Duke et al. 2003).

6.3 MECHANISMS OF GLYPHOSATE RESISTANCE IN WEEDS

The mechanisms that can confer herbicide resistance can be broadly grouped into two categories: (1) target-site-based resistance and (2) nontarget-site-based resistance. Target-site-based resistance involves a modification of the site of action (e.g., EPSP synthase) so that the herbicide has reduced affinity and no longer binds to the altered target enzyme. This modification results from a single nucleotide change (i.e., mutation) in the gene encoding the enzyme to which the herbicide binds (Devine and Shukla 2000; Preston and Mallory-Smith 2001). Target-site-based resistance also includes overproduction of the target enzyme, either by gene amplification, gene over-expression, and increased enzyme stability (Pline-Srnic 2006). Nontarget-site-based resistance involves the exclusion of the herbicide molecule (i.e., glyphosate) from the target site due to differential uptake and/or translocation, sequestration, or increased metabolic detoxification (Preston 2004; Yuan et al. 2006).

Glyphosate was used worldwide for over 20 years with no evidence of weeds evolving resistance to this herbicide (Dyer 1994; Holt et al. 1993). Genetic and biochemical constraints on the evolution of a potential mechanism of resistance to glyphosate would seem to exist in higher plants (Jasieniuk 1995). Also, the unique properties of glyphosate such as its mode of action, chemical structure, limited metabolism in plants, and lack of residual activity in soil were proposed as possible reasons for the lack of glyphosate resistance in weeds (Bradshaw et al. 1997). However, evolved resistance to glyphosate in weed species was first reported in 1996 in *Lolium rigidum* in Australia (Pratley et al. 1996). Today, evolved resistance to glyphosate has been identified in 18 weed species around the world (Heap 2010). The mechanisms of glyphosate resistance that are known in weed species will be discussed, which include alteration of the target enzyme (i.e., EPSP synthase) in *Eleusine indica* and *Lolium* spp., reduced glyphosate translocation in *Conyza* spp. and *Lolium* spp., and gene amplification and increased EPSP synthase expression in *Amaranthus palmeri*.

6.3.1 *Eleusine indica*

Eleusine indica L. is an annual, self-pollinated, highly prolific weed species that has evolved resistance to several herbicides, including glyphosate. GR biotypes of *E. indica* were collected from orchards in Jahor, Malaysia, that had been treated with glyphosate an average of eight times per year for at least 10 years (Tran et al. 1999). Glyphosate resistance was confirmed in dose–response experiments that determined that the resistant biotype was two- to fourfold more resistant to glyphosate than the susceptible biotype. In a different study, a GR biotype of *E. indica* that was collected in Teluk Intan, Malaysia, was found to be 8- to 12-fold more resistant to glyphosate than the susceptible biotype (Lee and Ngim 2000).

The GR Jahor biotype was characterized, and the mechanism of glyphosate resistance was investigated (Tran et al. 1999). After treatment with glyphosate, the susceptible biotype accumulated approximately twofold more shikimic acid than the resistant biotype. There were minor differences in absorption and translocation of ^{14}C-glyphosate between the resistant and the susceptible biotypes (in fact, the resistant biotype did absorb and translocate slightly more glyphosate), but the differences were not associated with the mechanism of resistance. No significant breakdown of ^{14}C-glyphosate was detected in either biotype, indicating that the metabolism of glyphosate was not involved. EPSP synthase activity levels between the two biotypes indicated that resistance was not associated with enzyme over-expression. However, it was found that EPSP synthase from the resistant biotype was fivefold less sensitive to glyphosate (IC_{50}, glyphosate concentrations required to cause a 50% reduction in EPSP synthase activities from the susceptible and resistant biotypes were 3.0 and 16.0 μM glyphosate, respectively), indicating that the mechanism of resistance was associated with the target site (Baerson et al. 2002a).

EPSP synthase coding regions from both biotypes were compared by sequence analysis, and two nucleotide changes (from cytosine to thymine) that resulted in the amino acid substitutions proline (CCA) to serine (TCA) at position 106 (Pro_{106} to Ser), and proline (CCG) to leucine (CTG) at position 381 (Pro_{381} to Leu) were identified. Recombinant enzymes were constructed via site-directed mutagenesis and expressed in *Escherichia coli* to determine the apparent affinity for glyphosate. The kinetic data suggested that only the Pro_{106} to Ser substitution contributed significantly to the reduced glyphosate sensitivity resulting in a 16-fold increase in $K_{i(app)}$ (glyphosate) (Baerson et al. 2002a). This same amino acid substitution was present in the GR *aroA* gene (encoding EPSP synthase) from a mutagenized strain of *Salmonella typhimurium* and was determined to be the basis of resistance (Comai et al. 1983; Stalker et al. 1985). In a similar experiment, the site-directed mutation Pro_{106} to Ser of petunia EPSP synthase increased $K_{i(app)}$ (glyphosate) by 7.5-fold, indicating reduced sensitivity to glyphosate (Padgette et al. 1991). In a recent study, the Pro_{106} to Ser was strongly correlated with the resistant phenotype in a GR population of *E. indica* from the Philippines (Kaundun et al. 2008). Additional studies with GR *E. indica* from Malaysia revealed the same Pro_{106} to Ser amino acid substitution in the EPSP synthase gene in two resistant biotypes (Bidor and Temerloh), and also identified Pro_{106} to threonine (ACA) (Pro_{106} to Thr) in another resistant biotype (Chaah) (Ng et al. 2003, 2004). These results indicate that at least two different amino acid substitutions at Pro_{106} of the EPSP synthase gene are found in GR *E. indica*.

6.3.2 *Conyza* spp.

Conyza canadensis L. Cronq. is a winter or summer annual, self-pollinated, wind-dispersed weed species that is native to North America and has evolved resistance to several herbicides, including glyphosate. No-till crop production systems (e.g., maize, soybean, and cotton rotations) have been widely adopted in the Midwest and mid-South regions of the United States, which favored the establishment and growth of *C. canadensis* populations (Main et al. 2004; VanGessel 2001). Glyphosate resistant *C. canadensis* was first documented in 2001, in a population collected from a field in Delaware, USA, that had continuous Roundup Ready® (Monsanto, St. Louis, MO) soybean for 3 years. The resistant population was 8- to 13-fold more resistant to glyphosate than the susceptible population (VanGessel 2001). In the U.S. states of Mississippi and Tennessee, four *C. canadensis* populations evolved glyphosate resistance in fields that were planted with Roundup Ready® cotton and soybean for 3 years, and fields that were in no-till for 6 years (Koger et al. 2004). These populations showed similar levels of resistance to the resistant population from Delaware, ranging from 8- to 12-fold. Today, GR *C. canadensis* has been reported in 17 states in the United States, and also in China, Brazil, Spain, and the Czech Republic (Heap 2010).

Feng et al. (2004) investigated the potential mechanisms of glyphosate resistance in *C. canadensis* examining shikimic acid accumulation, foliar retention, absorption, translocation, and metabolism of glyphosate in susceptible and resistant plants collected from Delaware, USA, using simulated field and drop applications of ^{14}C-glyphosate. Tissue from both resistant and susceptible plants showed increased levels of shikimic acid when treated with glyphosate, indicating that the EPSP synthase remained sensitive. Likewise, shikimic acid accumulated in resistant and susceptible populations from Tennessee; however, the shikimic acid concentration declined about 40% from 2 to 4 days after treatment (DAT) in the resistant population (Mueller et al. 2003). The lower shikimic acid-to-glyphosate ratio in the resistant population suggested that glyphosate may be partially excluded from the plastids. Foliar spray retention and ^{14}C-glyphosate absorption were similar and equally as variable in both susceptible and resistant plants and were not correlated with glyphosate resistance. No metabolic deactivation of ^{14}C-glyphosate was detected in resistant plants, demonstrating that metabolism did not contribute to glyphosate resistance either. However, results from drop and spray applications of ^{14}C-glyphosate indicated reduced glyphosate translocation in resistant plants. Foliar translocation to roots, which was apparently the strongest sink, was two times greater in susceptible plants (40%) than in resistant plants (20%) at 4 DAT (Feng et al. 2004).

Similar results in which the resistant plants showed no differences in glyphosate absorption, but reduced glyphosate translocation, were obtained by Koger and Reddy (2005) and by Dinelli et al. (2006). Absorption of ^{14}C-glyphosate was similar (from 47% to 54%) in both susceptible and resistant plants of *C. canadensis* from Arkansas, Delaware, Mississippi, and Tennessee. However, the resistant plants retained more ^{14}C-glyphosate in the treated leaves than did the susceptible plants. At 48 h after treatment (HAT), the amount of ^{14}C-glyphosate that translocated from the treated leaf to other parts of the plant (other leaves, crown, and roots) ranged from 28% to 31% in the susceptible plants and from 16% to 20% in the resistant plants (Koger and Reddy 2005). Resistant plants of *C. canadensis* from Arkansas, Delaware, Ohio, and Virginia showed similar ^{14}C-glyphosate absorption. However, less ^{14}C-glyphosate translocated downward from leaves to roots and more translocated upward from culm to leaves compared with the susceptible plants (Dinelli et al. 2006). Furthermore, the resistant populations showed two- to threefold higher EPSP synthase mRNA relative levels without treatment with glyphosate. The over-expression of EPSP synthase may be considered as a second factor that could be involved in the mechanism of glyphosate resistance (Dinelli et al. 2006).

Conyza bonariensis is a summer or winter annual weed species that is native to South America and is, as with *C. canadensis*, a predominant weed in no-tillage cropping systems. Glyphosate resistance in *C. bonariensis* has been identified in South Africa, Spain, Brazil, Israel, Colombia, and California in the United States, with the first report in South African vineyards and orchards in 2003 (Heap 2010). Four populations (R1, R2, R3, and R4) from olive groves in Spain were characterized and determined to have glyphosate resistance indexes ranging between 3.5 and 10.5 (Urbano et al. 2007). A limited translocation pattern of ^{14}C-glyphosate similar to GR *C. canadensis* was observed in the resistant populations of *C. bonariensis* (Dinelli et al. 2008). In the resistant plants, less glyphosate translocated downward from treated leaves to the culm and roots, ranging from 13% to 18% of absorbed ^{14}C-glyphosate, whereas in the susceptible plants, the amount of absorbed ^{14}C-glyphosate that translocated from treated leaves to other parts of the plant was 35%. Furthermore, more glyphosate was translocated upward (from culm to leaves) in the resistant plants compared with the susceptible plants. In addition, a twofold increase of basal EPSP synthase mRNA (before treatment with glyphosate) was observed in the most resistant populations (R3 and R4), which suggests that glyphosate resistance could be partially conferred by over-expression of EPSP synthase in these populations (Dinelli et al. 2008).

6.3.3 *Lolium* spp.

Lolium rigidum Gaudin is an annual, self-incompatible, cross-pollinated weed species occurring in cereals and orchards that has evolved resistance to glyphosate and other herbicides, including photosystem II (PSII), acetyl coenzyme A carboxylase (ACCase), and acetolactate synthase (ALS) inhibitors. Glyphosate resistance in *L. rigidum* was first reported in 1996 (Pratley et al. 1996) and then documented in 1998 and 1999 (Powles et al. 1998; Pratley et al. 1999). Two *L. rigidum* populations from Australia evolved resistance following 15 years of glyphosate use. The first population (designated NLR68 and then NLR70) originated from an apple orchard in New South Wales and was 7- to 11-fold more resistant to glyphosate, while the second population (designated 118a and then 48118a) was collected from a continuously cropped field (rotation of summer and winter crops) in Northern Victoria and was 10-fold more resistant to glyphosate when compared with the susceptible population. Both populations were found to be resistant to the herbicide diclofop-methyl, an ACCase inhibitor (Powles et al. 1998; Pratley et al. 1999). In addition, a GR population from Western Australia (designated WALR 50) had multiple resistance to ACCase and ALS-inhibiting herbicides (Neve et al. 2004).

The potential mechanisms of glyphosate resistance were explored in the two resistant *L. rigidum* populations (NLR70 and 48118a). After treatment with glyphosate, the susceptible biotype accumulated approximately twofold more shikimic acid compared with the resistant biotype (48118a) (Baerson et al. 2002b). Little to no metabolism of glyphosate was detected in either resistant or susceptible plants (Feng et al. 1999; Lorraine-Colwill et al. 1999, 2003). EPSP synthase from resistant (NLR70) and susceptible plants were equally sensitive to inhibition by glyphosate (IC$_{50}$ values of 1.4 ± 0.25 and

1.2 ± 0.25 mM, respectively). In addition, no significant differences in the level of expression of EPSP synthase between the resistant (NLR70) and the susceptible populations were found (Lorraine-Colwill et al. 1999, 2003). Nevertheless, a two- to threefold increase in basal EPSP synthase mRNA and enzyme activity levels were observed in the most resistant lines derived from 48118a, indicating that EPSP synthase over-expression may play a partial role in the resistance mechanism in this population (Baerson et al. 2002b).

Initially, no major differences were observed between the resistant and the susceptible plants in respect of ^{14}C-glyphosate uptake and translocation (Feng et al. 1999; Lorraine-Colwill et al. 1999). However, results of further studies determined that in the resistant population (NLR70), glyphosate resistance was directly correlated with increased translocation of glyphosate to the leaf tips. At 48 HAT, 50% of the absorbed ^{14}C-glyphosate accumulated above the application site in the leaf tips of the resistant plants, compared with only 15% in the susceptible plants. The major accumulation of ^{14}C-glyphosate in the susceptible plants occurred below the application site in the leaf bases (55%) and roots (20%), whereas the resistant plants accumulated only 33% and 6% in the leaf bases and roots, respectively (Lorraine-Colwill et al. 2003).

Translocation of glyphosate was further investigated in one *L. rigidum* population collected from a vineyard in South Australia (designated SLR76) and two populations that originated from chemical fallow, summer crop rotations in New South Wales (designated NLR71 and NLR72) (Wakelin et al. 2004). Based on dose–response experiments, these populations were between 4- and 10-fold more resistant to glyphosate than the susceptible populations. Similarly, all resistant plants translocated significantly more ^{14}C-glyphosate to the tip of the treated leaf than did the susceptible plants (27–42% for the resistant plants, and 7–12% for the susceptible plants). Susceptible plants translocated from 26% to 28% of the absorbed ^{14}C-glyphosate to the stem portion of the plant, whereas the resistant plants translocated significantly less, ranging from 15% to 19% (Wakelin et al. 2004).

Glyphosate resistance in three different *L. rigidum* populations is due to a target-site mechanism (altered EPSP synthase), where resistant plants have an amino acid substitution of the EPSP synthase gene. A *L. rigidum* population collected from a vineyard in South Australia (designated SLR77) was threefold more resistant to glyphosate and had a translocation pattern similar to that of the susceptible population (Wakelin and Preston 2006). The resistant population SLR77 accumulated less shikimic acid than the susceptible population when treated with glyphosate, but more than the resistant population NLR70, which has limited glyphosate translocation. The EPSP synthase gene in the population SLR77 had one nucleotide change, from cytosine to adenine, at the first position of codon 106 that resulted in the amino acid substitution Pro_{106} to Thr. This is the same mutation that was found in one GR population of *E. indica* from Malaysia (Ng et al. 2003, 2004).

A *L. rigidum* population that was collected from an almond orchard in California, USA, had been intensively treated with glyphosate for 20 years

(Simarmata et al. 2003). After glyphosate treatment, a 10-fold increase in shikimic acid accumulation was observed in the susceptible population compared with the resistant population. No significant differences were detected in foliar absorption and translocation of ^{14}C-glyphosate between susceptible and resistant plants. The movement of ^{14}C-glyphosate into *in vitro* and *in vivo* chloroplast was not significantly different for susceptible and resistant plants. Glyphosate metabolism was not observed in either the susceptible or the resistant population. However, EPSP synthase from the resistant population was 90-fold less sensitive to inhibition by glyphosate (IC$_{50}$ values of 1068 and 12 μM for the resistant and the susceptible population, respectively). In the resistant population, sequence analysis of the EPSP synthase gene revealed a single nucleotide mutation from cytosine to thymine in the first position of codon 106 that results in the amino acid substitution Pro$_{106}$ to Ser (Simarmata and Penner 2008), which corresponds to the same amino substitution of the first reported glyphosate resistant *E. indica* population from Malaysia (Baerson et al. 2002a).

Because *L. rigidum* is a self-incompatible, cross-pollinated species, it is not surprising that one population had both nontarget-site- and target-site-based mechanisms of glyphosate resistance. This was the case for a population collected from a farm in South Africa (designated AFLR2) where glyphosate was used for over 25 years (Yu et al. 2007). Dose–response experiments established that the resistant population was 14-fold more resistant to glyphosate. Leaf uptake of ^{14}C-glyphosate was similar in both resistant and susceptible plants. However, there was significantly less ^{14}C-glyphosate translocated from treated leaves to untreated leaves in resistant plants compared with susceptible plants. In the resistant plants, 40% of the applied ^{14}C-glyphosate remained in the treated leaves and stem and 2% was translocated to untreated leaves, whereas in the susceptible plants, 30% of the applied ^{14}C-glyphosate remained in the treated leaves and stem and 4% was translocated to untreated leaves. In addition, sequence analysis of the EPSP synthase gene revealed a single nucleotide change in resistant plants, from cytosine to guanine, in the first position of codon 106, resulting in the amino acid substitution Pro$_{106}$ to alanine (GCA) (Pro$_{106}$ to Ala). The population AFLR2 also had evolved resistance to paraquat and to ACCase-inhibiting herbicides.

Lolium multiflorum Lam, as *L. rigidum*, is an annual, self-incompatible, cross-pollinated, major weed species of cereals and orchards that has evolved resistance to glyphosate and other herbicides, including PSII, ACCase, and ALS inhibitors. Glyphosate resistance in *L. multiflorum* was first reported in Chile (Perez and Kogan 2003; Perez et al. 2004). Two populations suspected to be GR were collected from fruit orchards that received continuous applications of glyphosate for 10 years. Dose–response experiments confirmed that both populations were resistant to glyphosate, with a level of resistance that ranged between two- and sixfold with respect to the susceptible population (Perez and Kogan 2003). Later, glyphosate resistance in *L. multiflorum* was reported in Oregon, California, Arkansas, and Mississippi in the United States,

Brazil, Spain, and Argentina (Heap 2010). In Oregon, USA, a GR *L. multiflorum* population was collected from a filbert orchard where glyphosate was applied continuously for 15 years (Perez-Jones et al. 2005). Dose–response experiments indicated that the Oregon population was fivefold more resistant to glyphosate than the susceptible population. After glyphosate treatment, the susceptible population accumulated between three and five times more shikimic acid than the resistant population. In Mississippi, USA, two *L. multiflorum* populations were selected from fields where Roundup Ready® soybean and cotton were grown continuously for 4 years. Based on dose–response experiments, it was determined that both populations were threefold more resistant to glyphosate than the susceptible population (Nandula et al. 2007). In California, USA, four GR *L. multiflorum* populations were identified out of 118 populations that were collected from a diversity of crops, including orchards, vineyards, and field crops, and also from noncropped areas. The GR *L. multiflorum* populations differed in the level of resistance, which ranged from 2- to 15-fold (Jasieniuk et al. 2008).

The potential mechanisms of glyphosate resistance were first explored in two resistant *L. multiflorum* populations with different origins, Chile (designated SF) and Oregon, USA (designated OR) (Perez-Jones et al. 2007). The susceptible population accumulated two and three times more shikimic acid after glyphosate application than the resistant OR and SF populations, respectively. Leaf uptake of ^{14}C-glyphosate between the susceptible and the GR populations was not different. However, there was a difference in the proportion of ^{14}C-glyphosate translocated from the treated leaf to the untreated parts of the plant in the GR OR population compared with the susceptible and the GR SF populations. At 72 HAT, 41% of the absorbed ^{14}C-glyphosate translocated above the treated section to the tip of the leaf in the OR population, in contrast to 16% and 24% in the susceptible and GR SF population, respectively (Perez-Jones et al. 2007). In a similar study, the two GR populations from Mississippi (designated T1 and T2) retained significantly more of the absorbed ^{14}C-glyphosate in the treated leaf (from 65% to 67%) at 48 HAT compared with the susceptible population (45%) (Nandula et al. 2008). The T1 population also absorbed less ^{14}C-glyphosate (43%) than the susceptible population (59%). In addition to an altered translocation pattern of ^{14}C-glyphosate, lower spray retention and foliar uptake by the abaxial leaf surface were described as possible factors involved in the mechanism of glyphosate resistance in an *L. multiflorum* population collected from a wheat field in Chile (Michitte et al. 2007).

Altered target-site (EPSP synthase) resistance has also been explored in GR *L. multiflorum*. The SF population from Chile has a cytosine to thymine nucleotide substitution in the first position of codon 106, which results in a Pro_{106} to Ser amino acid substitution (Perez-Jones et al. 2007). Plants from all four GR populations from California exhibited point mutations at Pro_{106}. One population had the Pro_{106} to Ser mutation, whereas the other three populations had the Pro_{106} to Ala mutation (Jasieniuk et al. 2008).

6.3.4   *Amaranthus* spp.

Amaranthus tuberculatus var. *tuberculatus* and *Amaranthus rudis* (now considered *A. tuberculatus* var. *rudis*; Costea and Tardif 2003) are summer annual, dioecious, very prolific species that are considered among the most problematic weeds in maize and soybean in the midwestern region of the United States. *A. palmeri* is also a summer annual, dioecious, very prolific species, and is considered one of the most troublesome weeds of cotton, soybean, and maize in the southern region of the United States. Several *Amaranthus* spp. populations have been reported to have either differential responses or resistance to glyphosate in various states of the United States after continuous use of glyphosate in Roundup Ready® crops, including populations of *A. tuberculatus* var. *tuberculatus* (Volenberg et al. 2007; Zelaya and Owen 2005), *A. tuberculatus* var. *rudis* (Legleiter and Bradley 2008; Smith and Hallett 2006), and *A. palmeri* (Culpepper et al. 2006; Norsworthy et al. 2008a, 2008b; Steckel et al. 2008).

The mechanisms conferring glyphosate resistance were explored in one *A. palmeri* population from Georgia (Culpepper et al. 2006), which was six- to eightfold more resistant to glyphosate than the susceptible population. Shikimic acid accumulation was detected in leaf tissue from the susceptible population and increased linearly as glyphosate concentration increased, whereas in the resistant population, shikimic acid was not detected regardless of the glyphosate concentration. No differences in ^{14}C-glyphosate absorption were observed between the GR and susceptible populations at 48 HAT. In addition, translocation of ^{14}C-glyphosate out of the treated leaf and distribution throughout the plant were similar for both populations (Culpepper et al. 2006).

Further investigations to determine the molecular basis of the mechanism of glyphosate resistance in one population of *A. palmeri* from Georgia were conducted by Gaines et al. (2010). An amino acid change, from arginine to lysine, at codon 316 of the mature EPSP synthase was found in resistant plants, but it was not considered to be the cause of glyphosate resistance, since EPSP synthase from GR and -susceptible plants was equally inhibited by glyphosate. Quantitative real time polymerase chain reaction (RT-PCR) on cDNA was used to measure EPSP synthase transcripts abundance relative to the ALS gene. Compared with susceptible plants, GR plants had on average 35-fold higher EPSP synthase expression relative to ALS. Furthermore, DNA blot analysis using an EPSP synthase probe showed that gDNA from resistant plants had higher hybridization intensity, indicating a higher copy number of EPSP synthase than in susceptible plants. Quantitative PCR revealed that EPSP synthase genomic copy number was positively correlated with EPSP synthase expression. Thus, genomic copy number of the EPSP synthase gene relative to ALS ranged from 1.0 to 1.3 for susceptible plants, whereas the relative copy number for GR plants was much higher, ranging from 5 to more than 160. Fluorescent in situ hybridization (FISH) showed that multiple EPSP synthase genes in GR plants were dispersed throughout the genome, suggesting that gene amplification was not caused by unequal chromosome crossing-over or genome duplication, and could have been the result of an associated mobile genetic element (e.g., transposon) that activated and amplified the EPSP synthase gene (Gaines et al. 2010). The elevated EPSP synthase copy number was heritable and correlated with a higher expression level of EPSP synthase and glyphosate resistance. Thus, the molecular basis of resistance in *A. palmeri* is due to increased production of EPSP synthase due to gene amplification, which corresponds to a novel mechanism of glyphosate resistance in a weed population. The authors did not rule out the possibility that one or more copies have higher expression due to promoter changes, or have a target-site mutation that has not been detected.

6.4 SUMMARY

Currently, glyphosate resistance has evolved in at least 18 weed species around the world. In some of these weed populations, the mechanisms conferring glyphosate resistance have been determined. Some populations of *E. indica* from Malaysia and the Philippines, *L. multiflorum* from Chile and the United States, and *L. rigidum* from Australia, South Africa, and the United States exhibit target-site-based glyphosate resistance due to an amino acid change at Pro_{106} of the EPSP synthase gene (Table 6.1).

A single amino acid substitution at Pro_{106} in the EPSP synthase gene codes for an altered EPSP synthase that has reduced affinity for glyphosate (Baerson et al. 2002a; Stalker et al. 1985). Pro_{106} is not in the active site, hence is not directly involved in glyphosate binding, which explains the low levels of glyphosate resistance conferred by these target-site mutations (Sammons et al. 2007). According to the crystal structure of *E. coli* EPSP synthase, glyphosate directly forms a hydrogen bond with the main-chain nitrogen atom of glycine at position 101 (Gly_{101}) (Zhou et al. 2006). The critical role of Gly_{101} in glyphosate binding was previously confirmed in petunia using site-directed mutagenesis, in which the Gly_{101} to Ala amino acid substitution increased the $K_{i(app)}$ (glyphosate) 5000-fold (Padgette et al. 1991). In fact, CP4 EPSP synthase from *Agrobacterium* sp. strain CP4, which confers glyphosate resistance in Roundup Ready® crops, has an alanine residue at the equivalent position and was found to significantly reduce glyphosate sensitivity (Funke et al. 2006). However, the Gly_{101} to Ala mutation also increased the $K_{m(app)}$ (PEP) 40-fold, affecting the catalytic activity of EPSP synthase by decreasing PEP affinity. Amino acid substitutions at Pro_{106} might affect the conformation of an α-helix starting from Gly_{101}, or the orientation of arginine (Arg) at position 105, which results in a movement of Gly_{101} that would reorient the binding site and cause a reduction in glyphosate affinity (Schönbrunn et al. 2001; Stallings et al. 1991; Zhou et al. 2006).

On the other hand, some populations of *L. rigidum* from Australia, *L. multiflorum*, and *C. canadensis* from the United States, and *C. bonariensis* from

TABLE 6.1. Glyphosate-Resistant Weed Species and the Identified Mechanism of Glyphosate Resistance

Species	Country	Situation	Mechanism of Resistance	References
Amaranthus palmeri	United States	Glyphosate-resistant cotton	Amplification of EPSP synthase	Gaines et al. (2010)
Conyza bonariensis	Spain	Olive orchard	Impaired glyphosate translocation; vacuolar sequestration	Dinelli et al. (2008)
Conyza canadensis	United States	Glyphosate-resistant soybean and cotton	Impaired glyphosate translocation	Dinelli et al. (2006); Koger and Reddy (2005); Feng et al. (2004); Ge et al. (2010)
Eleusine indica	Malaysia	Fruit orchard	Decreased EPSP synthase sensitivity (Pro$_{106}$ to Ser; Pro$_{106}$ to Thr)	Baerson et al. (2002a); Ng et al. (2003, 2004)
	Philippines	Noncropped area		Kaundun et al. (2008)
Lolium multiflorum	Chile	Almond orchard	Decreased EPSP synthase sensitivity (Pro$_{106}$ to Ser; Pro$_{106}$ to Ala)	Perez-Jones et al. (2007)
	United States	Orchards and vineyards		Jasieniuk et al. (2008)
	United States	Filbert orchard	Impaired glyphosate translocation	Perez-Jones et al. (2007)
	United States	Glyphosate-resistant soybean and cotton		Nandula et al. (2008)
Lolium rigidum	Australia	Fruit orchard	Impaired glyphosate translocation	Lorraine-Colwill et al. (2003); Wakelin et al. (2004)
	Australia	Chemical fallow		
	Australia	Vineyard	Decreased EPSP synthase sensitivity (Pro$_{106}$ to Thr)	Wakelin and Preston (2006)
	South Africa	Vineyard	Decreased EPSP synthase sensitivity (Pro$_{106}$ to Ala)	Yu et al. (2007)
	United States	Almond orchard	Decreased EPSP synthase sensitivity (Pro$_{106}$ to Ser)	Simarmata and Penner (2008)

Spain exhibit a nontarget-site-based mechanism involving limited or reduced translocation of glyphosate to meristematic tissues (Table 6.1). Glyphosate is ambimobile and is able to move via apoplast and symplast (Franz et al. 1997). Glyphosate absorption by plant cells through the plasma membrane involves a passive diffusion mechanism at high concentrations (absorption increases when external glyphosate concentration increases), and also an active transport mechanism via a phosphate transport system at low concentrations (Sterling 1994). Once glyphosate enters the symplasmic system, to be effective it needs to translocate via phloem to meristematic tissues (i.e., growing points), following the same distribution pattern as photoassimilates (i.e., source to sink movement) (Franz et al. 1997). Using *Abutilon theophrasti* as a model, Feng et al. (2003) determined that roots and meristems, where there is a high expression of EPSP synthase, were the most sensitive plant tissues. Likewise, reproductive organs, where there is a high demand of carbohydrates and amino acids, and where there is a higher expression of EPSP synthase, are very sensitive to glyphosate (Pline et al. 2002). Therefore, it would be reasonable to associate limited translocation to meristematic sinks with glyphosate resistance.

In some GR populations of *Lolium* spp., it appears that glyphosate translocates via xylem with the transpiration stream and gets trapped in the tip of the leaf (Lorraine-Colwill et al. 2003; Perez-Jones et al. 2007; Wakelin et al. 2004). Similarly, in some GR populations of *Conyza* spp., glyphosate is initially loaded into the apoplast and moves toward the leaf tip along with the transpiration stream and does not readily load into the phloem; therefore, less glyphosate translocates out of the treated leaf (Dinelli et al. 2006, 2008; Feng et al. 2004; Koger and Reddy 2005). In the susceptible populations, glyphosate is loaded into the phloem and is exported out of the treated leaf to meristematic sinks (i.e., roots and meristem). The actual mechanism of glyphosate resistance is unknown, and it is still unclear if in the resistant populations glyphosate is trapped in the apoplast and fails to be loaded into the symplast, or if it is effluxed out of the symplast to the apoplast via an active carrier. At low glyphosate concentrations, isolated leaf tissue from susceptible plants of *L. multiflorum* (Perez-Jones et al. 2005) and *C. canadensis* (Koger et al. 2005) accumulated more shikimic acid than the resistant plants, indicating a greater inhibition of EPSP synthase. In contrast, no differences were observed between the susceptible and the resistant plants at high glyphosate concentrations. Based on these observations, Shaner (2009) proposed that if the external glyphosate concentration is low, when glyphosate is absorbed via an active transporter, there is impairment to glyphosate absorption at the leaf cell level that prevents the accumulation of glyphosate in the symplasmic system. This impairment, however, does not affect glyphosate absorption at high glyphosate concentrations, when glyphosate is absorbed via passive diffusion.

There could be several mechanisms by which glyphosate absorption by the plant cell is reduced, resulting in limited or reduced glyphosate translocation (Lorraine-Colwill et al. 2003; Preston and Wakelin 2008; Shaner 2009): first, an alteration of the phosphate transporter responsible for the active transport of glyphosate at low concentrations, resulting in a loss of loading efficiency; second, an efflux transporter (i.e., a cellular glyphosate pump) that pumps glyphosate out of the cell into the apoplast; third, the evolution of a new transporter that pumps and sequesters glyphosate into the vacuole; and fourth, the evolution of a transporter at the chloroplast envelope that pumps glyphosate out of the chloroplast into the cytoplasm. Any of the first three mechanisms could be associated with the impairment of glyphosate absorption at the leaf cell level that appears to exist in GR populations of *Lolium* spp. and *Conyza* spp. (Shaner 2009). Recently, ^{31}P nuclear magnetic resonance (NMR) experiments on GR and -susceptible *C. canadensis* showed significantly more glyphosate accumulation within the vacuole in the resistant plants (Ge et al. 2010). At 24 HAT, mature leaves from GR plants showed greater than 85% glyphosate fractional occupancy of the vacuole, compared with less than 15% in susceptible plants. Thus, glyphosate enters the cytoplasm of both GR and -susceptible *C. canadensis*. However, glyphosate is removed from the cytoplasm and sequestered within the vacuole in GR plants, which can explain the reduced translocation previously observed (Ge et al. 2010). The evolution of a transporter at the chloroplast envelope that pumps glyphosate out of the chloroplast seems less likely to exist in GR populations. First, because there should be no difference in the translocation pattern between the GR and the -susceptible plants, and second, because glyphosate has been shown to be active in the cytoplasm by inhibiting the import of the precursor to EPSP synthase (pEPSPS) into the chloroplast (Della-Cioppa and Kishore 1988; Della-Cioppa et al. 1986). Adenosine triphosphate (ATP)-binding cassette (ABC) transporters are membrane proteins (e.g., P-glycoprotein; AtPgp1) that are involved in a wide range of functions, including pumping molecules (e.g., xenobiotics and other toxic compounds) out of cells and sequestration in the vacuole by an ATP-dependent mechanism (i.e., active transport), and hence have been associated with pesticide resistance (Buss and Callaghan 2008). These transporters belong to one of the most diverse and largest protein family, with 129 putative members in *Arabidopsis thaliana* (Sánchez-Fernández et al. 2001). Unpublished experiments have indicated that glyphosate upregulates several ABC transporters in GR *C. canadensis*, which might be related to the mechanism of limited translocation that exists in the resistant plants (Yuan et al. 2006). Over-expression of AtPgp1 and psNTP9 (the garden pea apyrase) in *A. thaliana* was shown to confer resistance to several herbicides, including pendimethalin, oryzalin, dicamba, and monosodium methylarsonate (MSMA), but not to glyphosate (Windsor et al. 2003). Thus, more evidence is needed to establish a direct association between ABC transporters and limited glyphosate translocation in GR weed populations.

Over-expression of EPSP synthase might play a role in the mechanism(s) of glyphosate resistance in some populations of *L. rigidum* (Baerson et al. 2002b), *C. canadensis* (Dinelli et al. 2006), and *C. bonariensis* (Dinelli et al. 2008), where limited translocation seems to be the basis of the mechanism of resistance. On the other hand, the increased production of EPSP synthase due to gene amplification is the molecular basis of glyphosate resistance in *A. palmeri* (Gaines et al. 2010). Increased production of EPSP synthase could be the result of gene over-expression, reduced enzyme turnover, or gene amplification. Several studies using tissue culture selection for glyphosate resistance in plants reported the selection of GR lines due to either increased rates of transcription or stable amplification of the EPSP synthase genes. For example, a GR *Petunia hybrida* cell line was shown to overproduce EPSP synthase as a result of a 20-fold amplification of the gene (Shah et al. 1986). However, the stability of resistance in cell cultures in the absence of glyphosate varied considerably. Thus, gene amplification in plants that were derived from tissue culture selection was not genetically stable or heritable; therefore, plants did not retain high levels of glyphosate resistance, or even the resistance was lost (Pline-Srnic 2006). This mechanism of glyphosate resistance in *A. palmeri*, involving overproduction of EPSP synthase by gene amplification, corresponds to an entirely novel mechanism not yet seen for glyphosate or any other herbicide.

There are several GR weed populations (e.g., *Ambrosia artemisiifolia*, *Ambrosia trifida*, and *Sorghum halepense*) for which the mechanism(s) of resistance are still unknown. Recent data indicate that the mechanism of resistance in two GR *A. artemisiifolia* populations from Arkansas does not involve an insensitive EPSP synthase or reduced glyphosate absorption or translocation (Brewer and Oliver 2006). Investigating the biochemical and molecular basis of these mechanisms will be essential to better understand the evolution of glyphosate resistance in weed species.

REFERENCES

Alibhai, M. F. and W. C. Stalling. 2001. Closing down on glyphosate inhibition—with a new structure for drug discovery. *Proceedings of the National Academy of Sciences of the United States of America* 98:2944–2946.

Baerson, S. R., D. Rodriguez, M. Tran, Y. Feng, N. A. Biest, and G. M. Dill. 2002a. Glyphosate-resistant goosegrass. Identification of a mutation in the target enzyme 5-enolpyruvylshikimate-3-phosphate synthase. *Plant Physiology* 129:1265–1275.

Baerson, S. R., D. Rodriguez, N. A. Biest, M. Tran, J. You, R. W. Kreuger, G. M. Dill., J. E. Pratley, and K. J. Gruys. 2002b. Investigating the mechanism of glyphosate resistance in rigid ryegrass (*Lolium rigidum*). *Weed Science* 50:721–730.

Baylis, A. 2000. Why glyphosate is a global herbicide: strengths, weaknesses and prospects. *Pest Management Science* 56:299–308.

Bradshaw, L. D., S. R. Padgette, S. L. Kimball, and B. H. Wells. 1997. Perspectives on glyphosate resistance. *Weed Technology* 11:189–198.

Brewer, C. E. and L. R. Oliver. 2009. Confirmation and resistance mechanisms in glyphosate-resistant common ragweed (*Ambrosia artemisiifolia*) in Arkansas. *Weed Science* 57:567–573.

Bromilow, R. and K. Chamberlain. 2000. The herbicide glyphosate and related molecules: physicochemical and structural factors determining their mobility in the phloem. *Pest Management Science* 56:368–373.

Buss, D. S. and A. Callaghan. 2008. Interaction of pesticides with p-glycoprotein and other ABC proteins: a survey of the possible importance to insecticide, herbicide and fungicide resistance. *Pesticide Biochemistry and Physiology* 90:141–153.

Caseley, J. C. and D. Coupland. 1985. Environmental and plant factors affecting glyphosate uptake, movement and activity. In E. Grossbard and D. Atkinson, eds. *The Herbicide Glyphosate*. London: Butterworths, pp. 92–123.

Comai, L., L. C. Sen, and D. M. Stalker. 1983. An altered aroA gene product confers resistance to the herbicide glyphosate. *Science* 221:370–371.

Costea, M. and F. J. Tardif. 2003. Conspectus and notes on the genus *Amaranthus* in Canada. *Rhodora* 105:260–281.

Coupland, D. 1985. Metabolism of glyphosate in plants. In E. Grossbard and D. Atkinson, eds. *The Herbicide Glyphosate*. London: Butterworths, pp. 25–34.

Culpepper, A. S., T. L. Grey, W. K. Vencill, J. M. Kichler, T. M. Webster, S. M. Brown, A. C. York, J. W. Davis, and W. W. Hanna. 2006. Glyphosate-resistant palmer amaranth (*Amaranthus palmeri*) confirmed in Georgia. *Weed Science* 54:620–626.

Della-Cioppa, G., S. C. Bauer, B. K. Klein, D. M. Shah, R. T. Fraley, and G. M. Kishore. 1986. Translocation of the precursor of 5-enolpyruvylshikimate-3-phosphate synthase into chloroplast of higher plants in vitro. *Proceedings of the National Academy of Sciences of the United States of America* 83:6873–6877.

Della-Cioppa, G. and G. M. Kishore. 1988. Import of a precursor protein into chloroplast is inhibited by the herbicide glyphosate. *EMBO Journal* 7:1299–1305.

Devine, M. and A. Shukla. 2000. Altered target sites as a mechanism of herbicide resistance. *Crop Protection* 19:881–889.

Dill, G. M. 2005. Glyphosate-resistant crops: history, status and future. *Pest Management Science* 61:219–224.

Dill, G. M., C. A. CaJacob, and S. R. Padgette. 2008. Glyphosate-resistant crops: adoption, use and future considerations. *Pest Management Science* 64:326–331.

Dinelli, G., I. Marotti, A. Bonetti, P. Catizone, J. M. Urbano, and J. Barnes. 2008. Physiological and molecular bases of glyphosate resistance in *Conyza bonariensis* biotypes from Spain. *Weed Research* 48:257–265.

Dinelli, G., I. Marotti, A. Bonetti, M. Minelli, P. Catizone, and J. Barnes. 2006. Physiological and molecular insight on the mechanisms of resistance to glyphosate in *Conyza canadensis* (L.) Cronq. biotypes. *Pesticide Biochemistry and Physiology* 86:30–41.

Duke, S. O., S. R. Baerson, and A. M. Rimando. 2003. Glyphosate. In D. W. Gammon and N. R. Ragsdale, eds. *Encyclopedia of Agrochemicals Online*. John Wiley & Sons. http://www.mrw.interscience.wiley.com/emrw/9780471263630/home/ (accessed September 2008).

Dyer, W. E. 1994. Resistance to glyphosate. In S. B. Powles and J.A.M. Holtum, eds. *Herbicide Resistance in Plants: Biology and Biochemistry*. Boca Raton, FL: CRC Press, pp. 229–242.

Feng, P. C. C., T. Chiu, and R. D. Sammons. 2003. Glyphosate efficacy is contributed by its tissue concentration and sensitivity in velvetleaf (*Abutilon theophrasti*). *Pesticide Biochemistry and Physiology* 73:87–91.

Feng, P. C. C., J. E. Pratley, and J. Bohn. 1999. Resistance to glyphosate in *Lolium rigidum*. II. Uptake, translocation, and metabolism. *Weed Science* 47:412–415.

Feng, P. C. C., M. Tran, T. Chiu, R. D. Sammons, G. R. Heck, and C. A. Cajacob. 2004. Investigations into glyphosate-resistant horseweed (*Conyza canadensis*): retention, uptake, translocation, and metabolism. *Weed Science* 52:498–505.

Franz, J. E., M. K. Mao, and J. A. Sikorski. 1997. *Glyphosate: A Unique Global Herbicide*. ACS Monograph 189. Washington, DC: American Chemical Society.

Funke, T., H. Han, M. L. Healy-Fried, M. Fischer, and E. Schönbrunn. 2006. Molecular basis for the herbicide resistance of Roundup Ready® crops. *Proceedings of the National Academy of Sciences of the United States of America* 103:13010–13015.

Gaines, T. A., W. Zhang, D. Wang, B. Bukun, S. T. Chisholm, D. L. Shaner, S. J. Nissen, W. L. Patzoldt, P. J. Tranel, A. S. Culpepper, T. L. Grey, T. M. Webster, W. K. Vencill, R. D. Sammons, J. Jiang, C. Preston, J. E. Leach, and P. Westra. 2010. Gene amplification confers glyphosate resistance in *Amaranthus palmeri*. *Proceedings of the National Academy of Science of the United States of America* 107:1029–1034.

Ge, X., D. André d' Avignon, J. J. H. Ackerman, and R. D. Sammons. 2010. Rapid vacuolar sequestration: the horseweed glyphosate resistance mechanism. *Pest Management Science* 66:345–348.

Geiger, D. R. and M. A. Fuchs. 2002. Inhibitors of aromatic amino acid biosynthesis (glyphosate). In P. Böger, K. Wakabayashi, and K. Hirai, eds. *Herbicide Classes in Development*. Berlin: Springer-Verlag, pp. 59–85.

Gianessi, L. 2005. Economic and herbicide use impacts of glyphosate-resistant crops. *Pest Management Science* 61:241–245.

Gianessi, L. 2008. Economic impacts of glyphosate-resistant crops. *Pest Management Science* 64:346–352.

Gianessi, L. and N. Reigner. 2006. *Pesticide Use in U.S. Crop Production: 2002 with Comparison to 1992 and 1997*. Washington, DC: Crop Life Foundation.

Heap, I. M. 2010. International Survey of Herbicide-Resistant Weeds. http://www.weedscience.org (accessed April 8, 2010).

Herrmann, K. M. and L. M. Weaver. 1999. The shikimate pathway. *Annual Reviews of Plant Physiology and Plant Molecular Biology* 50:473–503.

Hess, F. D. 1985. Herbicide absorption and translocation and their relationship to plant tolerances and susceptibility. In S. O. Duke, ed. *Weed Physiology. Vol. II. Herbicide Physiology*. Boca Raton, FL: CRC Press, pp. 191–214.

Holt, J. S., S. B. Powles, and J. A. M. Holtum. 1993. Mechanisms and agronomic aspects of herbicide resistance. *Annual Reviews of Plant Physiology and Plant Molecular Biology* 44:203–229.

Jasieniuk, M. 1995. Constraints on the evolution of glyphosate resistance in weeds. *Resistant Pest Management* 7:31–32.

Jasieniuk, M., R. Ahmad, A. M. Sherwood, J. L. Firestone, A. Perez-Jones, W. T. Lanini, C. Mallory-Smith, and Z. Stednick. 2008. Glyphosate-resistant Italian ryegrass (*Lolium multiflorum*) in California: distribution, response to glyphosate, and molecular evidence for an altered target enzyme. *Weed Science* 56:496–502.

Kaundun, S. S., I. A. Zelaya, R. P. Dale, A. J. Lycett, P. Carter, K. R. Sharples, and E. McIndoe. 2008. Importance of the P106S target-site mutation in conferring resistance to glyphosate in a goosegrass (*Eleusine indica*) population from the Philippines. *Weed Science* 56:637–646.

Kishore, G. M. and D. M. Shah. 1988. Amino acid biosynthesis inhibitors as herbicides. *Annual Reviews of Biochemistry* 57:627–663.

Koger, C. H., D. H. Poston, R. M. Hayes, and R. F. Montgomery. 2004. Glyphosate-resistant horseweed in Mississippi. *Weed Technology* 18:820–825.

Koger, C. H. and K. N. Reddy. 2005. Role of absorption and translocation in the mechanism of glyphosate resistance in horseweed (*Conyza canadensis*). *Weed Science* 53:84–89.

Koger, C. H., D. L. Shaner, W. B. Henry, T. Nadler-Hassar, W. E. Thomas, and J. W. Wilcut. 2005. Assessment of two nondestructive assays for detecting glyphosate resistance in horseweed (*Conyza canadensis*). *Weed Science* 53:559–566.

Lee, L. J. and J. Ngim. 2000. A first report of glyphosate-resistant goosegrass (*Eleusine indica* (L) Gaertn) in Malaysia. *Pest Management Science* 56:336–339.

Legleiter, T. R. and K. W. Bradley. 2008. Glyphosate and multiple herbicide resistance in common waterhemp (*Amaranthus rudis*) populations from Missouri. *Weed Science* 56:582–587.

Lorraine-Colwill, D. F., T. R. Hawkes, P. H. Williams, S. A. J. Warner, P. B. Sutton, S. B. Powles, and C. Preston. 1999. Resistance to glyphosate in *Lolium rigidum*. *Pesticide Science* 55:486–503.

Lorraine-Colwill, D. F., S. B. Powles, T. R. Hawkes, P. H. Hollinshead, S. A. J. Warner, and C. Preston. 2003. Investigations into the mechanism of glyphosate resistance in *Lolium rigidum*. *Pesticide Biochemistry and Physiology* 74:62–72.

Main, C. L., T. C. Mueller, R. M. Hayes, and J. B. Wilkerson. 2004. Response of selected horseweed (*Conyza canadensis* (L.) Cronq.) populations to glyphosate. *Journal of Agricultural and Food Chemistry* 52:879–883.

Michitte, P., R. De Prado, N. Espinoza, J. P. Ruiz-Santaella, and C. Gauvrit. 2007. Mechanisms of resistance to glyphosate in a ryegrass (*Lolium multiflorum*) biotype from Chile. *Weed Science* 55:435–440.

Mueller, T. C., J. H. Massey, R. M. Hayes, C. L. Main, and C. N. Stewart. 2003. Shikimate accumulates in both glyphosate-sensitive and glyphosate-resistant horseweed (*Conyza canadensis* L. Cronq.). *Journal of Agricultural and Food Chemistry* 51:680–684.

Nandula, V. K., D. H. Poston, T. W. Eubank, C. H. Koger, and K. N. Reddy. 2007. Differential response to glyphosate in Italian ryegrass (*Lolium multiflorum*) populations from Mississippi. *Weed Technology* 21:477–482.

Nandula, V. K., K. N. Reddy, D. H. Poston, A. M. Rimando, and S. O. Duke. 2008. Glyphosate tolerance mechanism in Italian ryegrass (*Lolium multiflorum*) from Mississippi. *Weed Science* 56:344–349.

Neve, P., J. Sadler, and S. B. Powles. 2004. Multiple herbicide resistance in a glyphosate-resistant rigid ryegrass (*Lolium rigidum*) population. *Weed Science* 52:920–928.

Ng, C. H., R. Wickneswari, S. Salmijah, and B. S. Ismail. 2004. Glyphosate resistance in *Eleusine indica* (L.) Gaertn. from different origins and polymerase chain reaction amplification of specific alleles. *Australian Journal of Agricultural Research* 55:407–414.

Ng, C. H., R. Wickneswari, S. Salmijah, Y. T. Teng, and B. S. Ismail. 2003. Gene polymorphisms in glyphosate-resistant and -susceptible biotypes of *Eleusine indica* from Malaysia. *Weed Research* 43:108–115.

Norsworthy, J. K., G. M. Griffith, R. C. Scott, K. L. Smith, and L. R. Oliver. 2008a. Confirmation and control of glyphosate-resistant Palmer amaranth (*Amaranthus palmeri*) in Arkansas. *Weed Technology* 22:108–113.

Norsworthy, J. K., R. C. Scott, K. L. Smith, and L. R. Oliver. 2008b. Response of northeastern Arkansas Palmer amaranth (*Amaranthus palmeri*) accessions to glyphosate. *Weed Technology* 22:408–413.

Padgette, S. R., D. B. Re, C. S. Gasser, D. A. Eichholtz, R. B. Frazier, C. M. Hironaka, E. B. Levine, D. M. Shah, R. T. Fraley, and G. M. Kishore. 1991. Site-directed mutagenesis of a conserved region of the 5-enolpyruvylshikimate-3-phosphate synthase active site. *Journal of Biological Chemistry* 266:22364–22369.

Perez, A., C. Alister, and M. Kogan. 2004. Absorption, translocation, and allocation of glyphosate in resistant and susceptible Chilean biotypes of *Lolium multiflorum*. *Weed Biology and Management* 4:56–58.

Perez, A. and M. Kogan. 2003. Glyphosate-resistant *Lolium multiflorum* in Chilean orchards. *Weed Research* 43:12–19.

Perez-Jones, A., K. W. Park, J. Colquhoun, C. Mallory-Smith, and D. Shaner. 2005. Identification of glyphosate resistant Italian ryegrass (*Lolium multiflorum*) in Oregon. *Weed Science* 53:775–779.

Perez-Jones, A., K. W. Park, N. Polge, J. Colquhoun, and C. Mallory-Smith. 2007. Investigating the mechanisms of glyphosate resistance in *Lolium multiflorum*. *Planta* 226:395–404.

Pline, W., R. Viator, J. Wilcut, K. L. Edmisten, J. Thomas, and R. Wells. 2002. Reproductive abnormalities in glyphosate-resistant cotton caused by lower CP4-EPSPS levels in the male reproductive tissue. *Weed Science* 50:438–447.

Pline-Srnic, W. 2006. Physiological mechanisms of glyphosate resistance. *Weed Science* 20:290–300.

Powles, S. B. 2008. Evolved glyphosate-resistant weeds around the world: lessons to be learnt. *Pest Management Science* 64:360–365.

Powles, S. B., D. F. Lorraine-Colwill, J. J. Dellow, and C. Preston. 1998. Evolved resistance to glyphosate in rigid ryegrass (*Lolium rigidum*) in Australia. *Weed Science* 46:604–607.

Powles, S. B. and C. Preston. 2006. Evolved glyphosate resistance in plants: biochemical and genetic basis of resistance. *Weed Technology* 20:282–289.

Pratley, J., P. Baines, P. Eberbach, M. Incerti, and J. Broster. 1996. Glyphosate resistance in annual ryegrass. In J. Virgona and D. Michalk, eds. *Proceedings of the 11th Annual Conference of the Grassland Society of New South Wales*. Wagga Wagga, Australia: The Grassland Society of New South Wales, p. 126.

Pratley, J., N. Urwin, R. Stanton, P. Baines, J. Broster, K. Cullis, D. Schafer, J. Bohn, R. Krueger. 1999. Resistance to glyphosate in *Lolium rigidum*. I. Bioevaluation. *Weed Science* 47:405–411.

Preston, C. 2004. Herbicide resistance in weeds endowed by enhanced detoxification: complications for management. *Weed Science* 52:448–453.

Preston, C. and C. A. Mallory-Smith. 2001. Biochemical mechanisms, inheritance, and molecular genetics of herbicide resistance in weeds. In S. Powles and D. Shaner, eds. *Herbicide Resistance and World Grains*. Boca Raton, FL: CRC Press, pp. 23–60.

Preston, C. and A. M. Wakelin. 2008. Resistance to glyphosate from altered herbicide translocation patterns. *Pest Management Science* 64:372–376.

Reddy, K. N., A. M. Rimando, and S. O. Duke. 2004. Aminomethylphosphonic acid, a metabolite of glyphosate, causes injury in glyphosate-treated, glyphosate-resistant soybean. *Journal of Agricultural and Food Chemistry* 52:5139–5143.

Sammons, R. D., D. C. Heering, N. Dinicola, H. Glick, and G. A. Elmore. 2007. Sustainability and stewardship of glyphosate and glyphosate resistant crops. *Weed Technology* 21:347–354.

Sánchez-Fernández, R., T. G. E. Davies, J. O. D. Coleman, and P. A. Rea. 2001. The *Arabidopsis thaliana* ABC protein superfamily: a complete inventory. *Journal of Biological Chemistry* 276:30231–30244.

Schönbrunn, E., S. Eschenburg, W. A. Shuttleworth, J. V. Schloss, N. Amrhein, J. N. S. Evans, and W. Kabsch. 2001. Interaction of the herbicide glyphosate with its target enzyme 5-enolpyruvylshikimate 3-phosphate synthase in atomic detail. *Proceedings of the National Academy of Sciences of the United States of America* 98:1376–1380.

Shah, D. M., R. B. Horsch, H. J. Klee, G. M. Kishore, J. A. Winter, N. E. Tumer, C. M. Hironaka, P. R. Sanders, C. S. Gasser, S. Aykent, N. R. Siegel, S. G. Rogers, and R. T. Fraley. 1986. Engineering herbicide tolerance in transgenic plants. *Science* 233:478–481.

Shaner, D. L. 2009. Role of translocation as a mechanism of resistance to glyphosate. *Weed Science* 57:118–123.

Siehl, D. L. 1997. Inhibitors of EPSP synthase, glutamine synthetase and histidine synthesis. In R. M. Roe, J. D. Burton, and R. J. Kuhr, eds. *Herbicide Activity: Toxicology, Biochemistry and Molecular Biology*. Amsterdam: IOS Press, pp. 37–67.

Simarmata, M., J. E. Kaufmann, and D. Penner. 2003. Potential basis of glyphosate resistance in California rigid ryegrass (*Lolium rigidum*). *Weed Science* 51:678–682.

Simarmata, M. and D. Penner. 2008. The basis for glyphosate resistance in rigid ryegrass (*Lolium rigidum*) from California. *Weed Science* 56:181–188.

Smith, D. A. and S. G. Hallett. 2006. Variable response of common waterhemp (*Amaranthus rudis*) populations and individuals to glyphosate. *Weed Technology* 20:466–471.

Stalker, D. M., W. R. Hiatt, and L. Comai. 1985. A single amino acid substitution in the enzyme 5-enolpyruvylshikimate-3-phosphate synthase confers resistance to the herbicide glyphosate. *Journal of Biological Chemistry* 260:4724–4728.

Stallings, W. C., S. S. Abdel-Meguid, L. W. Lim, H. S. Shieh, H. E. Dayringer, N. K. Leimgruber, R. A. Stegeman, K. S. Anderson, J. A. Sikorski, S. R. Padgette, G. M. Kishore. 1991. Structure and topological symmetry of the glyphosate target 5-enol-pyruvylshikimate-3-phosphate synthase: a distinctive protein fold. *Proceedings of the National Academy of Sciences of the United States of America* 88:5046–5050.

Steckel, L. E., C. L. Main, A. T. Ellis, and T. C. Mueller. 2008. Palmer amaranth (*Amaranthus palmeri*) in Tennessee has low level glyphosate resistance. *Weed Technology* 22:119–123.

Steinrücken, H. and N. Amrhein. 1980. The herbicide glyphosate is a potent inhibitor of 5-enolpyruvylshikimic acid-3-phosphate synthase. *Biochemical and Biophysical Research Communications* 94:1207–1212.

Sterling, T. M. 1994. Mechanisms of herbicide absorption across plant membranes and accumulation in plant cells. *Weed Science* 42:263–276.

Tran, M., S. Baerson, R. Brinker, L. Casagrande, M. Faletti, Y. Feng, M. Nemeth, T. Reynolds, D. Rodriguez, D. Shaffer, D. Stalker, N. Taylor, Y. Teng, and G. Dill. 1999. Characterization of glyphosate resistant *Eleusine indica* biotypes from Malaysia. In *Proceedings 1 (B) of the 17th Asian-Pacific Weed Science Society Conference.* Bangkok, Thailand: The Asian-Pacific Weed Science Society, pp. 527–536.

Urbano, J. M., A. Borrego, V. Torres, J. M. Leon, C. Jimenez, G. Dinelli, and J. Barnes. 2007. Glyphosate-resistant hairy fleabane (*Conyza bonariensis*) in Spain. *Weed Technology* 21:396–401.

VanGessel, M. J. 2001. Glyphosate-resistant horseweed from Delaware. *Weed Science* 49:703–705.

Volenberg, D., W. L. Patzoldt, A. Hager, and P. J. Tranel. 2007. Responses of contemporary and historical waterhemp (*Amaranthus tuberculatus*) accessions to glyphosate. *Weed Science* 55:327–333.

Wakelin, A. M., D. F. Lorraine-Colwill, and C. Preston. 2004. Glyphosate resistance in four different population of *Lolium rigidum* is associated with reduced translocation of glyphosate to meristematic zones. *Weed Research* 44:453–459.

Wakelin, A. M. and C. Preston. 2006. A target-site mutation is present in a glyphosate-resistant *Lolium rigidum* population. *Weed Research* 46:432–440.

Windsor, B., S. J. Roux, and A. Lloyd. 2003. Multiherbicide tolerance conferred by AtPgp1 and apyrase overexpression in *Arabidopsis thaliana*. *Nature Biotechnology* 21:428–433.

Woodburn, A. T. 2000. Glyphosate: production, pricing and use worldwide. *Pest Management Science* 56:309–312.

Yu, Q., A. Cairns, and S. Powles. 2007. Glyphosate, paraquat and ACCase multiple herbicide resistance evolved in a *Lolium rigidum* biotype. *Planta* 225:499–513.

Yuan, J. S., P. J. Tranel, and N. Stewart. 2006. Non-target-site herbicide resistance: a family business. *Trends in Plant Science* 12:6–13.

Zelaya, I. A. and M. D. K. Owen. 2005. Differential response of *Amaranthus tuberculatus* (Moq ex DC) JD Sauer to glyphosate. *Pest Management Science* 61:936–950.

Zhou, M., H. Xu, X. Wei, Z. Ye, L. Wei, W. Gong, Y. Wang, and Z. Zhu. 2006. Identification of a glyphosate-resistant mutant of rice 5-enolpyruvylshikimate 3-phosphate synthase using a directed evolution strategy. *Plant Physiology* 140:184–195.

7

GLYPHOSATE RESISTANCE: GENETIC BASIS IN WEEDS

MICHAEL J. CHRISTOFFERS AND ARUNA V. VARANASI

7.1 OVERVIEW

The herbicidal activity of glyphosate is due to the inhibition of chloroplastic 5-enolpyruvylshikimate-3-phosphate synthase (EPSPS), a critical enzyme in the production of the aromatic amino acids, phenylalanine, tyrosine, and tryptophan. At least two mechanisms of resistance to glyphosate among weeds have been discovered. One mechanism is characterized by reduced translocation of glyphosate, for which the molecular genetic basis is yet to be understood. A second mechanism is associated with reduced EPSPS sensitivity to glyphosate due to nonsynonymous point mutations within a critical codon of the EPSPS coding region. Reduced translocation generally confers 8- to 12-fold resistance to glyphosate compared with susceptible biotypes, while the level of resistance conferred by target-site alteration tends to only be two- to fourfold above susceptibles (reviewed in Kaundun et al. 2008). A third potential mechanism for which there is recent evidence also involves the EPSPS target site, but is associated with gene amplification leading to increased expression of susceptible EPSPS (Gaines et al. 2009).

7.2 FREQUENCY AND FITNESS OF GLYPHOSATE RESISTANCE GENES AMONG WEEDS

The unselected frequency of glyphosate resistance genes in weed populations is generally considered to be relatively low, owing in part to the herbicide's

Glyphosate Resistance in Crops and Weeds: History, Development, and Management
Edited by Vijay K. Nandula
Copyright © 2010 John Wiley & Sons, Inc.

141

unique mode of action and limited metabolism in plants (Bradshaw et al. 1997). Hence, glyphosate-resistant (GR) weeds were not always considered to be a potential problem of economic significance. The low probability of glyphosate resistance resulting from a point mutation was supported by the findings of Jander et al. (2003), who screened 250,000 M_2 mouse-ear cress (*Arabidopsis thaliana* (L.) Heynh.) plants produced from ethyl methanesulfonate (EMS) mutagenesis and did not find GR mutants. Despite apparent constraints on the evolution of glyphosate resistance, the appearance of GR weeds has proven that variants exist at a high enough frequency to provide genetic variation upon which selection can act. The high frequency of glyphosate usage worldwide has no doubt contributed to the appearance of GR monocot and dicot weeds. As with other herbicides, there is evidence for independent mutation events leading to resistance within species (Jasieniuk et al. 2008).

The initial, unselected frequency of GR mutants is difficult to estimate due to their rarity and a lack of information on the variety of mutations that can confer glyphosate resistance. Weed populations that have not been exposed to glyphosate and have been genetically isolated from potentially selected populations are also not readily available. However, gene frequency is an important aspect of GR weed management and researchers using simulation models to assess management strategies need to estimate this frequency. Neve et al. (2003) estimated the initial frequency of glyphosate resistance genes in rigid ryegrass (*Lolium rigidum* Gaudin) to be between 1×10^{-8} and 1×10^{-6}, and Werth et al. (2008) also used these same estimates. However, based on modeling results, Neve (2008) suggested that an initial gene frequency estimate of 1×10^{-8} may be too high.

The pleiotropic effects of herbicide resistance genes are difficult to elucidate, in part due to the diverse genetic backgrounds of herbicide-resistant weeds. However, there is evidence that glyphosate resistance in weeds may be associated with reduced fitness, and that fitness costs may in part contribute to the relatively low frequency of glyphosate resistance prior to selection. Baucom and Mauricio (2004) found that glyphosate-tolerant tall morningglory (*Ipomoea purpurea* (L.) Roth) produced 35% fewer seeds compared with susceptible plants when not treated with glyphosate. Pedersen et al. (2007) did not observe differences in competitiveness between resistant and susceptible rigid ryegrass. However, the resistant phenotype did produce fewer, albeit larger, seeds compared with susceptible plants when grown in the absence of wheat (*Triticum aestivum* L.) or in low wheat densities.

Rigid ryegrass with the reduced translocation mechanism of resistance was also shown to become less frequent compared with susceptibles over the course of 3 years when glyphosate was not applied (Preston and Wakelin 2008). Wakelin and Preston (2006c, cited in Stanton et al. 2008) also observed that resistance frequency in rigid ryegrass decreased with time in the absence of glyphosate treatment. These studies suggest that reduced fitness associated with glyphosate resistance may be exploited in the management of some GR weeds (Preston et al. 2009).

7.3   INHERITANCE OF GLYPHOSATE RESISTANCE IN WEEDS

Studies investigating glyphosate resistance in weeds have typically revealed inheritance patterns consistent with an incompletely dominant single nuclear gene. This has been demonstrated in rigid ryegrass (Lorraine-Colwill et al. 2001), horseweed (*Conyza canadensis* (L.) Cronq.) (Zelaya et al. 2004, 2007), and goosegrass (*Eleusine indica* (L.) Gaertn.) (Ng et al. 2004a). Incompletely dominant resistance is typical of both reduced translocation and altered EPSPS mechanisms of resistance in weeds (reviewed in Powles and Preston 2006). However, single-gene resistance expression among populations of rigid ryegrass with reduced glyphosate translocation was found to vary from incompletely dominant to dominant (Preston et al. 2009; Preston and Wakelin 2008; Wakelin and Preston 2006b). The glyphosate response of suscepti-ble × resistant F_1 hybrids also suggested dominant resistance in horseweed (Feng et al. 2004).

Glyphosate resistance in weeds is not always a single-gene trait. Simarmata et al. (2005) found the inheritance of glyphosate resistance in rigid ryegrass from California to be consistent with two nuclear, incompletely dominant genes. Reduced glyphosate translocation, in addition to an EPSPS target-site mutation, was found to be responsible for glyphosate resistance in a rigid ryegrass biotype, indicating the presence of at least two genes (Yu et al. 2007). Variability in herbicide response among plants after recurrent selection suggested that decreased response to glyphosate might be a quantitative trait in tall waterhemp (*Amaranthus tuberculatus* (Moq.) Sauer), with multiple genes being one explanation for this observation (Zelaya and Owen 2005). Lorraine-Colwill et al. (2001) also suggested that additional minor genes may contribute to the resistance mainly provided by a single, major gene in rigid ryegrass. The existence of minor genes influencing response to glyphosate is supported by research in field bindweed (*Convolvulus arvensis* L.) (Duncan and Weller 1987).

7.4   MOLECULAR GENETICS OF GLYPHOSATE RESISTANCE IN WEEDS

Researchers have identified nonsynonymous point mutations within prokary-otic EPSPS that confer resistance to glyphosate (Comai et al. 1983; reviewed in Kaundun et al. 2008). However, similar point mutations among higher plant EPSPSs appear particularly prone to reduced fitness (Jander et al. 2003; Sammons et al. 2007). Despite the evolutionary constraints on altered target-site resistance to glyphosate, substitutions at amino acid position 106 have been identified among herbicide-resistant weeds (Table 7.1). Proline$_{106}$ is conserved among wild-type EPSPS gene sequences, but replacement of this amino acid with serine have been discovered among GR goosegrass (Baerson et al. 2002b; Kaundun et al. 2008; Ng et al. 2003, 2004b), Italian ryegrass

TABLE 7.1. Amino Acid Substitutions within EPSPS of Glyphosate-Resistant Weeds

Species	Amino Acid Position 106[a]	Country of Origin	Reference
Wild type	Proline	—	—
Goosegrass	Serine	Malaysia	Baerson et al. (2002b); Ng et al. (2003, 2004b)
		Philippines	Kaundun et al. (2008)
	Threonine	Malaysia	Ng et al. (2003, 2004b)
Italian	Alanine	United States	Jasieniuk et al. (2008)
ryegrass	Serine	Chile	Jasieniuk et al. (2008); Perez-Jones et al. (2007)
Rigid	Alanine	South Africa	Yu et al. (2007)
ryegrass	Serine	Australia	F. C. Dolman and C. Preston (unpublished data, cited in Preston et al. 2009)
		United States	Simarmata and Penner (2008)
	Threonine	Australia	F. C. Dolman and C. Preston (unpublished data, cited in Preston et al. 2009); Wakelin and Preston (2006a)

[a]Amino acid position number is based on Padgette et al. (1996) and Baerson et al. (2002b).

(*Lolium multiflorum* Lam.) (Jasieniuk et al. 2008; Perez-Jones et al. 2007), and rigid ryegrass (Simarmata and Penner 2008). Substitution of proline$_{106}$ with threonine has also been identified in goosegrass (Ng et al. 2003, 2004b) and rigid ryegrass (Wakelin and Preston 2006a). Additionally, proline$_{106}$-to-alanine substitutions have been identified in Italian ryegrass (Jasieniuk et al. 2008) and rigid ryegrass (Yu et al. 2007). Whether or not these mutations alone can fully account for field-selectable resistant biotypes is still unclear (Sammons et al. 2007). However, the growing number of reports identifying these mutations among herbicide-resistant weeds suggests that they play a notable role in the occurrence of GR weeds worldwide. Targeted mutagenesis in plants, although not yet readily achievable, shows promise for facilitating the assessment of specific mutation effects (Li et al. 2007), and will likely play an eventual role in GR weed research.

A few studies have compared EPSPS expression levels in resistant and susceptible weed biotypes in order to determine if target-site over-expression contributes to glyphosate resistance. Levels of EPSPS mRNA were found to be similar in resistant and susceptible rigid ryegrass (Lorraine-Colwill et al. 2003), but Baerson et al. (2002a) found untreated rigid ryegrass to have 2.5- to 3-fold higher levels of EPSPS mRNA in the most resistant biotypes compared with susceptibles or plants with intermediate resistance. GR horseweed also had levels of EPSPS mRNA that were 1.8- to 3.1-fold higher than susceptible biotypes (Dinelli et al. 2006). However, EPSPS over-expression did not seem to fully explain the level of resistance in these rigid ryegrass and horseweed biotypes.

Gaines et al. (2009) found EPSPS expression levels 30- to 40-fold higher in resistant Palmer amaranth (*Amaranthus palmeri* S. Wats.) compared with susceptible plants. An investigation of EPSPS gene copy number revealed gene amplification as a likely genetic mechanism of increased EPSPS expression and glyphosate resistance. Gene amplification has been implicated as a glyphosate resistance mechanism in cell culture of cultivated species (Widholm et al. 2001), but the Palmer amaranth research of Gaines et al. (2009) is the first report of such a mechanism among naturally occurring weeds.

7.5 FUTURE OUTLOOK

The diversity of gene mutations capable of conferring GR phenotypes among weeds have not been fully explored, especially for reduced glyphosate translocation. Further research is likely to reveal additional resistance mechanisms with diverse genetic causes, underscoring how little is known at present. Minor genes responsible for subtle variations in glyphosate response and their role in glyphosate resistance, both realized and potential, also seems to be underappreciated. Continued selection of glyphosate resistance in weeds will provide increased opportunity for minor genes to act in concert through additive effects. Weed populations will probably continue to evolve with continued selection, and it is likely that we have not yet seen the full capacity of weed adaptation to glyphosate.

While our current understanding of the genetics of glyphosate resistance among weeds is still rudimentary, it is apparent that a diversity of genes and alleles may be responsible for resistance to glyphosate worldwide. This suggests that research to investigate optimum management of GR weeds may not be universally applicable to all cases. Continued identification of the genetic causes of glyphosate resistance is necessary to better classify resistance types, thus improving weed control and glyphosate stewardship.

ACKNOWLEDGMENTS

This material is based on work supported by the Cooperative State Research, Education, and Extension Service, United States Department of Agriculture, under Agreement No. 2006-34361-16992; the Minnesota Soybean Research and Promotion Council; and the North Dakota Experimental Program to Stimulate Competitive Research.

REFERENCES

Baerson, S. R., D. J. Rodriguez, N. A. Biest, M. Tran, J. S. You, R. W. Kreuger, G. M. Dill, J. E. Pratley, and K. J. Gruys. 2002a. Investigating the mechanism of glyphosate resistance in rigid ryegrass (*Lolium ridigum*). *Weed Science* 50:721–730.

Baerson, S. R., D. J. Rodriguez, M. Tran, Y. M. Feng, N. A. Biest, and G. M. Dill. 2002b. Glyphosate-resistant goosegrass. Identification of a mutation in the target enzyme 5-enolpyruvylshikimate-3-phosphate synthase. *Plant Physiology* 129:1265–1275.

Baucom, R. S. and R. Mauricio. 2004. Fitness costs and benefits of novel herbicide tolerance in a noxious weed. *Proceedings of the National Academy of Sciences of the United States of America* 101:13386–13390.

Bradshaw, L. D., S. R. Padgette, S. L. Kimball, and B. H. Wells. 1997. Perspectives on glyphosate resistance. *Weed Technology* 11:189–198.

Comai, L., L. C. Sen, and D. M. Stalker. 1983. An altered aroA gene product confers resistance to the herbicide glyphosate. *Science* 221:370–371.

Dinelli, G., I. Marotti, A. Bonetti, M. Minelli, P. Catizone, and J. Barnes. 2006. Physiological and molecular insight on the mechanisms of resistance to glyphosate in *Conyza canadensis* (L.) Cronq. biotypes. *Pesticide Biochemistry and Physiology* 86:30–41.

Duncan, C. N. and S. C. Weller. 1987. Heritability of glyphosate susceptibility among biotypes of field bindweed. *Journal of Heredity* 78:257–260.

Feng, P. C. C., M. Tran, T. Chiu, R. D. Sammons, G. R. Heck, and C. A. CaJacob. 2004. Investigations into glyphosate-resistant horseweed (*Conyza canadensis*): retention, uptake, translocation, and metabolism. *Weed Science* 52:498–505.

Gaines, T., C. Preston, D. Shaner, J. Leach, S. Chisholm, B. Bukun, S. Ward, A. S. Culpepper, P. Tranel, and P. Westra. 2009. A novel mechanism of resistance to glyphosate in Palmer amaranth (*Amaranthus palmeri*). *Abstracts of the Weed Science Society of America* 49:368.

Jander, G., S. R. Baerson, J. A. Hudak, K. A. Gonzalez, K. J. Gruys, and R. L. Last. 2003. Ethylmethanesulfonate saturation mutagenesis in Arabidopsis to determine frequency of herbicide resistance. *Plant Physiology* 131:139–146.

Jasieniuk, M., R. Ahmad, A. M. Sherwood, J. L. Firestone, A. Perez-Jones, W. T. Lanini, C. Mallory-Smith, and Z. Stednick. 2008. Glyphosate-resistant Italian ryegrass (*Lolium multiflorum*) in California: distribution, response to glyphosate, and molecular evidence for an altered target enzyme. *Weed Science* 56:496–502.

Kaundun, S. S., I. A. Zelaya, R. P. Dale, A. J. Lycett, P. Carter, K. R. Sharples, and E. McIndoe. 2008. Importance of the P106S target-site mutation in conferring resistance to glyphosate in a goosegrass (*Eleusine indica*) population from the Philippines. *Weed Science* 56:637–646.

Li, J., A.-P. Hsia, and P. S. Schnable. 2007. Recent advances in plant recombination. *Current Opinion in Plant Biology* 10:131–135.

Lorraine-Colwill, D. F., S. B. Powles, T. R. Hawkes, P. H. Hollinshead, S. A. J. Warner, and C. Preston. 2003. Investigations into the mechanism of glyphosate resistance in *Lolium rigidum*. *Pesticide Biochemistry and Physiology* 74:62–72.

Lorraine-Colwill, D. F., S. B. Powles, T. R. Hawkes, and C. Preston. 2001. Inheritance of evolved glyphosate resistance in *Lolium rigidum* (Gaud.). *Theoretical and Applied Genetics* 102:545–550.

Neve, P. 2008. Simulation modelling to understand the evolution and management of glyphosate resistant in weeds. *Pest Management Science* 64:392–401.

Neve, P., A. J. Diggle, F. P. Smith, and S. B. Powles. 2003. Simulating evolution of glyphosate resistance in *Lolium rigidum*. I: population biology of a rare resistance trait. *Weed Research* 43:404–417.

Ng, C. H., W. Ratnam, S. Surif, and B. S. Ismail. 2004a. Inheritance of glyphosate resistance in goosegrass (*Eleusine indica*). *Weed Science* 52:564–570.

Ng, C. H., R. Wickneswari, S. Salmijah, Y. T. Teng, and B. S. Ismail. 2004b. Glyphosate resistance in *Eleusine indica* (L.) Gaertn. from different origins and polymerase chain reaction amplification of specific alleles. *Australian Journal of Agricultural Research* 55:407–414.

Ng, C. H., R. Wickneswari, S. Salmijah, Y. T. Teng, and B. S. Ismail. 2003. Gene polymorphisms in glyphosate-resistant and -susceptible biotypes of *Eleusine indica* from Malaysia. *Weed Research* 43:108–115.

Padgette, S. R., D. B. Re, G. F. Barry, D. E. Eichholtz, X. Delannay, R. L. Fuchs, G. M. Kishore, and R. T. Fraley. 1996. New weed control opportunities: development of soybeans with a Roundup Ready gene. In S. Duke, ed. *Herbicide-Resistant Crops: Agricultural, Economic, Environmental, Regulatory, and Technological Aspects*. Boca Raton, FL: CRC Press, pp. 53–84.

Pedersen, B. P., P. Neve, C. Andreasen, and S. B. Powles. 2007. Ecological fitness of a glyphosate-resistant *Lolium rigidum* population: growth and seed production along a competition gradient. *Basic and Applied Ecology* 8:258–268.

Perez-Jones, A., K. W. Park, N. Polge, J. Colquhoun, and C. A. Mallory-Smith. 2007. Investigating the mechanisms of glyphosate resistance in *Lolium multiflorum*. *Planta* 226:395–404.

Powles, S. B. and C. Preston. 2006. Evolved glyphosate resistance in plants: biochemical and genetic basis of resistance. *Weed Technology* 20:282–289.

Preston, C. and A. M. Wakelin. 2008. Resistance to glyphosate from altered herbicide translocation patterns. *Pest Management Science* 64:372–376.

Preston, C., A. M. Wakelin, F. C. Dolman, Y. Bostamam, and P. Boutsalis. 2009. A decade of glyphosate-resistant *Lolium* around the world: mechanisms, genes, fitness, and agronomic management. *Weed Science* 57:435–441.

Sammons, R. D., D. C. Heering, N. Dinicola, H. Glick, and G. A. Elmore. 2007. Sustainability and stewardship of glyphosate and glyphosate-resistant crops. *Weed Technology* 21:347–354.

Simarmata, M., S. Bughrara, and D. Penner. 2005. Inheritance of glyphosate resistance in rigid ryegrass (*Lolium rigidum*) from California. *Weed Science* 53:615–619.

Simarmata, M. and D. Penner. 2008. The basis for glyphosate resistance in rigid ryegrass (*Lolium rigidum*) from California. *Weed Science* 56:181–188.

Stanton, R. A., J. E. Pratley, D. Hudson, and G. M. Dill. 2008. A risk calculator for glyphosate resistance in *Lolium rigidum* (Gaud.). *Pest Management Science* 64:402–408.

Wakelin, A. M. and C. Preston. 2006a. A target-site mutation is present in a glyphosate-resistant *Lolium rigidum* population. *Weed Research* 46:432–440.

Wakelin, A. M. and C. Preston. 2006b. Inheritance of glyphosate resistance in several populations of rigid ryegrass (*Lolium rigidum*) from Australia. *Weed Science* 54:212–219.

Wakelin, A. and C. Preston. 2006c. The cost of glyphosate resistance: is there a fitness penalty associated with glyphosate resistance in annual ryegrass? In C. Preston, J. H. Watts, and N. D. Crossman, eds. *15th Australian Weeds Conference Proceedings: Managing Weeds in a Changing Climate.* Torrens Park, SA: Weed Management Society of South Australia, pp. 515–518.

Werth, J. A., C. Preston, I. N. Taylor, G. W. Charles, G. N. Roberts, and J. Baker. 2008. Managing the risk of glyphosate resistance in Australian glyphosate-resistant cotton production systems. *Pest Management Science* 64:417–421.

Widholm, J. M., A. R. Chinnala, J.-H. Ryu, H.-S. Song, T. Eggett, and J. E. Brotherton. 2001. Glyphosate selection of gene amplification in suspension cultures of 3 plant species. *Physiologia Plantarum* 112:540–545.

Yu, Q., A. Cairns, and S. Powles. 2007. Glyphosate, paraquat and ACCase multiple herbicide resistance evolved in a *Lolium rigidum* biotype. *Planta* 225:499–513.

Zelaya, I. A. and M. D. K. Owen. 2005. Differential response of *Amaranthus tuberculatus* (Moq ex DC) JD Sauer to glyphosate. *Pest Management Science* 61:936–950.

Zelaya, I. A., M. D. K. Owen, and M. J. VanGessel. 2004. Inheritance of evolved glyphosate resistance in *Conyza canadensis* (L.) Cronq. *Theoretical and Applied Genetics* 110:58–70.

Zelaya, I. A., M. D. K. Owen, and M. J. VanGessel. 2007. Transfer of glyphosate resistance: evidence of hybridization in *Conyza* (Asteraceae). *American Journal of Botany* 94:660–673.

8

GENOMICS OF GLYPHOSATE RESISTANCE

C. Neal Stewart, Jr., Yanhui Peng, Laura G. Abercrombie,
Matthew D. Halfhill, Murali R. Rao, Priya Ranjan,
Jun Hu, R. Douglas Sammons, Gregory R. Heck,
Patrick J. Tranel, and Joshua S. Yuan

8.1 INTRODUCTION

Genomics is a subject area composed of molecular biology and bioinformatics tools concerned about the structure and function of genes and genomes. There are several other "omics" technologies that are often placed within the realm of genomics, including transcriptomics and proteomics, which, taken together, are often referred to as systems biology (Yuan et al. 2008a). Systems biology, genomics included, is studied using sophisticated methodologies and instrumentation for high-throughput data collection and analysis. These data are manipulated and parsed using bioinformatics tools that are equally sophisticated. Led by tremendous funding and breakthroughs in genomics pioneered first in biomedical fields, the methods, equipment, and computational resources have trickled down to the world of plant biology and agriculture, where they are now relatively inexpensive (per datum) and becoming routinely used in a few laboratories. Weed biology and other much applied agricultural areas have been slow to embrace genomics, however (Stewart et al. 2009). Discussions about using genomics to answer research problems in weed biology have ensued just recently (Basu et al. 2004; Chao et al. 2005), and in fact very little weed genomics has been accomplished to date. Much like the weather, it is something that we have enjoyed talking about, but nobody did anything about

Glyphosate Resistance in Crops and Weeds: History, Development, and Management
Edited by Vijay K. Nandula
Copyright © 2010 John Wiley & Sons, Inc.

it! That situation is changing rapidly (Stewart 2009; Stewart et al. 2009). The transcriptomes of several weedy species are being sequenced to answer questions of weed biology relevance, including glyphosate resistance—we know of at least four species (*Conyza canadensis, Amaranthus rudis, Centaurea solstitialis*, and *Orobanche* sp.) in which next-generation sequencing is being used. Therefore, we expect to see dramatic increases of available sequence data for weeds, which should also spur greater interest in weedy plant genomics.

8.2 TARGET-SITE GENOMICS

Herbicide resistance can be the result of either of two mechanisms: changes in the target site, that is, within the sequences of the gene encoding the enzyme that is inhibited by the herbicide, and changes in the nontarget site, for example, within other genes that might impact the ability of the herbicide to reach or affect its target enzyme. Target-site resistance typically requires little to no genomics since it invokes a single gene cause—either a mutation in the coding sequences that alters herbicide binding or function or a mutation that alters expression of the targeted enzyme such that it is made in sufficient quantities that normal levels of herbicide cannot prevent enzymatic action.

Upon discovering a new case of glyphosate resistance, an intuitive first step would be to test for alterations within the target site. For example, mutations in 5-enolypyruvylshikimate-3-phosphate synthase (EPSPS), such as a change from proline at the 106th position to alanine, serine, or threonine, are known to confer glyphosate resistance (Alibhai et al. 2004). Such target-site mutations have been found to be responsible for conferring glyphosate resistance in goosegrass, *Eleusine indica* (Baerson et al. 2002; Ng et al. 2003), and the ryegrasses, *Lolium rigidum* and *Lolium multiflorum* (Perez-Jones et al. 2007; Simarmata and Penner 2008; Wakelin and Preston 2006). These data can be obtained by conventional sequencing technologies following amplification or cloning for the target gene from resistant and susceptible biotypes.

Another possible mechanism for target-site mutations that could confer glyphosate resistance is increased expression of the EPSPS gene. Increased expression could result from mutations with the promoter that result in binding of additional transcription factors, modifications in intron sequences that result in more efficient splicing, or modifications in the coding or noncoding sequences in the mRNA that increase the stability or translation of the message. It is also feasible, and indeed demonstrated, that glyphosate resistance, albeit in tissue cultured plant cells, resulted from amplification of EPSPS genes (Widholm et al. 2001). Increased expression in intact plants has been suggested as a mechanism for glyphosate resistance in horseweed, *C. canadensis* and hairy fleabane, *Conyza bonariensis*. Researchers reported that EPSPS transcripts were higher in resistant biotypes compared with susceptible biotypes (Dinelli et al. 2006, 2008). However, in both of these studies, a "semiquantitative" reverse transcriptase polymerase chain reaction (RT-

PCR) study was used (Nebenfuhr and Lomax 1998), and the transcription of only one of the transcribed EPSPS genes (in these species there are two active genes and one pseudo-gene) was examined. While there was apparent increased transcription in both species' EPSPS genes, 1.5- to 3.0-fold higher in resistant *C. canadensis* (Dinelli et al. 2006) and twofold higher in resistant *C. bonariensis* (Dinelli et al. 2008), the suboptimal assay choice coupled with examining the transcription of just one of the EPSPS genes indicates that the picture is far from complete with regard to target gene expression in resistant biotypes in *Conyza*. Therefore, no conclusive data exist that show a clear cause and effect of increased gene expression and glyphosate resistance in the refereed literature. That said, EPSPS gene duplication resulting in increased expression and glyphosate resistance was recently reported to have evolved in wild populations of Palmer amaranth (*Amaranthus palmeri*). Gaines et al. (2010) found between 2 to over 100 copies of the EPSPS gene in resistant Palmer amaranth. This increased copy number correlated with increased gene expression and EPSPS protein accumulation. It is yet unknown if EPSPS gene duplication is the result of unequal crossovers between randomly duplicated genes at a single locus, or association of the EPSPS gene with an active transposable element that resulted in multiple copies of the EPSPS gene.

Investigating target-site resistance is typically straightforward; however, as the above discussions illustrate, target-site resistance may involve more than a simple point mutation in the DNA coding sequence and characterizing the cause of resistance might require advanced analysis. The *Conyza* cases illustrate the importance of assay choice, and the Palmer amaranth case begs for follow-up research. As weed genomics becomes a reality, it will be increasingly important to choose the best assays to sort out complicated data. Technologies such as RT-PCR should replace older technologies such as Northern and Southern blotting and semiquantitative RT-PCR with the caveat that proper controls and statistical methods are used (Yuan et al. 2006, 2008b). It is also important to establish that alterations in gene expression or copy number result in increased accumulation and/or activity of the encoded protein since numerous mechanisms might alter mRNA stability, translation, or protein function and thus limit the correlation between gene expression and resistance. Thus, confirmation of over-expression should include enzyme-linked immunosorbent assay (ELISA) or Western blotting to confirm protein accumulation, as well as enzyme assays.

As we segue from analyzing single genes and proteins using relatively simple and mature procedures to genomics, where analyzing the sequence and expression of thousands of genes simultaneously are performed, there are many scientific and technical hurdles that will need to be overcome. Indeed, interrelated problems are that genomics is currently still very expensive—as measured by price per experiment and equipment requirements, and weed biologists largely lack training in genomics and molecular biology (Stewart et al. 2009). This situation begets other technical and scientific issues when practicing weed genomics; but these issues are largely tractable.

8.3 NONTARGET-SITE GENOMICS

Whereas target-site gene analysis typically requires relatively simple gene cloning, Sanger sequencing, and basic gene expression analysis, nontarget-site changes are likened to the proverbial needle in the haystack, and this is assuming that there is only one needle! Once a target-site mechanism is eliminated as a possibility, then the strategy becomes "1 down and 29,999 to go"—that is to say, finding a nontarget-site resistance molecular mechanism is much more complex and requires genomics-based techniques that allow one to analyze thousands of genes simultaneously. This complexity coupled with the lack of genomic information and resources in weed genomics makes for difficult work indeed (Stewart et al. 2009).

Based on the known number of genes in several plant and animal species, it is likely that the average plant genome contains between 30,000 and 50,000 genes. Although studying this many genes may seem daunting, few are likely targets for resistance. Yuan et al. (2007) laid out a theoretical framework of nontarget-site resistance mechanisms, by examining the problem from known biochemical and metabolic flux data, which is summarized below. In addition, plant physiological mechanisms for known nontarget-site resistance mesh well with this theoretical framework (Yuan et al. 2007).

Nontarget resistance could result from the possible modification of glyphosate, thereby detoxifying it. In addition, the detoxified, or even toxic, versions could be sequestered into subcellular compartments, which are important considerations if herbicide is not excluded from cells—another possible mechanism. Although it is possible that some other undetermined physiological or developmental mechanism might result in resistance, studies on antibiotic resistance in bacteria suggest modification, transport, and/or sequestration are the most likely. Therefore, we should be able to narrow our gene space to just a few families or classes of genes that could confer resistance; only a few thousand genes have the potential to result in resistance. We believe that genomics offers our only hope to delineate the cause of nontarget-site resistance.

8.3.1 Detoxification and Conjugation

In plant cells, detoxification of herbicides is often accomplished via oxidation by cytochrome P450 monooxygenases, peroxidases, and other compounds. One non-P450 enzymate example is the detoxification and oxidation of glyphosate by the *gox* gene, which encodes glyphosate oxidoreductase (GOX) that degrades glyphosate to glyoxalate (Tan et al. 2006). Most data from herbicide oxidation implicates P450s as the chief responsible enzyme family that could play a role in the evolution of herbicide detoxification. The plant P450 gene family is large, with 246 genes among 44 subfamilies that have been identified in *Arabidopsis thaliana* (Nelson et al. 2004). Other plant species seem to have more cytochrome P450 genes, for example, over 550 each in rice and poplar (http://drnelson.utmem.edu/CytochromeP450.html). This diversification has arisen through gene duplications and divergence, especially multiple local tandem duplications. While there are no cases of naturally evolved glyphosate resistance in weeds via altered P450 activity, transgenic experiments have demonstrated that modified P450 expression can increase glyphosate resistance (Yuan et al. 2007). Over-expression studies of a single P450 gene in transgenic plants endowed its host with resistance to up to 13 different herbicides (Hirose et al. 2005).

Another gene family that has been shown to participate in herbicide resistance is the glutathione S-transferase (GST) gene family. Plant GSTs catalyze the conjugation of glutathione (γ-glutamyl-cysteinyl-glycine) to various substrates (R-X) to form a polar S-glutothionylated product (R-SG), thereby serving to detoxify compounds (Yuan et al. 2007). Detoxification can occur in conjunction with R-SG products being sequestered into vacuoles by transporters such as adenosine triphosphate (ATP)-binding cassette (ABC) transporters (Dixon et al. 2002; Reade et al. 2004). While no GSTs have been shown to be active in glyphosate resistance, they play a role in the nontarget resistance to other herbicide chemistries (Yuan et al. 2007).

8.3.2 Transporters

Various membrane-associated proteins are responsible for passive and active transport of compounds across membranes. A large gene family of active transport proteins is the ABC transporters. Subdivided among nine subfamilies, there are 129 members in *A. thaliana* (Sánchez-Fernández et al. 2001; Verrier et al. 2008). Plants seem to have greater ABC transporter diversity than any other type of organism. ABC transporters are of interest with regard to glyphosate resistance because of their diverse substrates. Hence, they are good candidates to screen against sequestration of herbicides and their metabolites into subcellular compartments. ABC transporters are responsible for a wide range of functions in plants including export of toxins, sequestration of plant secondary metabolites, translocation of fatty acids and phospholipids, and cell homeostasis (Schulz and Kolukisaglu 2006). Also of interest is transport of xenobiotics and wobble among substrates (see, e.g., Mentewab and Stewart 2005). ABC transporters can be targeted to any component of the endomembrane system, but of particular interest here is tonoplast targeting; glyphosate pumped into vacuoles would be rendered harmless and such a mechanism corresponds with described physiological effects. Even though little glyphosate resistance work has been published in this gene family, there exist physiological studies showing affinity to some other herbicides. For example, *Arabidopsis* AtMRP1 was shown to be the first ABC transporter able to transport the glutathione S (GS)-conjugated herbicide metolachlor (Lu et al. 1997). Likewise, multiple plant ABC transporters have been shown to transport different herbicides and herbicide metabolites (Klein et al. 2006; Liu et al. 2001; Schulz and Kolukisaglu 2006).

8.4   THE BIOCHEMISTRY AND PHYSIOLOGY CONNECTION

In cases of glyphosate resistance in which target-site mutations have been eliminated or deemed not critical as mechanisms, reduced translocation of herbicide has been implicated as the physiological mechanism for glyphosate resistance (reviewed by Shaner 2009). These cases include several important weeds, such as *C. canadensis* (Feng et al. 2004; Koger and Reddy 2005), *C. bonariensis* (Dinelli et al. 2008), and the ryegrasses *L. rigidum* (Wakelin and Preston 2006) and *L. multiflorum* (Perez-Jones et al. 2007). These studies indicate a similar pattern of lack of transport from source leaves that receive glyphosate application to other parts of the plant. In resistant horseweed, serious injury and death occurs on sprayed tissue, but meristems seem to be protected from a lethal dose and regrow after a few weeks. Indeed, the physiological data indicate that glyphosate is largely sequestered in source leaves, preventing its transport to other organs (Ge et al. 2010). Thus, a priori, there is reason to search for a cellular mechanism that allows sequestration to vacuoles or another organelle within cells, with the exception of chloroplasts, the site of EPSPS inhibition. The most effective means to search among candidate gene families is using genomics methods, of which two types are illustrated in our own research described next.

8.4.1   *C. canadensis* Case Study: The Search for Transporters

Horseweed was the first dicot that evolved glyphosate resistance; hence, we adopted it as a weedy genomics model for glyphosate resistance. It has a small genome of about 335 Mb, from our own studies using genome size estimation based on flow cytometry. This is a tractably sized genome, only slightly more than two times the size of *A. thaliana*. It is self-fertile and is relatively easy to maintain in low-light growth rooms until plants bolt, when they quickly outgrow light rack spacing. In addition, the nontarget mechanism appears to have a relatively simple Mendelian inheritance, indicating that a single locus was responsible for conferring resistance, perhaps endowed by a single gene (Halfhill et al. 2007; Zelaya et al. 2004). Thus, the most parsimonious genomic search would be for genes that were upregulated by glyphosate in a resistant biotype compared with an isogenic susceptible genotype with a narrowed search for those genes involved with transport. It seems obvious that to perform genomics research, genomics resources should be available to facilitate success. When we started the research, however, very few genomics tools were available for weeds (Basu et al. 2004). On the other hand, we were routinely performing microarray experiments using the long oligo *A. thaliana* arrays from the University of Arizona in other research projects. Thus, we performed experiments to hybridize horseweed cDNAs with the *Arabidopsis* two-color arrays.

8.4.1.1   Microarray Analysis   Microarrays are glass slides or chips that contain parts of genes, consisting of representative DNA. The cDNA of the transcriptomes of interest are then hybridized to microarrays, thereby allowing the simultaneous transcriptomic analysis of thousands of genes at once. Ideally, a microarray will contain gene sequences of the species of interest, but few plant species have whole genome arrays that are commercially available. We first demonstrated successful hybridization of horseweed leaf cDNAs with an *A. thaliana* microarray in a proof-of-concept study. The arrays were obtained from the University of Arizona, which contained a spotted Qiagen Operon (Huntsville, AL) *A. thaliana* Genome Oligo Set, Version 1.0 (26,000 unique DNA sequences) on glass slides for hybridization. This near-whole genome oligonucleotide array was probed with *Arabidopsis* cDNAs (with one fluorescent dye) and horseweed cDNAs (with a second fluorescence dye), using a dye-swap protocol. The results demonstrated that this approach was feasible since approximately 15,000–18,000 spots yielded positive hybridization at over 50% background (± 2 SD) among the four array replicates. Indeed, there is precedence for using a heterologous approach for microarray studies, even with species that are more divergent from *Arabidopsis* than horseweed such as *Avena fatua* and *Euphorbia esula* (Horvath et al. 2003). The sequence information available for *A. thaliana* can clearly be applied to elucidate potential molecular mechanisms for plants such as horseweed and other weedy species in which little genomic information is known (Horvath et al. 2003).

We next conducted a preliminary microarray experiment to probe the response to glyphosate treatment of a single Tennessee (TN) accession of glyphosate-resistant horseweed. Horseweed plants were grown on soil in a growth room under a 16-h photoperiod at ambient temperatures. Young leaves and meristematic tissue from 3- to 4-month-old plants were used as source materials for harvested tissue. Microarray analysis was performed on a comparison of two treatment groups: (1) resistant horseweed from western Tennessee (Mueller et al. 2003) sprayed with field rate of glyphosate ($0.84 \, \text{kg a.e.ha}^{-1}$ Roundup WeatherMAX, Monsanto Company, St. Louis, MO) and (2) a water-sprayed respective control, harvested 24 h post treatment. Both glyphosate and water applications were performed with a backpack sprayer.

Total RNA was extracted from three biological replicates of each treatment group (eight plants per replicate) utilizing TriReagent according to the manufacturer's protocol (MRC, Cincinnati, OH). After purification with RNeasy columns (QIAGEN, Valencia, CA), mRNA was isolated using the Oligotex mRNA purification kit (QIAGEN). Transcripts were labeled using the SuperScript Plus Direct cDNA Labeling System for DNA microarray (Invitrogen, Carlsbad, CA).

A total of six hybridized chips including dye-swap technical replication were used. Slides were scanned using GenePix 4000B microarray scanner (Axon Instruments, Union City, CA) and analyzed using GenePix Pro 4.1 software for spot detection and intensity determination. Bad spots were flagged and removed. Microarray data was normalized with the Loess transformation and $\log_2$ transformed using SAS software version 9.2 (SAS Institute, Inc., Cary, NC).

Microarray data was analyzed using rank products (Breitling et al. 2004). Bioconductor (www.bioconductor.org) RankProd package was used to perform the rank product analysis (Gentleman et al. 2004; Hong et al. 2006). The false discovery rate (FDR) value obtained was based on 10,000 random permutations. Since 10,000 random permutations was very computer intensive, 1000 random permutations were performed 10 different times, each time starting with a different random seed number, and the average FDR value was used for further analysis. The genes that had FDR values less than or equal to 0.10 were considered as differentially expressed.

Increased gene expression from the glyphosate treatment of resistant horseweed is reported on a linear scale (fold change) in Table 8.1. The top 20 upregulated genes are displayed with four ABC transporter genes on the list, with the most upregulated gene being an ABC transporter (Table 8.1). In addition to the ABC transporters, also identified was a mitochondrial phosphate transporter. Phosphate transporters are also reasonable candidates for nontarget glyphosate resistance (Shaner 2009). Thus, 5 of 12 of the most upregulated genes from the glyphosate treatment code for proteins whose functions are to transport compounds across membranes.

Typically, RT-PCR on individual candidate genes is performed as a validation of microarray results (Yuan et al. 2006). However, since horseweed sequences were not known, this important step has not been possible. Therefore, we must either use known *Arabidopsis* gene sequence to assist in cloning and sequencing individual horseweed orthologs, a laborious exercise, especially since ABC transporter genes are often very large, or we can wait to obtain horseweed sequences from a large-scale sequencing project. We have chosen the latter and proceeded to a sequencing effort.

8.4.1.2   454 Transcriptome Sequencing   For nonmodel plants, such as weed species, the traditional method of obtaining genome or transcriptome data has been library construction, repeated rounds of normalization/subtraction, and then Sanger sequencing. Although Sanger sequencing technology has been progressively improved over the past decade, the application of this relatively expensive approach to large-scale sequencing has remained beyond the typical grant-funded investigators. Recently, the Sanger method has been partially supplanted by several "next-generation" sequencing technologies that offer dramatic increases in cost-effective sequence throughput (Morozova and Marra 2008). The next-generation technologies that are commercially available and been most used today include the Roche-FLX 454 (Branford, CT) and the Illumina-Solexa (San Diego, CA) systems, which have had a tremendous impact on genomic research for increasing sequencing depth and coverage while reducing time, labor, and cost (Morozova and Marra 2008; Rothberg and Leamon 2008). Roche-FLX 454 technology, the first next-generation sequencing technology that was released onto the market in October 2005 (Margulies et al. 2005), has been the most widely published next-generation technology. It has been used in more than 330 peer-reviewed research publica-

TABLE 8.1. Microarray Analysis of Differentially Expressed and Upregulated Genes after Glyphosate Treatment of Resistant Horseweed

Gene Name	ID	Fold Change	FDR	p-Value
ABC transporter family protein	**At3g13080**	29.64	0.000	0.000
Cryptochrome 1 apoprotein (CRY1)/flavin-type blue-light photoreceptor (HY4)	**At4g08920**	13.59	0.000	0.000
Expressed protein	**At1g05060**	7.34	0.000	0.000
ABC transporter family protein similar to PDR5-like ABC transporter GI:1514643 from (*Spirodela polyrhiza*)	**At3g16340**	6.66	0.000	0.000
Proline-rich extensin-like family protein	**At5g06640**	6.24	0.000	0.000
Mitochondrial phosphate transporter	**At3g48850**	5.80	0.000	0.000
Proline-rich extensin-like family protein	**At3g54580**	5.70	0.000	0.000
ABC transporter family protein	**At4g15233**	5.55	0.000	0.000
Glutamine-dependent asparagine synthetase 1 (ASN1)	**At3g47340**	5.24	0.000	0.000
Terpene synthase/cyclase family protein 5-epi-aristolochene synthase	**At4g20200**	5.10	0.000	0.000
Elongation factor 1-alpha/ EF-1-alpha	**At5g60390**	5.03	0.000	0.000
Multidrug-resistant (MDR) ABC transporter	**At3g62150**	4.67	0.000	0.000
AAA-type ATPase family protein contains Pfam profile: ATPase family PF00004	**At5g40010**	4.66	0.000	0.000
Proline-rich extensin-like family protein	**At3g54590**	4.57	0.000	0.000
tRNA synthetase-related/tRNA ligase-related	**At5g10880**	4.53	0.000	0.000
Lysyl-tRNA synthetase, putative/ lysine—tRNA ligase	**At3g11710**	4.23	0.000	0.000
Caffeoyl-CoA 3-O-methyltransferase	**At4g34050**	4.22	0.000	0.000
Proline-rich extensin-like family protein	**At5g49080**	4.16	0.000	0.000
Protease inhibitor/seed storage/lipid transfer protein (LTP) family protein	**At4g15160**	4.12	0.000	0.000
tRNA synthetase class II (G, H, P, and S) family protein	**At3g62120**	4.07	0.000	0.000

Top 20 heterologous *Arabidopsis* genes identified as upregulated ($p < 0.05$) in horseweed 24 h post glyphosate treatment are displayed. p-Values and false discovery rate ([FDR] all less than 0.0001) were calculated using a permutational analysis.

tions (http://www.454.com) and used for standard sequencing applications, such as de novo genome sequencing and resequencing, and for novel applications previously unexplored by Sanger sequencing (Droege and Hill 2008). The 454 technology avoids expensive cloning-based library construction by taking advantage of a highly efficient *in vitro* DNA amplification method known as emulsion PCR (Margulies et al. 2005). Followed by pyrosequencing (Nyren et al. 1993; Ronaghi et al. 1996), the FLX-454 standard system is capable of generating an average of 100 Mb of 250 base reads per 7.5-h run (http://www.454.com). Compared with similar dollar expenditure by Sanger sequencing, it yields redundant coverage for many more genes. Also, even though the read length is shorter than Sanger sequencing, the lower error level (<0.5%) associated with 454 technology is beneficial for sufficient coverage depth to allow assembly of overlapping reads (Droege and Hill 2008). There is even less of a concern for transcriptome sequencing and analysis as transcriptomes are smaller than the genomes from which they are derived, and typically contain less repetitive DNA. Although next-generation technologies allow genome sequencing to become more efficient, the sequencing of complex genomes remains expensive, often prohibitively so. Therefore, we choose horseweed transcriptome sequencing by using the 454 technique to acquire candidate sequences for functional genomics analysis.

To obtain comprehensive transcriptome data, total RNA was isolated from meristematic tissues of a horseweed sample from Knoxville, TN, that was glyphosate susceptible and those from a TN-resistant biotype from western Tennessee that was glyphosate-treated and water-treated as above (Mueller et al. 2003). These samples were subsequently pooled to generate double-stranded cDNA using SMART™ cDNA Library Construction Kit (Clontech, Mountain View, CA). The cDNA sample was then sheared into smaller pieces (300–500 bp) and fractionated by size. Sheared DNAs of the appropriate size were subsequently polished (short stretches of nonpaired nucleotides resulting from the shearing processes were removed from the ends of each clone). Short adaptors were then ligated on to each resulting fragment, which provide priming sequences for both emulsion PCR amplification and pyrosequencing. This process resulted in a single-stranded template library suitable for pyrosequencing Three sequencing runs were performed on a Roche-FLX sequencer at the University of Illinois Keck Center using methods previously described (Margulies et al. 2005; Poinar et al. 2006).

The sequence run yielded 411,962 raw reads. The average length of each read was 233 bp, and 79.2% of them were distributed between 200 and 300 bp, and the total data size was 95.8 Mb (Fig. 8.1). Initial quality filtering of the 454 reads was performed at the machine level before base calling. These sequences were subsequently trimmed using Lucy (Chou and Holmes 2001) under default settings. In addition, poly (A/T) tails and adapters used during cDNA synthesis were removed from raw 454 sequences with EGassembler (Human Genome Center, University of Tokyo) (Masoudi-Nejad et al. 2006) using default settings. Finally, the returned 379,152 high-quality clean sequences were assem-

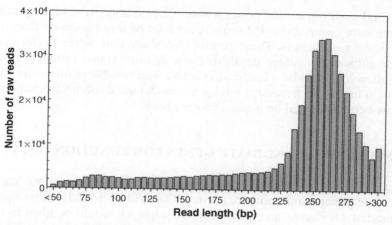

Figure 8.1. Raw data distribution. The 454 horseweed transcriptome data set contained 411,962 raw reads, 79.2% of them between 200 and 300 bp. The average length of each read was 233 bp, and the total data size is 95.8 Mb.

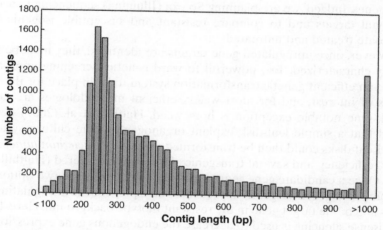

Figure 8.2. Contig distribution. Of the sequence reads obtained, 379,152 were usable sequences. In the data set, there were 31,783 unique sequences, consisting of 16,102 contigs and 15,681 singletons. The total data included 10.35 Mb of total nonredundant and usable sequence.

bled using CAP3 (Huang and Madan 1999) and EGassembler (Masoudi-Nejad et al. 2006) to generate unigenes using default settings. This returned 31,783 unique sequences, including 16,102 contigs and 15,681 singletons. After assembling, ~55% of contigs (8817) were longer than 300 bp, and 19.5% of contigs (3145) were longer than 600 bp (Fig. 8.2). The average coverage depth for each contig and each nucleotide position was ~22-fold and ~12-fold, respectively. That measure ensured our 454 sequences could be more accurate than traditional Sanger sequences. These results also show that, when performed to provide sufficient coverage depth to allow *de novo* transcriptome assembly, 454 sequencing could be a fast, cost-effective, and reliable platform for development of functional genomic tools for nonmodel weed species. Sequence data are now being analyzed on a gene-by-gene basis.

8.5   SCREENING CANDIDATE GENES FOR FUNCTIONALITY

In order to elucidate nontarget glyphosate resistance genomics, we see that several components and tools are needed. The most profound tool is sequence information. Of course, an entire genomic sequence would be ideal but cost prohibitive, but having sequence to the majority of transcribed genes in a weed is very helpful. As sequencing technologies continue to advance, not only will whole genome sequencing be economically feasible, but sequencing might entirely replace microarray analysis as a tool for whole transcriptome screening for biologically interesting genes, in this case nontarget glyphosate resistance genes. Indeed, we are planning Solexa (Illumina) sequencing to acquire additional targets and to compare resistant and susceptible isogenic lines: glyphosate treated and untreated.

However, once upregulated gene targets are identified, they must be functionally characterized. Two powerful forward genetics screening approaches require an efficient genetic transformation system to be in place for the weed species of interest, and for most weeds, efficient methodologies are largely absent. One notable exception is horseweed. Halfhill et al. (2007) demonstrated that a simple leaf-disk explant organogenic tissue culture system is facile. Leaf disks could then be transformed using *Agrobacterium tumefaciens* at 90% efficiency, and several transgenic plants were recovered (Halfhill et al. 2007). Once a candidate gene sequence is identified, there are two options for screening using a transgenic approach. One option is downregulating the expression of the target gene in the resistant biotype background. Here, RNAi or antisense silencing is used to decrease the endogenous gene expression and then analyzing the phenotype. The advantage of this approach is that only a partial sequence is required. A positive result would be the conversion of a resistant biotype into a transgenic susceptible biotype. One potentially negative result would be pleiotropic effects altering the endogenous functions of the gene product that are independent of glyphosate resistance, which could confound the conclusions. Thus, a second screening strategy is the overexpression of a candidate gene in a transgenic plant, potentially converting a susceptible biotype into a transgenic resistant biotype, which would lend significant evidence that the candidate gene is responsible for endowing resistance. Indeed, while a full-length cDNA is needed to perform this experiment, a positive result would be a plant that is highly resistant to glyphosate. This would only be the expectation, however, if the trait is a Mendelian dominant trait, such as is the case with glyphosate resistance in horseweed.

8.6 CONCLUSIONS

Genomics should help unravel not only the mechanisms of target and non-target herbicide resistance, but should also be useful in understanding the evolution of resistance and its spread. We have discussed how transcriptome sequences are useful for determining which genes over-express in functional assays. Transcriptome sequencing (a combination of 454 and short, deeper sequencing such as Solexa) will also reveal single nucleotide polymorphisms that could be helpful should the coding region of nontarget genes be responsible for conferring resistance instead of gene regulation changes. Whole genome, de novo sequencing should be done on economically important weeds, and a weed with a small genome, such as horseweed, could be accomplished in 2009 with approximately only $100,000.

ACKNOWLEDGMENTS

We appreciate funding from the University of Tennessee, the United States Department of Agriculture (USDA), and Monsanto. Thanks to David Horvath for making very helpful suggestions on an earlier draft and the constructive conversations with many scientists in the area of weed genomics.

REFERENCES

Alibhai, M. F., C. Cajacob, P. C. C. Feng, G. R. Heck, Y. Qi, S. Flasinski, and W. C. Stallings, inventors; Monsanto Technology LLC, assignee. 2004. Glyphosate resistant class I 5-enolpyruvylshikimate-3-phosphate synthase (EPSPS). U.S. Patent 2004/004636, filed September 2, 2004.

Baerson, S. R., D. J. Rodriguez, M. Tran, Y. Feng, N. A. Biest, and G. M. Dill. 2002. Glyphosate-resistant goosegrass. Identification of a mutation in the target enzyme 5-enolpyruvylshikimate-3-phosphate synthase. *Plant Physiology* 129:1265–1275.

Basu, C., M. D. Halfhill, T. C. Mueller, and C. N. Stewart, Jr. 2004. Weed genomics: new tools to understand weed biology. *Trends in Plant Science* 9:391–398.

Breitling, R., P. Armengaud, A. Amtmann, and P. Herzyk. 2004. Rank products: a simple, yet powerful, new method to detect differentially regulated genes in replicated microarray experiments. *FEBS Letters* 573:83–92.

Chao, W. S., D. P. Horvath, J. V. Anderson, and M. Foley. 2005. Potential model weeds to study genomics, ecology, and physiology in the 21st century. *Weed Science* 53:927–937.

Chou, H. H. and M. H. Holmes. 2001. DNA sequence quality trimming and vector removal. *Bioinformatics* 17:1093–1104.

Dinelli, G., I. Marotti, A. Bonetti, P. Catizone, J. M. Urbano, and J. Barnes. 2008. Physiological and molecular bases of glyphosate resistance in *Conyza bonariensis* biotypes from Spain. *Weed Research* 48:257–265.

Dinelli, G., I. Marotti, A. Bonetti, M. Minelli, P. Catizone, and J. Barnes. 2006. Physiological and molecular insight on the mechanisms of resistance to glyphosate in *Conyza canadensis* (L.) Cronq. biotypes. *Pesticide Biochemistry and Physiology* 86:30–41.

Dixon, D. P., A. Lapthorn, and R. Edwards. 2002. Plant glutathione transferases. *Genome Biology* 3:reviews 3004.1–3005.10.

Droege, M. and B. Hill. 2008. The genome sequencer FLX™ system-longer reads, more applications, straight forward bioinformatics and more complete datasets. *Journal of Biotechnology* 136:3–10.

Feng, P. C. C., M. Tran, T. Chiu, R. D. Sammons, G. R. Heck, and C. A. Jacob. 2004. Investigations into glyphosate-resistant horseweed (*Conyza canadensis*): retention, uptake, translocation, and metabolism. *Weed Science* 52:498–505.

Gaines, T. A., W. Zhang, D. Wang, B. Bukun, S. T. Chisholm, D. L. Shaner, S. J. Nissen, W. L. Patzoldt, P. J. Tranel, A. S. Culpepper, T. L. Grey, T. M. Webster, W. K. Vencill, R. D. Sammons, J. Jiang, C. Preston, J. E. Leach, and P. Westra. 2010. Gene amplification confers glyphosate resistance in *Amaranthus palmeri*. *Proceedings of the National Academy of Science of the USA* 107:1029–1034.

Ge, X., D. A. A. d'Avignon, J. J. H. Ackerman, and R. D. Sammons. 2010. Rapid vacuoler sequestration: the horseweed glyphosate resistance mechanism. *Pest Management Science* 66:345–348.

Gentleman, R. C., V. J. Carey, D. M. Bates, B. Bolstad, M. Dettling, S. Dudoit, B. Ellis, L. Gautier, Y. C. Ge, J. Gentry, K. Hornik, T. Hothorn, W. Huber, S. Iacus, R. Irizarry, F. Leisch, C. Li, M. Maechler, A. J. Rossini, G. Sawitzki, C. Smith, G. Smyth, L. Tierney, J. Y. H. Yang, and J. H. Zhang. 2004. Bioconductor: open software development for computational biology and bioinformatics. *Genome Biology* 5:R80.

Halfhill, M. D., L. L. Good, C. Basu, J. Burris, C. L. Main, T. C. Mueller, and C. N. Stewart, Jr. 2007. Transformation and segregation of GFP fluorescence and glyphosate resistance in horseweed (*Conyza canadensis*) hybrids. *Plant Cell Reports* 26:303–311.

Hirose, S., H. Kawahigashi, K. Ozawa, N. Shiota, H. Inui, H. Ohkawa, and Y. Ohkawa. 2005. Transgenic rice containing human CYP2B6 detoxifies various classes of herbicides. *Journal of Agricultural and Food Chemistry* 53:3461–3467.

Hong, F. X., R. Breitling, C. W. McEntee, B. S. Wittner, J. L. Nemhauser, and J. Chory. 2006. RankProd: a bioconductor package for detecting differentially expressed genes in meta-analysis. *Bioinformatics* 22:2825–2827.

Horvath, D. P., R. Schaffer, M. West, and E. Wisman. 2003. *Arabidopsis* microarrays identify conserved and differentially expressed genes involved in shoot growth and development from distantly related plant species. *The Plant Journal* 34:125–134.

Huang, X. and A. Madan. 1999. CAP3: a DNA sequence assembly program. *Genome Research* 9:868–877.

Klein, M., B. Burla, and E. Martinoia. 2006. The multidrug resistance-associated protein (MRP/ABCC) subfamily of ATP-binding cassette transporters in plants. *FEBS Letters* 580:1112–1122.

Koger, C. H. and K. N. Reddy. 2005. Role of absorption and translocation in the mechanism of glyphosate resistance in horseweed (*Conyza canadensis*). *Weed Science* 53:84–89.

Liu, G., R. Sanchez-Fernandez, Z. S. Li, and P. A. Rea. 2001. Enhanced multispecificity of *Arabidopsis* vacuolar multidrug resistance-associated protein-type ATP-binding cassette transporter, AtMRP2. *Journal of Biological Chemistry* 276:8648–8656.

Lu, Y. P., Z. S. Li, and P. A. Rea. 1997. AtMRP1 gene of *Arabidopsis* encodes a glutathione S-conjugate pump: isolation and functional definition of a plant ATP-binding cassette transporter gene. *Proceedings of the National Academy of Sciences of the United States of America* 94:8243–8248.

Margulies, M., E. Egholm, W. E. Altman, et al. 2005. Genome sequencing in microfabricated high-density picolitre reactors. *Nature* 437:376–380.

Masoudi-Nejad, A., K. Tonomura, S. Kawashima, Y. Moriya, M. Suzuki, M. Itoh, M. Kanehisa, T. Endo, and S. Goto. 2006. EGassembler: online bioinformatics service for large-scale processing, clustering and assembling ESTs and genomic DNA fragments. *Nucleic Acids Research* 34:459–462.

Mentewab, A. and C. N. Stewart, Jr. 2005. Overexpression of an *Arabidopsis thaliana* ABC transporter confers kanamycin resistance to transgenic plants. *Nature Biotechnology* 23:1177–1180.

Morozova, O. and M. A. Marra. 2008. Applications of next-generation sequencing technologies in functional genomics. *Genomics* 92:255–264.

Mueller, T. C., J. E. Massey, R. M. Hayes, C. L. Main, and C. N. Stewart, Jr. 2003. Shikimate accumulates in both glyphosate-sensitive and glyphosate-resistant horseweed (*Conyza canadensis* L. Cronq.). *Journal of Agricultural and Food Chemistry* 51:680–684.

Nebenfuhr, A. and T. L. Lomax. 1998. Multiplex titration RT-PCR: rapid determination of gene expression patterns for a large number of genes. *Plant Molecular Biology Reporter* 16:323–339.

Nelson, D. R., M. A. Schuler, S. M. Paquette, D. Werck-Reichhart, and S. Bak. 2004. Comparative genomics of rice and *Arabidopsis*. Analysis of 727 cytochrome genes and pseudogenes from a monocot and a dicot. *Plant Physiology* 135:756–772.

Ng, C. H., R. Wickneswari, S. Salmijah, Y. T. Teng, and B. S. Ismail. 2003. Gene polymorphisms in glyphosate-resistant and -susceptible biotypes of *Eleusine indica* from Malaysia. *Weed Research* 43:108–115.

Nyren, P., B. Pettersson, and M. Uhlen. 1993. Solid phase DNA minisequencing by an enzymatic luminometric inorganic pyrophosphate detection assay. *Analytical Biochemistry* 208:171–175.

Perez-Jones, A., K. W. Park, N. Polge, J. Colquhoun, and C. A. Mallory-Smith. 2007. Investigating the mechanisms of glyphosate resistance in *Lolium multiflorum*. *Planta* 226:395–404.

Poinar, H. N., C. Schwarz, J. Qi, et al. 2006. Metagenomics to paleogenomics: large-scale sequencing of mammoth DNA. *Science* 311:392–394.

Reade, J. P. H., L. J. Milner, and A. H. Cobb. 2004. A role for glutathione S-transferases in resistance to herbicides in grasses. *Weed Science* 52:468–474.

Ronaghi, M., S. Karamohamed, B. Pettersson, M. Uhlen, and P. Nyren. 1996. Real-time DNA sequencing using detection of pyrophosphate release. *Analytical Biochemistry* 242:84–89.

Rothberg, J. M. and J. H. Leamon. 2008. The development and impact of 454 sequencing. *Nature Biotechnology* 26:1117–1124.

Sánchez-Fernández, R., T. G. E. Davies, J. O. D. Coleman, and P. A. Rea. 2001. The *Arabidopsis thaliana* ABC protein superfamily: a complete inventory. *Journal of Biological Chemistry* 276:30231–30244.

Simarmata, M. and D. Penner. 2008. The basis for glyphosate resistance in rigid ryegrass (*Lolium rigidum*) from California. *Weed Science* 56:181–188.

Shaner, D. 2009. Role of translocation as a mechanism of resistance to glyphosate. *Weed Science* 57:118–123.

Schulz, B. and H. U. Kolukisaglu. 2006. Genomics of plant ABC transporters: the alphabet of photosynthetic life forms or just holes in membranes? *FEBS Letters* 580:1010–1016.

Stewart, C. N., Jr., ed. 2009. *Weedy and Invasive Plant Genomics*. Ames, IA: Blackwell Scientific Publishing.

Stewart, C. N., Jr., P. J. Tranel, D. P. Horvath, J. V. Anderson, L. H. Rieseberg, J. H. Westwood, C. A. Mallory-Smith, M. L. Zapiola, and K. M. Dlugosch. 2009. Evolution of weediness and invasiveness: charting the course for weed genomics. *Weed Science* 57:451–462.

Tan, S., R. Evans, and B. Singh. 2006. Herbicidal inhibitors of amino acid biosynthesis and herbicide-tolerant crops. *Amino Acids* 30:195–204.

Verrier, P. J., D. Bird, B. Burla, E. Dassa, C. Forestier, M. Geisler, M. Klein, U. Kolukisaoglu, Y. Lee, E. Martinoia, A. Murphy, P. A. Rea, L. Samuels, B. Schulz, E. Spalding, K. Yazaki, and F. L. Theodoulou. 2008. Plant ABC proteins a unified nomenclature and updated inventory. *Trends in Plant Science* 13:151–159.

Wakelin, A. M. and C. Preston. 2006. A target-site mutation is present in a glyphosate-resistant *Lolium rigidum* population. *Weed Research* 46:432–440.

Widholm, J. M., A. R. Chinnala, J. H. Ryu, H. S. Song, T. Eggett, and J. E. Brotherton. 2001. Glyphosate selection of gene amplification in suspension cultures of 3 plant species. *Physiologia Plantarum* 112:540–545.

Yuan, J. S., D. W. Galbraith, S. Y. Dai, P. Griffin, and C. N. Stewart, Jr. 2008a. Plant systems biology comes of age. *Trends in Plant Science* 13:165–171.

Yuan, J., D. Wang, and C. N. Stewart, Jr. 2008b. Statistical methods for efficiency adjusted real-time PCR analysis. *Biotechnology Journal* 3:112–123.

Yuan, J. S., A. Reed, F. Chen, and C. N. Stewart, Jr. 2006. Statistical analysis of real-time PCR data. *BMC Bioinformatics* 7:85.

Yuan, J. S., P. J. Tranel, and C. N. Stewart, Jr. 2007. Non-target site herbicide resistance: a family business. *Trends in Plant Science* 12:6–13.

Zelaya, I. A., M.D.K. Owen, and M. J. VanGessel. 2004. Inheritance of evolved glyphosate resistance in *Conyza canadensis* (L.) Cronq. *Theoretical and Applied Genetics* 110:58–57.

9

GLYPHOSATE-RESISTANT CROP PRODUCTION SYSTEMS: IMPACT ON WEED SPECIES SHIFTS

Krishna N. Reddy and Jason K. Norsworthy

9.1 OVERVIEW

The era of glyphosate-resistant crops (GRCs) began with the commercial launch of glyphosate-resistant (GR) canola and GR soybean in 1996 (see Table 9.1 for common and scientific names). Within 2 years, two more GRCs (cotton and corn) were commercialized in the United States. GRCs provide the flexibility to apply glyphosate (a nonselective herbicide) post emergence to control emerged weeds without concern for crop damage. The consistent weed control and economic benefits of GRCs have encouraged rapid adoption by U.S. farmers. The remarkable commercial success of GRCs has impacted herbicide use. Glyphosate use has increased tremendously, with a concomitant increase in selection pressure that promotes weed species shifts. Farmers' widespread adoption of GRC technology is being challenged by the looming threat of weed species shifts in GR cropping systems. Weed species shifts, a relative change in weed abundance and species diversity in response to continuous use of glyphosate in GRCs, are inevitable and are rapidly increasing (e.g., Benghal dayflower in GR cotton). This chapter will summarize the impact of GR corn, GR cotton, and GR soybean cropping systems on weed species shifts as well as late-season weed problems in the United States.

Glyphosate Resistance in Crops and Weeds: History, Development, and Management
Edited by Vijay K. Nandula

TABLE 9.1. Common and Scientific Names of Crops and Weeds Used in This Chapter

Plant Species	Scientific Name
Crop	
Canola	*Brassica napus* L.
Corn	*Zea mays* L.
Cotton	*Gossypium hirsutum* L.
Soybean	*Glycine max* (L.) Merr.
Spring wheat	*Triticum aestivum* L.
Sugarbeet	*Beta vulgaris* L.
Weed	
Arrowleaf sida	*Sida rhombifolia* L.
Asiatic dayflower	*Commelina communis* L.
Barnyardgrass	*Echinochloa crus-galli* (L.) Beauv.
Benghal dayflower	*Commelina benghalensis* L.
Bermudagrass	*Cynodon dactylon* (L.) Pers.
Birdsfoot trefoil	*Lotus corniculatus* L.
Brazil pusley	*Richardia brasiliensis* (Moq.) Gomez
Bristly starbur	*Acanthospermum hispidum* DC.
Browntop millet	*Urochloa ramosa* (L.) Nguyen
Broadleaf signalgrass	*Urochloa platyphylla* (Nash) R.D. Webster
Carpetweed	*Mollugo verticillata* L.
Chinese foldwing	*Dicliptera chinensis* (L.) Juss.
Coffee senna	*Senna occidentalis* (L.) Link
Common cocklebur	*Xanthium strumarium* L.
Common evening primrose	*Oenothera biennis* L.
Common lambsquarters	*Chenopodium album* L.
Common milkweed	*Asclepias syriaca* L.
Common pokeweed	*Phytolacca americana* L.
Common purslane	*Portulaca oleracea* L.
Common waterhemp	*Amaranthus rudis* Sauer
Crowfootgrass	*Dactyloctenium aegyptium* (L.) Willd.
Cutleaf evening primrose	*Oenothera laciniata* Hill
Eastern black nightshade	*Solanum ptychanthum* Dunal
Entireleaf morningglory	*Ipomoea hederacea* var. *integriuscula* Gray
Fall panicum	*Panicum dichotomiflorum* Michx.
Florida beggarweed	*Desmodium tortuosum* (Sw.) DC.
Florida pusley	*Richardia scabra* L.
Field bindweed	*Convolvulus arvensis* L.
Field horsetail	*Equisetum arvense* L.
Giant ragweed	*Ambrosia trifida* L.
Goosegrass	*Eleusine indica* (L.) Gaertn.
Hairy indigo	*Indigofera hirsuta* Harvey
Hairy nightshade	*Solanum physalifolium* Rusby
Hemp dogbane	*Apocynum cannabinum* L.
Hemp sesbania	*Sesbania herbacea* (P. Mill.) McVaugh
Hophornbeam copperleaf	*Acalypha ostryifolia* Riddell

TABLE 9.1. *Continued*

Plant Species	Scientific Name
Horseweed	*Conyza canadensis* (L.) Cronq.
Italian ryegrass	*Lolium perenne* L. ssp. *multiflorum* (Lam.) Husnot
Ivyleaf morningglory	*Ipomoea hederacea* Jacq.
Johnsongrass	*Sorghum halepense* (L.) Pers.
Jungle rice	*Echinochloa colona* (L.) Link
Kochia	*Kochia scoparia* (L.) Schrad.
Large crabgrass	*Digitaria sanguinalis* (L.) Scop.
Longspine sandbur	*Cenchrus longispinus* (Hack.) Fern.
Palmer amaranth	*Amaranthus palmeri* S. Wats.
Pennsylvania smartweed	*Polygonum pensylvanicum* L.
Pitted morningglory	*Ipomoea lacunosa* L.
Prickly sida	*Sida spinosa* L.
Red sprangletop	*Leptochloa panicea* (Retz.) Ohwi
Redvine	*Brunnichia ovata* (Walt.) Shinners
Rigid ryegrass	*Lolium rigidum* Gaudin
Shattercane	*Sorghum bicolor* (L.) Moench ssp. *arundinaceum* (Desv.) de Wet & Harlan
Sicklepod	*Senna obtusifolia* (L.) H.S. Irwin & Barneby
Smallflower morningglory	*Jacquemontia tamnifolia* (L.) Griseb.
Smooth pigweed	*Amaranthus hybridus* L.
Spurred anoda	*Anoda cristata* (L.) Schlecht.
Tall morningglory	*Ipomoea purpurea* (L.) Roth
Texas millet	*Urochloa texana* (Buckl.) R. Webster
Tropic croton	*Croton glandulosus* var. *septentrionalis* Muell.-Arg.
Trumpetcreeper	*Campsis radicans* (L.) Seem. ex Bureau
Velvetleaf	*Abutilon theophrasti* Medik.
Yellow nutsedge	*Cyperus esculentus* L.
Wild buckwheat	*Polygonum convolvulus* L.
Wild parsnip	*Pastinaca sativa* L.
Wild poinsettia	*Euphorbia heterophylla* L.
Wild proso millet	*Panicum miliaceum* L.
Wild radish	*Raphanus raphanistrum* L.

9.2 COMMERCIALIZATION AND ADOPTION OF GRCs

In 1996, transgenic GR canola and GR soybean containing a bacterial gene that imparts resistance to glyphosate were commercialized in the United States (Duke 2005; Reddy and Koger 2006). Later, GR cotton (1997) and GR corn (1998) were commercialized for planting in the United States (Duke 2005; Reddy and Koger 2006). GRCs enabled in-crop postemergence application of glyphosate. The effectiveness of glyphosate on a wide spectrum of weeds, simplicity and flexibility in application, lower herbicide cost, and

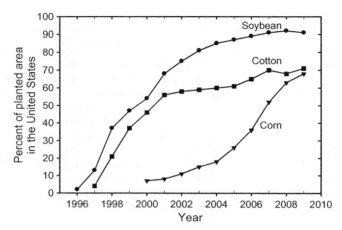

Figure 9.1. Increased adoption of glyphosate-resistant (GR) corn (▼), GR cotton (■), and GR soybean (●) in the United States, 1996–2009 (Gianessi et al. 2002; USDA 2009). GR soybean, GR cotton, and GR corn were commercialized in the United States in 1996, 1997, and 1998, respectively.

freedom to rotate crops have encouraged a rapid adoption by U.S. farmers (Gianessi 2005, 2008; Reddy and Whiting 2000). Because of efficient and consistent weed control and economic benefits, U.S. farmers have continued to plant more area with GRCs each year. GR soybean, GR cotton, and GR corn are dominant among all other GRCs grown commercially in there. In the United States, the soybean area planted with GR soybean cultivars has increased from 2% in 1996 to 91% in 2009 (Fig. 9.1). The area planted with GR cotton cultivars has increased from 4% in 1997 to 71% in 2009, but the percentage of area planted with GR cotton cultivars was influenced by the region. In the southeastern and mid-southern United States, at least 98% of the cotton planted is GR; however, use of GR cotton cultivars in Texas and the southwestern United States has been limited due to a lack of availability of high-yielding, adapted cultivars. In corn, area planted with GR corn hybrids has increased from 7% in 2000 to 68% in 2009. However, it should be stressed that the area reported for corn includes all herbicide-resistant (single and stacked gene) hybrids (USDA 2009). Currently, corn hybrids with the GR trait alone or stacked with glufosinate-resistant trait are commercially available in the United States. Due to their remarkable success, GR crops have dominated the U.S. seed market; thus, the area planted with glufosinate-resistant crops is negligible. In 2008, GR corn, GR cotton, and GR soybean crops were grown on 53 of the total 69 million ha of the total corn, cotton, and soybean planted in the United States. On a global perspective, this country ranked first in adoption of all transgenic crops (both insect and herbicide traits), with 57.7 million ha, which accounted for 50% of the global transgenic area in 2007 (ISAAA 2008).

9.3   GLYPHOSATE USE

The unprecedented commercial success of GRCs has impacted herbicide use patterns. Glyphosate is the predominant and often only herbicide used for managing weeds in GRCs. Glyphosate use has increased rapidly with a concomitant decrease in the use of other herbicides in the United States (Reddy 2001; Norsworthy 2003; Norsworthy et al. 2007; Shaner 2000; Young 2006). Glyphosate applied preplant, in-crop, and post harvest has effectively controlled a wide range of weeds. The use of soil-applied residual herbicides has declined; thus, weed control systems are increasingly total post emergence and are glyphosate-based in the United States. For example, the total active ingredient of glyphosate use in corn has increased from 0.6 million kg year^{-1} in 1997 (the year before GR corn was commercialized) to 5.6 million kg year^{-1} in 2003. Similarly, in cotton, glyphosate use has increased from 0.5 million kg year^{-1} in 1996 (the year before GR cotton was commercialized) to 5.8 million kg year^{-1} in 2003. Glyphosate use in soybean has increased from 2.9 million kg year^{-1} in 1995 (the year before GR soybean was commercialized) to 41.7 million kg year^{-1} in 2006 (USDA 2008). This represents a 9- to 14-fold increased use of glyphosate in these three crops alone since commercialization of GR technology.

9.4   IMPACT OF GRCs ON WEEDS

The total reliance on glyphosate and lack of herbicide diversity in weed-management tactics in GRCs has increased selection pressure that led to evolution of GR weeds and to weed shifts toward difficult-to-control species. The change in relative frequency of weeds within a species (e.g., GR weed abundance) or among species (diversity) in agricultural systems in response to weed management tactics is referred to as *weed species shift*. Weed species shifts occur because of natural tolerance to the primary herbicide used for weed control and/or elimination of competition from other weed species controlled by the primary herbicide. In GRCs, the weeds that escape control because of large weed size at treatment and/or high tolerance to glyphosate or those that emerge after a glyphosate application can fill ecological niches vacated by the weeds that were effectively controlled by glyphosate. Furthermore, elimination of competition from early-season weeds creates a favorable environment for late-season weeds. The sole use of glyphosate without a residual herbicide has resulted in late-season weeds becoming a major problem in GRCs.

9.4.1   Evolution of GR Weeds

Increased intensity of glyphosate use has led to the evolution of GR weeds, regardless of cropping systems. Twenty-four years after the introduction of glyphosate, the first evidence of evolved resistance to glyphosate was reported in a population of rigid ryegrass from an orchard in Australia following two to three applications of glyphosate for 15 consecutive years (Powles et al. 1998). In GR cropping systems, the first evolved glyphosate resistance was reported in horseweed (VanGessel 2001). Horseweed evolved resistance to glyphosate within 3 years of using only glyphosate in GR soybean. To date, a total of 18 weed species have evolved resistance to glyphosate under various cropping systems (Heap 2010; Nandula et al. 2005; Powles 2008). Among these 18 weed species, 11 weeds have evolved resistance to glyphosate in GRCs. In just the past 4 years (since 2005), eight weeds have evolved resistance to glyphosate in GRCs. Thus, there is a high selection pressure for the evolution of GR weeds in GRCs. Clearly, the rate of evolution of GR weeds is high in GRCs. Although, GR weeds are an integral part of weed species shifts, they will not be discussed in this chapter. The evolution of specific GR weeds and mechanisms of glyphosate resistance are the topics of several chapters published in this book.

9.4.2   Weed Species Shifts

Weed shifts in GRCs are no different from the shifts associated with use of other herbicides. Weed shifts are location specific and are a result of repeated use of the same herbicide mode of action or sole reliance on any particular production practice (e.g., continuous no-till). Webster and Coble (1997) have documented changes in weed species composition by analyzing data from annual surveys of most troublesome weeds in several major crops conducted by the Southern Weed Science Society since 1971. In cotton, johnsongrass, which ranked first among the most troublesome weeds in 1983, had dropped to being the ninth most troublesome weed by 1995. Morningglories (*Ipomoea* spp.), ranked as the fourth troublesome weed in 1983, became the most troublesome (first rank) weed by 1995. In soybean, morningglories ranked first among troublesome weeds in 1983, but had dropped to second rank among troublesome weeds by 1995. Sicklepod was ranked as the fourth troublesome weed in 1983, but became the most (first rank) troublesome weed in 1995. In corn, johnsongrass remained the most troublesome (first rank) weed in the 1974, 1983, and 1994 surveys.

A wide array of herbicides, with different modes of action, has been commercialized since the 1970s (Appleby 2005). For example, protoporphyrinogen oxidase inhibitors (1970s), acetyl-coenzyme A carboxylase inhibitors (1970–1980s), chloroacetamides (1970s), acetolactate synthase inhibitors (1980s), and pigment inhibitors (1980s) greatly expanded weed control options in crops. The newer herbicides are more specific and more active than previous herbicides. As a result, rankings of individual weed species changed during a two-decade period because the new herbicides effectively controlled some troublesome weeds (Webster and Coble 1997). For example, acetyl-coenzyme A carboxylase inhibitors (graminicides) selectively controlled most annual and perennial grasses in cotton and soybean, which led to a decrease in johnsongrass as a troublesome weed in these crops. However, johnsongrass continued to be a troublesome weed in corn over the same period as there were no herbicides developed that selectively controlled johnsongrass in corn.

Some weed species are inherently more tolerant to glyphosate than other weeds. For example, a naturally occurring GR biotype of field bindweed has been reported with no history of glyphosate use (DeGennaro and Weller 1984). A biotype of birdsfoot trefoil resistant to labeled glyphosate use rates was identified by Boerboom et al. (1990). Benghal dayflower (Culpepper et al. 2004), common lambsquarters (Westhoven et al. 2008), Asiatic dayflower, Chinese foldwing, common evening primrose, field horsetail, giant ragweed, common pokeweed, velvetleaf, and wild parsnip (Owen 2008) have various levels of natural tolerance to glyphosate. The species with high tolerance to glyphosate have adapted to glyphosate-based weed management systems and occupied ecological niches vacated by other weed species in GRCs.

The Southern Weed Science Society publishes annually the weed survey of southern U.S. states for several major crops. The participating extension weed scientists of each state provide the list of the 10 most troublesome weeds for each crop, and the list for each crop in each state is published in the annual proceedings of the Southern Weed Science Society. The information is not always clear because weed species are grouped as pigweed spp. (*Amaranthus* spp.), morningglories spp., nutsedge spp. (*Cyperus* spp.), and spurge spp. (*Chamaesyce* spp.). We made an attempt to highlight for each crop the new weed species that were not present on surveys before commercialization of GRCs but are listed in the most recent survey (Dowler 1995, 1997, 1998; Webster 2005, 2008). For example, the 10 most troublesome weeds in Georgia cotton, soybean, and corn cropping systems that appeared in the most recent surveys and around commercialization of GRCs are shown in Table 9.2.

9.4.2.1 *Cotton*

In Georgia, Benghal dayflower was not among the 10 most troublesome weeds of cotton until 1998, a year after GR cotton introduction (Dowler 1998). Three years later, it was ranked as the ninth most troublesome weed (Webster 2001), and by 2005, Benghal dayflower had become the most troublesome weed in Georgia cotton (Webster 2005). Benghal dayflower is relatively tolerant to glyphosate. In GR cotton in Georgia, glyphosate at 0.84 kg a.e. ha^{-1} applied over the top or post directed controlled Benghal dayflower 53–70% at 21 days after treatment (Culpepper et al. 2004). They concluded that the rapid shift toward Benghal dayflower was mainly due to inadequate control by glyphosate in GR cotton. Furthermore, Asiatic dayflower, Florida pusley, and Palmer amaranth were among the 10 most troublesome weeds in 2005 but not in 1998 (Dowler 1998; Webster 2005). In Florida cotton, Benghal dayflower was not on the list of the 10 most troublesome weeds until 1998, and by 2005, it was the most troublesome weed (Dowler 1998; Webster 2005).

TABLE 9.2. Ten Most Troublesome Weeds of Cotton, Soybean, and Corn in Georgia That Appeared on Annual Weed Surveys Conducted by the Southern Weed Science Society

Rank	Cotton		Soybean		Corn	
	1998	2005	1995	2005	1997	2008
1	Nutsedge spp.	Benghal dayflower	Sicklepod	Sicklepod	Texas panicum	Texas millet
2	Sicklepod	Palmer amaranth	Morningglory spp.	Morningglory spp.	Morningglories	Crabgrass
3	Pigweed spp.	Morningglory spp.	Pigweed spp.	Palmer amaranth	Sicklepod	Morningglories
4	Texas panicum	Florida pusley	Coffee senna	Nutsedge spp.	Cocklebur	Pigweeds
5	Morningglory spp.	Nutsedge spp.	Cocklebur	Florida pusley	Johnsongrass	Sicklepod
6	Bristly starbur	Asiatic dayflower	Johnsongrass	Benghal dayflower	Pigweed	Nutsedge spp.
7	Wild poinsettia	Smallflower morningglory	Crabgrass	Texas panicum	Crabgrass	Johnsongrass
8	Cocklebur	Texas panicum	Texas panicum	Common cocklebur	Bermudagrass	Annual ryegrass
9	Bermudagrass	Wild poinsettia	Florida beggarweed	Florida beggarweed	Wild radish	Pennsylvania smartweed
10	Coffee senna	Bermudagrass	Yellow nutsedge	Roundup Ready cotton	Florida beggarweed	Benghal dayflower

The information was extracted from annual weed surveys published in the *Proceedings of the Southern Weed Science Society* for the years before and after commercialization of glyphosate-resistant crops (Dowler 1995, 1997, 1998; Webster 2005, 2008). Weeds are listed in this table as they appeared in the Southern Weed Science Society surveys. Weed names are not always clear because species were grouped as pigweed spp., morningglory spp., nutsedge spp., and so on.

In North Carolina cotton, Benghal dayflower, Florida pusley, and goosegrass, and in South Carolina cotton, Florida pusley and spurred anoda were not among the 10 troublesome weeds in 1998, but by 2005, they were among the 10 most troublesome weeds (Dowler 1998; Webster 2005). In Tennessee cotton, horseweed was not in the list of 10 most troublesome weeds in 1998, but was the most troublesome weed in 2005 (Dowler 1998; Webster 2005). Horseweed became a widespread problem because of its windblown seed and ability to germinate under a wide range of environmental conditions (Nandula et al. 2006), which allowed glyphosate resistance to be rapidly spread over a broad region.

9.4.2.2   Soybean   In Florida soybean, Asiatic dayflower, cutleaf evening primrose, hairy indigo, and tropic croton were among the 10 most troublesome weeds in 2005 but not in 1995 (Dowler 1995; Webster 2005). In Georgia soybean, Florida pusley and Benghal dayflower were among the 10 most troublesome weeds in 2005 but not in 1995. In North Carolina soybean, bermudagrass, common milkweed, and hemp dogbane were among the 10 most troublesome weeds in 2005 but not in 1995. In Tennessee soybean, horseweed, hophornbeam copperleaf, Palmer amaranth, and trumpetcreeper were among the most troublesome weeds in 2005 but not in 1995. In South Carolina soybean, arrowleaf sida, common cocklebur, coffee senna, Florida pusley, prickly sida, and Texas millet were among the most troublesome in 2005 but not in 1995. Pusley (*Richardia* spp.) can survive multiple glyphosate applications and produce more than 2400 seeds m^{-2}, evidence of its tolerance to glyphosate (Jha et al. 2008).

Infestations of Asiatic dayflower, common lambsquarters, and wild buckwheat, which exhibit a high level of tolerance to glyphosate, are increasing in occurrence in Iowa soybean fields as a result of extensive glyphosate use (Owen and Zelaya 2005). Also, in Pennsylvania, wild buckwheat is an increasingly problematic weed in GR soybean (Curran et al. 2002). Additionally, in research plots in Pennsylvania, volunteer GR corn is a problem in the subsequent GR soybean crop. In soybean research plots along a transect from Minnesota to Louisiana, intensity of glyphosate use was found to be related to the number of weed escapes, leading to the conclusion that species shifts were more likely in intensively managed glyphosate-only systems (Scursoni et al. 2006). Common lambsquarters, eastern black nightshade, and *Amaranthus* species were the most prevalent weeds in plots prior to soybean harvest.

In a survey conducted in Mississippi in 2000, prickly sida (40%) was the most common weed followed by pitted morningglory (34%) and entireleaf morningglory (29%) in soybean fields (Rankins et al. 2005). Broadleaf signalgrass and barnyardgrass were the most common annual grasses, and yellow nutsedge was the most common sedge observed. Trumpetcreeper and redvine were the most common perennial vines.

9.4.2.3   Corn   In Florida corn, common bermudagrass, Benghal dayflower, cutleaf evening primrose, hemp sesbania, and wild radish were among the 10 most troublesome weeds in 2008 but not in 1997 (Dowler 1997; Webster 2008). In Georgia corn, annual ryegrass, Benghal dayflower, and Pennsylvania smartweed were among the 10 most troublesome weeds in 2008 but not in 1997. In North Carolina corn, johnsongrass, Palmer amaranth, Pennsylvania smartweed, trumpetcreeper, and yellow nutsedge were among the 10 most troublesome weeds in 2008 but not in 1997. In Tennessee corn, fall panicum, goosegrass, Italian ryegrass, large crabgrass, morningglories, Palmer amaranth, redvine, and smooth pigweed were among the 10 most troublesome weeds in 2008 but not in 1997. These states collectively produce less than 2% of the U.S. corn crop. Because of limited corn acreage and initial grower reluctance to adopt GR corn, the weed species shifts should have been less evident in corn than in soybean and cotton. This is in agreement with the recent survey reported by Culpepper (2006). Culpepper (2006) surveyed 12 weed scientists from 11 states across the United States to assess weed shifts in GR corn, cotton, and soybean. *Amaranthus*, *Commelina*, *Ipomoea*, and *Cyperus* species in GR cotton and *Ipomoea* and *Commelina* species in GR soybean are becoming problem weeds. However, no weed shifts were observed in GR corn. This is attributed to diversity of herbicides used in corn. Atrazine in combination with several residual herbicides provide cost-effective season-long weed control in most situations (Gianessi 2005, 2008).

9.4.3   Late-Season Emergence: Lack of Glyphosate Exposure

Tolerance to glyphosate or even evolution of GR weeds is not the sole cause of weed species shifts in glyphosate-only production systems. Weeds that germinate over a long period of time or emerge late in the cropping season after the final glyphosate application also contribute to species shifts in glyphosate-only systems. Glyphosate avoidance (nonexposure) through late-season emergence is a mechanism by which some species are increasing in prominence. Examples of weeds that have natural tolerance to glyphosate and avoid glyphosate due to late-season and/or continual emergence in GR cropping systems are shown in Table 9.3.

Emergence of weeds late in the cropping season is becoming a problem in GR crops. This is due to elimination of competition from early-season weeds controlled by glyphosate, the absence of residual control with glyphosate alone, and the decision by producers not to use a residual herbicide that will extend weed control into the later portions of the cropping season. It is these late-emerging weeds, sometimes after crop maturity, that are able to complete their life cycle in a short period, causing appreciable increases in the soil weed seedbank. In a 4-year study in South Carolina evaluating the effect of tillage intensity and GR soybean and corn grown alone versus traditional programs involving residual herbicides, crowfootgrass rapidly colonized the glyphosate-only system due to late-season emergence (Norsworthy 2008). Furthermore, carpetweed, a relatively noncompetitive late-season weed, became the predominant broad-leaved weed in plots where residual herbicides were excluded.

TABLE 9.3. Examples of Weeds That Have Natural Tolerance to Glyphosate and Avoid Glyphosate Due to Late-Season and/or Continual Emergence in Glyphosate-Resistant Cropping Systems

Natural Tolerance to Glyphosate		Glyphosate Avoidance by Late-Season/ Continual Emergence	
Weed Species	Reference	Weed Species	Reference
Asiatic dayflower	Owen (2008)	Barnyardgrass	Payne and Oliver (2000)
Benghal dayflower	Culpepper et al. (2004)	Benghal dayflower	Webster et al. (2005)
Brazil pusley	Jha et al. (2008)	Broadleaf signalgrass	D. O. Stephenson IV (pers. comm.)
Chinese foldwing	Owen (2008)	Browntop millet and jungle rice	D. O. Stephenson IV (pers. comm.) and K. N. Reddy (unpublished observations)
Common evening primrose	Owen (2008)	Carpetweed	Norsworthy (2008)
Common lambsquarters	Owen (2008); Owen and Zelaya (2005)	Common lambsquarters	Scursoni et al. (2007)
Common pokeweed	Owen (2008)	Crowfootgrass	Norsworthy (2008)
Florida pusley	Jha et al. (2008)	Eastern black nightshade	Scursoni et al. (2007)
Field horsetail	Owen (2008)	Ivyleaf morningglory	Hilgenfeld et al. (2004a)
Hemp sesbania	Norsworthy et al. (2001)	Pitted morningglory	Norsworthy and Oliveira (2007)
Giant ragweed	Owen (2008)	Shattercane	Hilgenfeld et al. (2004a)
Pitted morningglory	Norsworthy et al. (2001); Burke et al. (2009)	Sicklepod	Norsworthy and Oliveira (2007)
Velvetleaf	Owen (2008)	Red sprangletop	L. R. Oliver (pers. comm.)
Wild buckwheat	Owen and Zelaya (2005)	Tall morningglory	P. Jha (pers. comm.)
Wild parsnip	Owen (2008)	Texas millet	D. O. Stephenson IV (pers. comm.)

In Nebraska, shattercane and ivyleaf morningglory emerged from late April through mid-August in soybean, with some cohorts of both weeds avoiding glyphosate applications (Hilgenfeld et al. 2004a). Additionally, the projected seedbanks of shattercane, ivyleaf morningglory, and five other weed species were estimated, leading to the conclusion that shattercane and ivyleaf morningglory would become problematic in subsequent years because of increases in the soil weed seedbank following sequential glyphosate applications, whereas the other weeds would remain constant or decline over time in the soil seedbank (Hilgenfeld et al. 2004b). Glyphosate is highly effective in controlling shattercane; hence, increases of shattercane in the soil seedbank are largely a function of lack of glyphosate exposure to late-emerging cohorts. In other research, ivyleaf morningglory populations increased in a glyphosate-alone program that was evaluated in a corn–soybean rotation in Kansas (Marshall et al. 2000). The authors attributed the increases in the ivyleaf morningglory population to tolerance of the weed to glyphosate. However, tall morningglory, a closely related species, emerges in the spring, summer, and early fall when soybean begins to senesce and when glyphosate applications have ceased (P. Jha and J. K. Norsworthy, unpublished data). Although ivyleaf morningglory and other morningglories have a degree of tolerance to glyphosate, both tolerance and continual or late-season emergence may explain why morningglories have continued to be among the most problematic weeds of GRCs. Thus, glyphosate nonexposure because of late-season emergence probably plays a partial role in the continued problem with morningglories in soybean.

Late-season emergence of other weeds has been noted. Common lambsquarters and eastern black nightshade in soybean in Minnesota routinely escape control due to emergence following the last glyphosate application (Scursoni et al. 2007). Late-season barnyardgrass emergence in Arkansas following sequential glyphosate applications reduced control compared with programs having a residual herbicide (Payne and Oliver 2000). The most effective means of reducing late-season seed production of barnyardgrass in GR soybean was to apply glyphosate at two-trifoliolate leaf stage followed by 10-day sequential glyphosate applications beginning at barnyardgrass flowering, which reduced seed production up to 97% (Walker and Oliver 2008). In Argentina after four years, Puricelli and Tuesca (2005) reported a reduction in density of early-season emerging weeds and a reduction in species richness in crops that were regularly treated with glyphosate. However, less-competitive late-emerging annual broadleaf weeds increased in these systems, presumably because of emergence after the final glyphosate application.

Seed production by weed escapes is the cause of replenishment of the soil weed seedbank and dictates the prominence of species in a subsequent year's crop (Hartzler 1996; Schweizer and Zimdahl 1984). Hence, prevention of weed seed production is vital to minimizing the risks of buildup of a particular weed species over time. Late-season glyphosate applications have been effectively used to reduce seed production of early-season weed escapes and control of those weeds that emerge later in the year. For example, pitted morningglory, a common weed of the southern United States with partial tolerance to glyphosate, emerges over an approximate 7-month period (Norsworthy and Oliveira 2007). Early-emerging cohorts often survive glyphosate treatment and produce viable seed (Norsworthy and Oliver 2002b), and later-emerging cohorts are seldom treated with glyphosate. In Arkansas, three sequential glyphosate applications in soybean, with the latest applied at flowering reduced pitted morningglory seed production compared with a standard treatment of two early-season glyphosate applications (Norsworthy and Oliver 2002a). The third glyphosate application was sufficient to eliminate hemp sesbania seed production, a weed also difficult to control with glyphosate. In other research, pitted morningglory seed production was reduced 98% when an application of glyphosate at two-trifoliolate leaf stage of soybean was followed by sequential applications of glyphosate beginning at initial flowering of pitted morningglory (Walker and Oliver 2008).

Another weed that emerges over an extended period and can be problematic in a glyphosate-only system is sicklepod (Norsworthy and Oliveira 2007). A single early-season application of glyphosate alone was not effective in preventing sicklepod seed production in soybean, but a subsequent late-season application prevented seed production (Norsworthy et al. 2007). This strategy of using a late-season glyphosate application to reduce or eliminate seed production is effective on those weeds that have tolerance to glyphosate and often survive early-season applications. For instance, a mixture of Brazil and Florida pusley, two weeds with a high tolerance to glyphosate, were prevented from producing seed when two early-season glyphosate applications in soybean were followed by a third late-season application (Jha et al. 2008).

9.4.4 Crop and Herbicide Rotations

Crop rotation is considered an effective strategy to manage weed species associated with a monoculture system. Rotating crops breaks weed population buildup that may be detrimental to long-term management of a particular field. When crops are rotated, the change in production practices and herbicides could create an unfavorable environment for a specific weed species. Thus, the weed species that has dominance under a monoculture system can be prevented from becoming unmanageable by rotating with another crop. However, rotations among GR crops (e.g., GR soybean/GR corn, GR cotton/GR corn) without diversity in herbicide use will probably have little positive impact on reducing the potential for weed species shifts.

Glyphosate, the most dominant herbicide worldwide, provides flexible, efficient, economical, and environmentally safe weed control in GRCs (Duke and Powles 2008; Flint et al. 2005; Reddy and Whiting 2000). In a 3-year study, weed species decreased over time with the continued use of glyphosate in cotton and soybean (Flint et al. 2005). Several published reports on rotation studies with GRCs did not reveal any weed shifts in continuous GRC plots. In a 4-year ultranarrow-row cotton-soybean rotation, the purple nutsedge population markedly increased with a nonglyphosate-based program compared with a glyphosate-based program in continuous cotton (Bryson et al. 2003). In fact, the purple nutsedge population was reduced with the glyphosate-based program in continuous GR cotton, continuous GR soybean, or GR cotton/GR soybean rotation compared with continuous cotton with a nonglyphosate-based program. After a 3-year study, common purslane, sicklepod, and yellow nutsedge densities were higher in continuous bromoxynil-resistant cotton than bromoxynil-resistant cotton grown in rotation with GR cotton or continuous GR cotton (Reddy 2004).

In a 6-year cotton–corn rotation study, control of yellow nutsedge decreased in continuous non-GR cotton compared with rotated non-GR cotton and GR cotton (Reddy et al. 2006). In a 4-year GR and non-GR soybean rotation, yellow nutsedge densities were higher in non-GR than in GR soybean in 2 of 4 years (Heatherly et al. 2005). In a 6-year GR corn rotation with GR sugarbeet and GR spring wheat, common lambsquarters density averaged over glyphosate and nonglyphosate treatments increased compared with continuous corn (Wilson et al. 2007). This increase was likely because sugarbeet and spring wheat are less competitive with weeds than corn is. Wilson et al. (2007) have also demonstrated the role of glyphosate use rate in weed shifts. The density of common lambsquarters was higher with two applications of glyphosate at $0.4\,kg\,ha^{-1}$ per year compared with two applications of glyphosate at $0.8\,kg\,ha^{-1}$. However, common lambsquarters density remained similar over a 6-year period when glyphosate was used at $0.8\,kg\,ha^{-1}$ twice each year. Apparently, under the conditions of the above studies, no weed shifts were detected as a result of repeated glyphosate applications in continuous GR cropping systems, although this would be expected. Perhaps the glyphosate selection pressure in these studies was not sufficient to force weed species shifts. Scursoni et al. (2006) have observed that limited use of glyphosate in GR soybean had no profound effect on weed diversity in several field studies conducted across Arkansas, Iowa, Louisiana, Minnesota, and Missouri.

In 2005–2006, a survey of growers from six states (Illinois, Indiana, Iowa, Mississippi, Nebraska, and North Carolina) was conducted to discern the impact of GRCs on crop rotations, weed pressure, tillage practices, herbicide use, and GR weeds. The survey results reported in three publications are summarized briefly. No growers using a GR cropping system for more than 5 years reported heavy weed pressure (Kruger et al. 2009). In five cropping systems (continuous GR soybean, continuous GR cotton, GR corn/GR soybean, GR soybean/non-GR crop, and GR corn/non-GR crop), only 0–7% of growers reported greater weed pressure after implementing rotations using GR crops. *Amaranthus* spp. (pigweeds), *Ipomoea* spp. (morningglories), johnsongrass, *Ambrosia* spp. (ragweeds), *Setaria* spp. (foxtails), and velvetleaf were the most problematic weeds, depending on cropping system. Overall, systems using GR crops improved weed management compared with methods used prior to adoption of GR crops (Kruger et al. 2009). A GR soybean/non-GR crop rotation was more widely used than a GR soybean/GR corn rotation system (Shaw et al. 2009). Most corn and soybean growers reported using some type of crop rotation system, whereas very few cotton growers rotated out of cotton. Overall, rotations were more common in midwestern states than in southern states (Shaw et al. 2009). A high percentage of growers in crop rotations using a GR crop have made one to three applications of glyphosate (Givens et al. 2009). GR corn, GR cotton, and non-GRCs had the highest percentage of growers applying nonglyphosate herbicides compared to GR soybean during the 2005 growing season.

Recently, a survey was also conducted by Foresman and Glasgow (2008) to determine U.S. producer awareness, perceptions, attitudes, and experiences with GR weeds. General comparisons were made between a group of nine northern states and eight southern states. Growers from northern states planted an average of 112 ha per farm to continuous GR crops compared with 750 ha in southern states. Only 27% of farmers in northern states had used GR technology continuously for 3–5 years compared with 67% of southern farmers. The greater continuous use of GR technology in southern states is partially because of a monoculture production system for cotton.

9.5 MANAGEMENT OF WEED SPECIES SHIFTS

Weed species shifts in GRCs can be prevented with prudent use of glyphosate in combination with other weed management tactics. Farmers should not completely rely on glyphosate in GRCs but, rather, should diversify weed management systems, plant crops on seedbeds maintained weed-free by pre-plant tillage or "burndown" herbicides, and use preemergence herbicides on fields with a history of heavy weed pressure. Preemergence herbicides not only reduce detrimental early-season weed interference, but also widen the window of application for glyphosate. The wider window for glyphosate application can benefit farmers during extended rainy periods and also those farmers who have limited farm equipment. Additionally, glyphosate should be applied at a rate that targets the most difficult-to-control weed species present in the field and should be combined with other herbicides (selective and/or residual) either as a tank-mix or as sequential applications as needed.

To reduce the risks of weed species shifts, it is imperative to know the emergence pattern of the most difficult-to-control species. Crop rotation and use of herbicides with different modes of action reduce the likelihood of a weed species shift. However, use of two modes of action prior to emergence of a particular species may not prevent seed production or an increase in the population in subsequent years, especially in the absence of a residual herbicide. Rotating GRCs with non-GRCs (e.g., GR soybean/non-GR soybean, GR soybean/non-GR corn) and GRCs with glufosinate-resistant crops could aid in rotating herbicides with different modes of action. However, rotations among GRCs (e.g., GR soybean/GR corn, GR cotton/GR corn) with glypho-sate-only weed control program may only increase selection pressure. Fields with chronic weed problems must be monitored to control weed escapes and to prevent seed set. Diversity in weed management tactics involving chemical (herbicide rates, tank mixtures, application timing, different modes of action, soil-residual) and nonchemical (preplant tillage, cultivation, crop rotations, narrow row spacing) methods is critical to prevent weed species shifts and for sustainability of GRCs.

Multiple herbicide-resistant crops that combine traits such as resistance to glyphosate with resistance to other herbicides increase herbicide options to manage weeds with many modes of action. Corn and soybean with a high level of metabolic resistance to glyphosate combined with acetolactate synthase herbicide resistance (Optimum® GAT®, Pioneer Hi-Bred, Johnston, IA) are in the final stage of development (Green et al. 2008). Additionally, Monsanto (St. Louis, MO) is developing soybean resistant to both glyphosate and dicamba (auxinic herbicide), and Dow AgroSciences (Indianapolis, IN) is developing corn, cotton, and soybean resistant to glyphosate with (2,4-dichlorophenoxy) acetic acid (broad-leaved herbicide) and aryloxyphenoxypropionates (grass-specific herbicides). These crops are currently approved for field testing and will likely be available to growers in the near future (Duke 2005; Duke and Cerdeira 2010; Green et al. 2008). These multiple-resistant crops will provide a wide range of foliar- and soil-applied herbicide options to manage weeds that survive glyphosate. Crops resistant to other herbicides, such as glufosinate, could provide an alternative to glyphosate to diversify weed management options. Glyphosate- and glufosinate-resistant (stacked gene) corn, glufosinate-resistant cotton, and glufosinate-resistant soybean are commercially available. The GRCs with stacked traits could provide a tool to manage some GR weeds and reduce weed species shifts.

9.6 CONCLUSIONS

The widespread adoption of GRCs coupled with a spectacular increase in glyphosate use has exerted tremendous selection pressure on weed communities. Overreliance on glyphosate and inadequate diversity in weed management tactics in GRCs have resulted in weed species shifts. Weed species shifts refers to a relative change in weed population (abundance) or species (diversity) as well as late-season weed emergence in an agricultural system in response to weed management tactics. Weed species shifts in GRCs are a result of weeds that have escaped control because of a high level of tolerance to glyphosate or glyphosate avoidance from late-emerging cohorts. Common lambsquarters, johnsongrass, Italian ryegrass, *Amaranthus*, *Ambrosia*, *Commelina*, *Cyperus*, *Ipomoea*, and *Setaria* species are becoming problematic weeds in GRCs. Shifts in weed species can be prevented, and shifts when they occur can be managed with prudent selection of weed control methods. Multiple herbicide-resistant (stacked traits) crops that combine glyphosate resistance with resistance to other herbicides and/or inclusion of residual herbicides can provide a wide range of foliar- and soil-applied herbicide options to manage weeds that survive glyphosate. Diversity in weed management systems is critical to reduce weed species shifts and to maintain sustainability of GRCs as an effective weed management tool.

REFERENCES

Appleby, A. P. 2005. A history of weed control in the United States and Canada—a sequel. *Weed Science* 53:762–768.

Boerboom, C. M., D. L. Wyse, and D. A. Somers. 1990. Mechanism of glyphosate tolerance in birdsfoot trefoil (*Lotus corniculatus*). *Weed Science* 38:463–467.

Bryson, C. T., K. N. Reddy, and W. T. Molin. 2003. Purple nutsedge (*Cyperus rotundus*) population dynamics in narrow row transgenic cotton (*Gossypium hirsutum*) and soybean (*Glycine max*) rotation. *Weed Technology* 17:805–810.

Burke, I. C., K. N. Reddy, and C. T. Bryson. 2009. Pitted and hybrid morningglory accessions have variable tolerance to glyphosate. *Weed Technology* 23:592–598.

Culpepper, A. S. 2006. Glyphosate-induced weed shifts. *Weed Technology* 20:277–281.

Culpepper, A. S., J. T. Flanders, A. C. York, and T. M. Webster. 2004. Tropical spiderwort (*Commelina benghalensis*) control in glyphosate-resistant cotton. *Weed Technology* 18:432–436.

Curran, W. S., K. Handwerk, and D. D. Lingenfelter. 2002. Temporal weed dynamics as influenced by corn and soybean herbicides. *Abstracts of the Weed Science Society of America* 42:2.

DeGennaro, F. P. and S. C. Weller. 1984. Differential sensitivity of field bindweed (*Convolvulus arvensis*) biotypes to glyphosate. *Weed Science* 32:472–476.

Dowler, C. C. 1995. Weed survey—southern states. *Proceedings of the Southern Weed Science Society* 48:290–302.

Dowler, C. C. 1997. Weed survey—southern states. *Proceedings of the Southern Weed Science Society* 50:227–246.

Dowler, C. C. 1998. Weed survey—southern states. *Proceedings of the Southern Weed Science Society* 51:299–313.

Duke, S. O. 2005. Taking stock of herbicide-resistant crops ten years after introduction. *Pest Management Science* 61:211–218.

Duke, S. O and A. L. Cerdeira. 2010. Transgenic crops for herbicide resistance. In C. Kole, C. H. Michler, A. G. Abbott, and T. C. Hall, eds. *Transgenic Crop Plants, Volume 2: Utilization and Biosafety*. Berlin: Springer, pp. 133–166.

Duke, S. O. and S. B. Powles. 2008. Glyphosate: a once-in-a-century herbicide. *Pest Management Science* 64:379–325.

Flint, S. G., D. R. Shaw, F. S. Kelley, and J. C. Holloway. 2005. Effect of herbicide systems on weed shifts in soybean and cotton. *Weed Technology* 19:266–273.

Foresman, C. and L. Glasgow. 2008. US grower perceptions and experiences with glyphosate-resistant weeds. *Pest Management Science* 64:388–391.

Gianessi, L. P. 2005. Economic and herbicide use impacts of glyphosate-resistant crops. *Pest Management Science* 61:241–245.

Gianessi, L. P. 2008. Economic impacts of glyphosate-resistant crops. *Pest Management Science* 64:346–352.

Gianessi, L. P., C. S. Silvers, S. Sankula, and J. E. Carpenter. 2002. Plant Biotechnology: Current and Potential Impact for Improving Pest Management in U.S. Agriculture: An Analysis of 40 Case Studies. Washington, DC: National Center for Food and Agricultural Policy. http://www.ncfap.org/40casestudies.html (accessed October 14, 2008).

Givens, W. A., D. R. Shaw, W. G. Johnson, S. C. Weller, B. G. Young, R. G. Wilson, M. D. K. Owen, and D. L. Jordan. 2009. A grower survey of herbicide use patterns in glyphosate-resistant cropping systems. *Weed Technology* 23:156–161.

Green, J. M., C. B. Hazel, D. R. Forney, and L. M. Pugh. 2008. New multiple-herbicide crop resistance and formulation technology to augment the utility of glyphosate. *Pest Management Science* 64:332–339.

Hartzler, R. G. 1996. Velvetleaf (*Abutilon theophrasti*) population dynamics following a single year's seed rain. *Weed Technology* 10:581–586.

Heap, I. 2010. The International Survey of Herbicide Resistant Weeds. http://www.weedscience.com (accessed April 8, 2010).

Heatherly, L. G., K. N. Reddy, and S. R. Spurlock. 2005. Weed management in glyphosate-resistant and non-glyphosate resistant soybean grown continuously and in rotation. *Agronomy Journal* 97:568–577.

Hilgenfeld, K. L., A. R. Martin, D. A. Mortensen, and S. A. Mason. 2004a. Weed management in glyphosate resistant soybean: weed emergence patterns in relation to glyphosate treatment timing. *Weed Technology* 18:277–283.

Hilgenfeld, K. L., A. R. Martin, D. A. Mortensen, and S. A. Mason. 2004b. Weed management in a glyphosate resistant soybean system: weed species shifts. *Weed Technology* 18:284–291.

International Service for the Acquisition of Agri-Biotech Applications (ISAAA). 2008. ISAAA Brief 37—2007: Executive Summary. http://www.isaaa.org/resources/publications/briefs/37/executivesummary/default.html. (accessed October 8, 2008).

Jha, P., J. K. Norsworthy, W. Bridges, Jr., and M. B. Riley. 2008. Influence of glyphosate timing and row width on Palmer amaranth (*Amaranthus palmeri*) and pusley (*Richardia* spp.) demographics in glyphosate-resistant soybean. *Weed Science* 56:408–415.

Kruger, G. R., W. G. Johnson, S. C. Weller, M. D. K. Owen, D. R. Shaw, J. W. Wilcut, D. L. Jordan, R. G. Wilson, M. L. Bernards, and B. G. Young. 2009. U.S. grower views on problematic weeds and changes in weed pressure in glyphosate-resistant corn, cotton, and soybean cropping systems. *Weed Technology* 23:162–166.

Marshall, M. W., K. Al-Khatib, and L. Maddux. 2000. Impact of continuous glyphosate use on weed populations in a corn-soybean rotation. *Abstracts of the Weed Science Society of America* 40:21–22.

Nandula, V. K., T. W. Eubank, D. H. Poston, C. H. Koger, and K. N. Reddy. 2006. Factors affecting germination of horseweed (*Conyza canadensis*). *Weed Science* 54:898–902.

Nandula, V. K., K. N. Reddy, S. O. Duke, and D. H. Poston. 2005. Glyphosate-resistant weeds: current status and future outlook. *Outlooks on Pest Management* 16:183–187.

Norsworthy, J. K. 2003. Use of soybean production surveys to determine weed management needs of South Carolina farmers. *Weed Technology* 17:195–201.

Norsworthy, J. K. 2008. Effect of tillage intensity and herbicide programs on changes in weed species density and composition in the southeastern coastal plains of the United States. *Crop Protection* 27:151–160.

Norsworthy, J. K., N. R. Burgos, and L. R. Oliver. 2001. *Differences* in weed tolerance to glyphosate involve different mechanisms. *Weed Technology* 15:725–731.

Norsworthy, J. K., P. Jha, and W. Bridges, Jr. 2007. Sicklepod (*Senna obtusifolia*) survival and fecundity in wide- and narrow-row glyphosate-resistant soybean. *Weed Science* 55:252–259.

Norsworthy, J. K. and M. J. Oliveira. 2007. Effect of tillage and soybean on *Ipomoea lacunosa* and *Senna obtusifolia* emergence. *Weed Research* 47:499–508.

Norsworthy, J. K. and L. R. Oliver. 2002a. Effect of irrigation, soybean (*Glycine max*) density, and glyphosate on hemp sesbania (*Sesbania exaltata*) and pitted morning-glory (*Ipomoea lacunosa*) interference in soybean. *Weed Technology* 16:7–17.

Norsworthy, J. K. and L. R. Oliver. 2002b. Pitted morningglory interference in drill-seeded glyphosate-resistant soybean. *Weed Science* 50:26–33.

Norsworthy, J. K., K. L. Smith, R. C. Scott, and E. E. Gbur. 2007. Consultant perspectives on weed management needs in Arkansas cotton. *Weed Technology* 21:825–831.

Owen, M. D. K. 2008. Weed species shifts in glyphosate-resistant crops. *Pest Management Science* 64:377–387.

Owen, M. D. K. and I. A. Zelaya. 2005. Herbicide-resistant crops and weed resistance to herbicides. *Pest Management Science* 61:301–311.

Payne, S. A. and L. R. Oliver. 2000. Weed control programs in drilled glyphosate-resistant soybean. *Weed Technology* 14:413–422.

Powles, S. B. 2008. Evolved glyphosate-resistant weeds around the world: lessons to be learnt. *Pest Management Science* 64:360–365.

Powles, S. B., D. F. Lorraine-Colwill, J. J. Dellow, and C. Preston. 1998. Evolved resistance to glyphosate in rigid ryegrass (*Lolium rigidum*) in Australia. *Weed Science* 46:604–607.

Puricelli, E. and D. Tuesca. 2005. Weed density and diversity under glyphosate-resistant crop sequences. *Crop Protection* 24:533–542.

Rankins, A. Jr., J. D. Byrd, Jr., D. B. Mask, J. W. Barnett, and P. D. Gerard. 2005. Survey of soybean weeds in Mississippi. *Weed Technology* 19:492–498.

Reddy, K. N. 2001. Glyphosate-resistant soybean as weed management tool: opportunities and challenges. *Weed Biology and Management* 1:193–202.

Reddy, K. N. 2004. Weed control and species shifts in bromoxynil- and glyphosate-resistant cotton (*Gossypium hirsutum*) rotation systems. *Weed Technology* 18:131–139.

Reddy, K. N. and C. H. Koger. 2006. Herbicide-resistant crops and weed management. In H. P. Singh, D. R. Batish, and R. K. Kohli, eds. *Handbook of Sustainable Weed Management*. New York: Food Products Press, an Imprint of the Haworth Press, Inc., pp. 549–580.

Reddy, K. N., M. A. Locke, C. H. Koger, R. M. Zablotowicz, and L. J. Krutz. 2006. Cotton and corn rotation under reduced tillage management: impacts on soil properties, weed control, yield, and net return. *Weed Science* 54:768–774.

Reddy, K. N. and K. Whiting. 2000. Weed control and economic comparisons of glyphosate-resistant, sulfonylurea-tolerant, and conventional soybean (*Glycine max*) systems. *Weed Technology* 14:204–211.

Schweizer, E. E. and R. L. Zimdahl. 1984. Weed seed decline in irrigated soil after six years of continuous corn (*Zea mays*) production and herbicides. *Weed Science* 32:76–83.

Scursoni, J. A., F. Forcella, and J. Gunsolus. 2007. Weed escapes and delayed weed emergence in glyphosate-resistant soybean. *Crop Protection* 26:212–218.

Scursoni, J., F. Forcella, J. Gunsolus, M. Owen, R. Oliver, R. Smeda, and R. Vidrine. 2006. Weed diversity and soybean yield with glyphosate management along a north-south transect in the United States. *Weed Science* 54:713–719.

Shaner, D. L. 2000. The impact of glyphosate-tolerant crops on the use of other herbicides and on resistance management. *Pest Management Science* 56:320–326.

Shaw, D. R., W. A. Givens, L. A. Farno, P. D. Gerard, D. L. Jordan, W. G. Johnson, S. C. Weller, B. G. Young, R. G. Wilson, and M. D. K. Owen. 2009. Using a grower survey to assess the benefits and challenges of glyphosate-resistant cropping systems for weed management in U.S. corn, cotton, and soybean. *Weed Technology* 23:134–149.

United States Department of Agriculture (USDA). 2008. National Agricultural Statistics Service. Agricultural Chemical Database. http://www.pestmanagement. info/nass/ (accessed October 14, 2008).

United States Department of Agriculture (USDA). 2009. National Agricultural Statistics Service. Acreage. http://usda.mannlib.cornell.edu/MannUsda/viewDocumentInfo. do?documentID=1000 (accessed July 1, 2009).

VanGessel, M. J. 2001. Glyphosate-resistant horseweed from Delaware. *Weed Science* 49:703–705.

Walker, E. R. and L. R. Oliver. 2008. Weed seed production as influenced by glyphosate applications at flowering across a weed complex. *Weed Technology* 22:318–325.

Webster, T. M. 2001. Weed survey—southern states. *Proceedings of the Southern Weed Science Society* 54:244–259.

Webster, T. M. 2005. Weed survey—southern states. *Proceedings of the Southern Weed Science Society* 58:291–306.

Webster, T. M. 2008. Weed survey—southern states. *Proceedings of the Southern Weed Science Society* 61:224–243.

Webster, T. M. and H. D. Coble. 1997. Changes in the weed species composition of the southern United States: 1974 to 1995. *Weed Technology* 11:308–317.

Webster, T., A. Culpepper, J. Flanders, and T. Grey. 2005. Planting date affects critical tropical spiderwort (*Commelina benghalensis*)-free interval in cotton. *Proceedings of the Beltwide Cotton Conferences* 2842–2843.

Westhoven, A. M., J. M. Stachler, M. M. Loux, and W. G. Johnson. 2008. Management of glyphosate-tolerant common lambsquarters (*Chenopodium album*) in glyphosate-resistant soybean. *Weed Technology* 22:628–634.

Wilson, R. G., S. D. Miller, P. Westra, A. R. Kniss, P. W. Stahlman, G. W. Wicks, and S. D. Kachman. 2007. Glyphosate-induced weed shifts in glyphosate-resistant corn or a rotation of glyphosate-resistant corn, sugarbeet, and spring wheat. *Weed Technology* 21:900–909.

Young, B. G. 2006. Changes in herbicide use patterns and production practices resulting from glyphosate-resistant crops. *Weed Technology* 20:301–307.

10

GLYPHOSATE-RESISTANT HORSEWEED IN THE UNITED STATES

LAWRENCE E. STECKEL, CHRISTOPHER L. MAIN, AND
THOMAS C. MUELLER

10.1 BACKGROUND ON ECOLOGY OF *CONYZA CANADENSIS*

Conyza canadensis (L.) Cronq., a member of the Asteraceae or sunflower family, is known by several common names, including Canada fleabane, horseweed, and mare's tail. The term "horseweed" will be used in reference to this species throughout the remainder of this chapter.

Horseweed is a winter or summer annual native to North America (Weaver 2001) and commonly found in agricultural habitats, wherever tillage has been reduced or eliminated as well as nonagricultural sites with minimal disturbance. Most seedlings emerge from late August through October in the Midwest and form rosettes that overwinter. In general, in states south of the Ohio River, germination can occur during most months of the year. The seeds do not appear to have dormancy mechanisms and can germinate promptly under favorable conditions of temperature and humidity (Lazaroto et al. 2008). Seeds emerged under highly variable conditions, but only emerged from the soil surface; no seedlings emerged from seeds placed at a depth of 0.5 cm or greater (Nandula et al. 2006). Horseweed emerged mainly during April and September in Tennessee when average daytime temperatures fluctuated between 10 and 15°C, as long as adequate moisture was available at the soil

Glyphosate Resistance in Crops and Weeds: History, Development, and Management
Edited by Vijay K. Nandula
Copyright © 2010 John Wiley & Sons, Inc.

surface (Main et al. 2006). In Iowa, most of the horseweed emerged in the fall (Buhler and Owen 1997). Winter survival of fall-emerged seedlings range from 59% to 91%, indicating substantial self-thinning of the population. Additional emergence occurred in the spring at this location, and horseweed emerged well into the growing season. Reports from Indiana also indicated substantial winter mortality (up to 80%) of fall-emerged horseweed (Davis et al. 2008). Crop residue present at the time of horseweed germination affected plant density (Main et al. 2006). Residue from a previous corn crop reduced horse-weed emergence compared with soybean and cotton residues in a no-tillage production system.

10.2   PREVIOUS HERBICIDE RESISTANCE IN HORSEWEED

Horseweed has a history of developing herbicide-resistant biotypes under limited amount of selection pressure. The resistance of horseweed to paraquat application has been extensively studied in both Europe and North America (Lehoczki et al. 1992; Smisek et al. 1998; Szigeti et al. 1996; Varadi et al. 2000). Horseweed has also developed resistance to atrazine (Darko et al. 1996), and some populations have multiple resistance to both paraquat and atrazine (Polos et al. 1988). Cross-resistance of paraquat to a relatively similar herbi-cide diquat and to atrazine was also reported (Szigeti et al. 1994). Horseweed has also been reported to have resistance to cloransulam-methyl, an herbicide that inhibits the enzyme acetolactate synthase (Trainer et al. 2005). These findings indicate that horseweed is a pervasive species that has large genetic variability from which herbicide resistance can evolve under a broad range of agricultural production systems.

10.3   HISTORICAL REVIEW OF GLYPHOSATE RESISTANCE

At the initiation of the development of Roundup Ready® (Monsanto Company, St. Louis, MO) crops, some weed scientists believed the probability of weed resistance developing to glyphosate was low (Bradshaw et al. 1997). As we close the first decade of glyphosate-resistant (GR) crops planted over millions of acres, it is clear that weeds have developed resistance to glyphosate. Other chapters in this book discuss this topic, so we make only a few brief comments. Prior to glyphosate resistance in horseweed, several plant species with glyphosate resistance were documented (Powles and Preston 2006). Goosegrass from Malaysia, rigid ryegrass from Australia, and Italian ryegrass from Chile have glyphosate resistance based on an altered form of the 5-enolpyruvylshikimate-3-phosphate synthase (EPSPS) enzyme, the site of action for glyphosate. Still, in the years around 2000, many farmers in the United States were producing crops with glyphosate being the only herbicide used (Young 2006). It should not have been a surprise that resistance has developed.

10.4 GLYPHOSATE RESISTANCE IN HORSEWEED

The landmark paper first reporting GR horseweed was published by Mark VanGessel in 2001. It clearly demonstrated that selection pressure from no-till soybean production systems in which glyphosate applications were the only mechanism used for weed management resulted in horseweed populations not being controlled, thus evolving GR horseweed. This is in contrast to his previous experience where horseweed had been previously controlled by the same treatments. There was a clear demarcation between susceptible and resistant biotypes.

The next group to investigate the GR horseweed was in Tennessee (Mueller et al. 2003). Resistance was confirmed and no-tillage cotton production systems and subsequent studies showed that it had spread throughout the mid-South region (Koger et al. 2004; Main et al. 2004). This work also indicated that there was a range of sensitivities in the "resistant" populations. The work also conclusively demonstrated that the target site for glyphosate, EPSPS, was still sensitive to glyphosate and being inhibited since shikimate accumulated in both susceptible and resistant species (Mueller et al. 2003). Since this time, GR horseweed has spread over a large geographic area (Grantz et al. 2008; Moreira et al. 2007).

The agronomic importance of GR horseweed would be difficult to overstate. While there were other GR weeds documented prior to horseweed, this was the first weed that forced large-acreage, row crop farmers to adjust their production systems to control an herbicide-resistant weed. Farmers who had enjoyed the extraordinarily simple, completely effective, and economical weed control with glyphosate-only systems now had to readjust their management levels. In Tennessee, there were many instances where farmers had to revert to tillage to control escaped horseweed in their fields.

With the widespread adoption of Roundup Ready crops and the subsequent decline in herbicide sales, some basic herbicide manufacturing companies had reduced their research efforts in herbicide discovery. With the widespread occurrence of a weed that glyphosate would not kill, several companies began to pursue herbicide active ingredients to meet this new market opportunity. Many of these products are just now coming to market.

10.5 PHYSIOLOGICAL BASIS FOR GLYPHOSATE RESISTANCE IN HORSEWEED

The underlying nature and cause of GR horseweed has been studied by several researchers. It is clear that shikimate accumulates in both susceptible and resistant biotypes, indicating that the active site is still being inhibited (Mueller et al. 2003).

In the Mueller et al. (2003) study, shikimic acid levels in shoot tissue of GR horseweed increased (45×) more rapidly by 48 hours after treatment (HAT) compared with glyphosate-susceptible (GS) horseweed (28×), while shikimic acid concentration in root tissue increased to similar levels by 72 HAT for both GR (21×) and GS (22×) horseweed. However, shikimic acid levels begin to decrease by 96 HAT in shoot and root tissue of GR horseweed, while shikimic acid levels continued to increase with time in GS horseweed. These data are consistent with Feng et al. (2004) in that these results can be explained by reduced phloem loading of glyphosate in GR horseweed, since shikimic acid levels stop increasing in root tissue 72 HAT. However, shikimic acid concentration in GR horseweed root tissue prior to 72 HAT is similar to the levels found in GS horseweed, so it is evident that glyphosate is moving into the phloem in GR horseweed (Fig. 10.1). This research indicates that 72 HAT shikimic acid levels stop increasing and begin to decrease in shoot and root tissue of GR horseweed.

Researchers with Monsanto have performed and published the landmark paper on this topic (Feng et al. 2004). They reported reduced root translocation

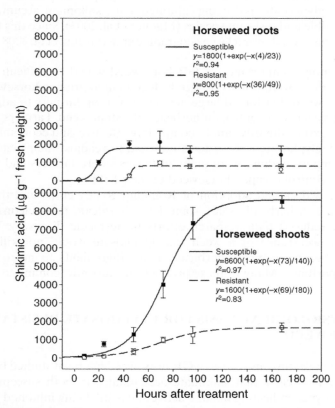

Figure 10.1. Shikimic acid response in glyphosate-resistant and glyphosate-susceptible horseweed shoot and root tissue over time following treatment with glyphosate (0.84 kg a.e. ha^{-1}).

of glyphosate in resistant biotypes, and that glyphosate loading into the apoplast and phloem was delayed and reduced in the resistant biotypes. Their explanation for glyphosate resistance was impaired glyphosate translocation, and that the resistance is likely due to altered cellular distribution that impaired phloem loading and plastidic import of glyphosate, resulting in reduced overall translocation as well as inhibition of the enzyme at the active sight. Dinelli et al. (2006) also reported altered glyphosate movement in resistant horseweed plants. They reported that impaired translocation of the herbicide and increased EPSPS transcript levels may account for glyphosate resistance in horseweed. Additional work examining the inheritance of glyphosate resistance in *Conyza* species suggested that glyphosate resistance was governed by an incompletely dominant, single-locus gene located in the nuclear genome (Zelaya et al. 2004). These researchers predicted a rapid increase in frequency in the resistance allele under continuous glyphosate selection pressures. In related research, they also reported that this resistance could be disseminated via hybrids of other *Conyza* species (Zelaya et al. 2007).

10.6 GR HORSEWEED DISSEMINATION

Seldom has a plant that has developed herbicide resistance grown as rapidly as has GR horseweed. It exists from California (Grantz et al. 2008) to Delaware (VanGessel 2001) in many agricultural areas. Researchers in Indiana have probably done the most detailed survey project investigating GR horseweed (Davis et al. 2008). In this research, GR horseweed populations were found in all regions of Indiana; however, the highest frequencies were in the southeastern region of the state, and very low (<5%) frequencies were noted throughout the rest of Indiana. Once a colonizing population of GR horseweed is established in an area, it frequently spreads to many adjacent farms.

Field studies examining horseweed dissemination have illustrated how easily seeds from horseweed spread (Dauer et al. 2007). Horseweed plants produce large numbers of very small seeds that easily move under field conditions. The seed regularly disbursed at least 500 m from source populations, although 99% of the seed was found within 100 m of the source (Dauer et al. 2007). Other field studies using remote-controlled model airplanes collected horseweed seeds as they floated on air currents above the ground as they entered the planetary boundary layer, where long-range transport of aerial biota frequently occurs (Shields et al. 2006). The results of this research indicated that horseweed seed dispersal could exceed 500 km in a single dispersal event. Wind-dispersed plant seeds such as this challenge the common practice of single-field management as a viable management option for herbicide-resistant weeds (Dauer et al. 2007). The unfortunate reality was that some farmers may have attempted to proactively manage their fields to avoid GR horseweed, yet because of the dispersal mechanisms previously described many still had their fields subsequently infested. This discourages farmers from using proactive resistance avoidance strategies since they are going to get resistance anyway from their neighbors' fields.

10.7   GR HORSEWEED MANAGEMENT STRATEGIES

Delaware growers were surveyed, and the most frequent change (66% of growers) in response to GR horseweed infestations was the application of another herbicide before planting with a different mode of action (Scott and VanGessel 2007). Several authors reported that the use of auxin-type herbicides such as 2,4-dichlorophenoxyacetic acid or dicamba were effective in controlling GR horseweed (Eubank et al. 2008; Steckel et al. 2006; Wiese et al. 1995), although planting too soon after the application of the auxin herbicide could result in crop injury (Thompson et al. 2007). Spring-applied, residual herbicide systems are often the most effective at reducing season-long horseweed densities and protecting crop yield (Davis et al. 2007). Another potential technology to control GR horseweed is the use of glufosinate in LibertyLink® (Bayer CropScience, Research Triangle Park, NC) cropping systems. Glufosinate can provide excellent control of horseweed if air temperatures are not below 70°F during the day (Steckel et al. 2006). This makes glufosinate a very good control option in crops. However, burndown application of glufosinate prior to planting in the spring typically provides less consistent control of horseweed due to cooler temperatures at that time of year.

There are many advantages to no-tillage production systems, but horseweed control is much easier when the crop seed bed is tilled. Disking in the spring controlled horseweed for the subsequent cotton crop (Brown and Whitwell 1988). There were many farmers in Tennessee who resorted to spring tillage when no other means of controlling GR horseweed were available.

GR horseweed was the first of many weeds that forced farmers to change their weed management, and as such, it has greatly affected weed control in this century. Given horseweed's intrinsic genetic variability, ability to produce many seeds, the ease and great distance that these seeds are dispersed, and the lack of dormancy mechanisms, it is quite reasonable to see the common occurrence of GR horseweed in many production systems.

REFERENCES

Bradshaw, L. D., S. R. Padgette, S. L. Kimball, and B. H. Wells. 1997. Perspectives on glyphosate resistance. *Weed Technology* 11:189–198.

Brown, S. M. and R. Whitwell. 1988. Influence of tillage on horseweed, *Conyza canadensis. Weed Technology* 2:269–270.

Buhler D. D. and M. D. K. Owen. 1997. Emergence and survival of horseweed (*Conyza canadensis*). *Weed Science* 45:98–101.

Darko, E., G. Varadi, S. Dulai, and E. Lehoczki. 1996. Atrazine-resistant biotypes of *Conyza canadensis* have altered fluorescence quenching and xanthophyll cycle pattern. *Plant Physiology and Biochemistry* 34:843–852.

Dauer, J. T., D. A. Mortensen, and M. J. VanGessel. 2007. Temporal and spatial dynamics of long-distance *Conyza canadensis* seed dispersal. *Journal of Applied Ecology* 44:105–114.

Davis, V. M., K. D. Gibson, T. T. Bauman, S. C. Weller, and W. G. Johnson. 2007. Influence of weed management practices and crop rotation on glyphosate-resistant horseweed population dynamics and crop yield. *Weed Science* 55:508–516.

Davis, V. M., K. D. Gibson, and W. G. Johnson. 2008. A field survey to determine distribution and frequency of glyphosate-resistant horseweed (*Conyza canadensis*) in Indiana. *Weed Technology* 22:331–338.

Dinelli, G., I. Marotti, A. Bonnetti, M. Minelli, P. Catizone, and J. Barnes. 2006. Physiological and molecular insight on the mechanisms of resistance to glyphosate in *Conyza canadensis* biotypes. *Pesticide Biochemistry and Physiology* 86:30–41.

Eubank, T. W., D. H. Poston, V. K. Nandula, C. H. Koger, D. R. Shaw, and D. B. Reynolds. 2008. Glyphosate-resistant horseweed (*Conyza canadensis*) control using glyphosate-, paraquat-, and glufosinate-based herbicide programs. *Weed Technology* 22:16–21.

Feng, P. C. C., M. Tran, T. Chiu, R. D. Sammons, G. R. Heck, and C. A. Jacob. 2004. Investigations into glyphosate-resistant horseweed (*Conyza canadensis*): retention, uptake, translocation, and metabolism. *Weed Science* 52:498–505.

Grantz, D. A., A. Shrestha, and H. B. Vu. 2008. Early vigor and ozone response in horseweed (*Conyza canadensis*) biotypes differing in glyphosate resistance. *Weed Science* 56:224–230.

Koger, C. H., D. H. Poston, R. M. Hayes, and R. F. Montgomery. 2004. Glyphosate-resistant horseweed (*Conyza canadensis*) in Mississippi. *Weed Technology* 18:820–825.

Lazaroto, C. A., N. G. Fleck, and R. A. Vidal. 2008. Biology and ecophysiology of hairy fleabane. *Ciencia Rural* 83:852–860.

Lehoczki, E., G. Laskay, I. Gaal, and Z. Szigeti. 1992. Mode of action of paraquat in leaves of paraquat-resistant *Conyza canadensis*. *Plant, Cell, & Environment* 15:531–539.

Main, C. L., T. C. Mueller, R. M. Hayes, and J. B. Wilkerson. 2004. Response of selected horseweed (*Conyza canadensis* (L.) Cronq.) populations to glyphosate. *Journal of Agricultural and Food Chemistry* 52:879–883.

Main, C. L., L. E. Steckel, R. M. Hayes, and T. C. Mueller. 2006. Biotic and abiotic factors influence horseweed emergence. *Weed Science* 54:1101–1105.

Moreira, M. S., M. Nicolai, S. J. P. Carvalho, and P. J. Christoffoleti. 2007. Glyphosate-resistance in *Conyza canadensis* and *C. bonariensis*. *Planta Daninha* 25:157–164.

Mueller, T. C., J. H. Massey, R. M. Hayes, C. L. Main, and C. N. Stewart, Jr. 2003. Shikimate accumulates in both glyphosate-sensitive and glyphosate-resistant horseweed (*Conyza canadensis* L. Cronq.). *Journal of Agricultural and Food Chemistry* 51:680–684.

Nandula, V. K., T. W. Eubank, D. H. Poston, C. H. Koger, and K. N. Reddy. 2006. Factors affecting germination of horseweed (*Conyza canadensis*). *Weed Science* 54:898–902.

Polos, E., J. Mikulas, Z. Szigetia, B. Matkovics, D. Q. Hai, A. Parducz, and E. Lehoczki. 1988. Paraquat and atrazine co-resistance in *Conyza canadensis*. *Pesticide Biochemistry and Physiology* 30:142–154.

Powles, S. B. and C. Preston. 2006. Evolved glyphosate resistance in plants: biochemical and genetic basis of resistance. *Weed Technology* 20:282–289.

Scott, B. A. and M. J. VanGessel. 2007. Delaware soybean grower survey on glyphosate-resistant horseweed (*Conyza canadensis*). *Weed Technology* 21:270–274.

Shields, E. J., J. T. Dauer, M. J. VanGessel, and G. Neumann. 2006. Horseweed (*Conyza canadensis*) seed collected in the planetary boundary layer. *Weed Science* 54:1063–1067.

Smisek, A., C. Doucer, M. Jones, and S. Weaver. 1998. Paraquat resistance in horseweed (*Conyza canadensis*) and Virginia pepperweed (*Lepidium virginicum*) from Essex. *Weed Science* 46:200–204.

Steckel, L. E., C. C. Craig, and R. M. Hayes. 2006. Glyphosate-resistant horseweed (*Conyza canadensis*) control with glufosinate prior to planting no-till cotton (*Gossypium hirsutum*). *Weed Technology* 20:1047–1051.

Szigeti, Z., E. Darko, E. Nagy, and E. Lehoczki. 1994. Diquat resistance of different paraquat-resistant *Conyza canadensis* biotypes. *Journal of Plant Physiology* 144:686–690.

Szigeti, Z., I. Racz, E. Darko, D. Lasztity, and E. Lehoczki. 1996. Are either SOD and catalyst or the polyamines involved in the paraquat resistance of *Conyza canadensis*. *Journal of Environmental Science and Health, Part B* 31:599–604.

Thompson, M. A., L. E. Steckel, A. T. Ellis, and T. C. Mueller. 2007. Soybean tolerance to early preplant applications of 2, 4-D amine and dicamba. *Weed Technology* 21:882–995.

Trainer, G. D., M. M. Loux, S. K. Harrison, and E. Regnier. 2005. Response of horseweed biotypes to foliar applications of cloransulam-methyl and glyphosate. *Weed Technology* 19:231–236.

VanGessel, M. J. 2001. Glyphosate-resistant horseweed from Delaware. *Weed Science* 49:703–705.

Varadi, G., E. Darko, and E. Lehoczki. 2000. Changes in the xanthophylls cycle and fluorescence quenching indicate light-dependent early events in the action of paraquat and the mechanism of resistance to paraquat in *Erigeron canadensis*. *Plant Physiology* 123:1459–1469.

Weaver, S. E. 2001. The biology of Canadian weeds. 115. *Conyza canadensis*. *Canadian Journal of Plant Science* 81:867–875.

Wiese, A. L., C. D. Salisbury, and B. W. Bean. 1995. Downy brome (*Bromus tectorum*), jointed goatgrass (*Aegilops cylindrical*) and horseweed (*Conyza canadensis*) control in fallow. *Weed Technology* 9:249–254.

Young, B. G. 2006. Changes in herbicide use patterns and production practices resulting from glyphosate-resistant crops. *Weed Technology* 20:301–307.

Zelaya, I. A., M. D. K. Owen, and M. J. VanGessel. 2004. Inheritance of evolved glyphosate resistance in *Conyza canadensis*. *Theoretical and Applied Genetics* 110:58–70.

Zelaya, I. A., M. D. K. Owen, and M. J. VanGessel. 2007. Transfer of glyphosate resistance: evidence of hybridization in *Conyza* (Asteraceae). *American Journal of Botany* 94:660–673.

11

GLYPHOSATE-RESISTANT PALMER AMARANTH IN THE UNITED STATES

A. STANLEY CULPEPPER, THEODORE M. WEBSTER,
LYNN M. SOSNOSKIE, AND ALAN C. YORK

11.1 INTRODUCTION

Palmer amaranth (*Amaranthus palmeri* S. Watson; subgenus *Acnida*, subsection *Saueranthus*) is one of the most common and problematic weeds in agronomic crop production throughout the southern United States (Webster 2005). Since the mid-1990s, when glyphosate-resistant (GR) crops were commercialized, glyphosate has been extensively used to effectively and economically manage Palmer amaranth (Culpepper and York 1998; Grichar et al. 2004). Many cotton (*Gossypium hirsutum*) growers transitioned away from soil-applied residual herbicides and cultivation for weed control to production systems that relied heavily on glyphosate applied multiple times throughout the season as the sole means of managing Palmer amaranth and other troublesome weeds. The adoption of GR technology also provided growers with the capabilities needed to rapidly adopt conservation tillage production systems, further increasing their dependence on herbicides, especially glyphosate. The selection pressure arising from this unprecedented use of glyphosate over space and time subsequently led to the evolution of GR biotypes in Palmer amaranth (Culpepper et al. 2006; Heap 2010). Beginning in 2005, many Georgia growers were forced to abandon their cotton crops due to the inability to

Glyphosate Resistance in Crops and Weeds: History, Development, and Management
Edited by Vijay K. Nandula
Copyright © 2010 John Wiley & Sons, Inc.

manage GR Palmer amaranth using herbicide programs that had previously provided excellent weed control. These growers now spend up to $130 ha^{-1} on herbicides, but herbicides applied alone may not adequately control GR Palmer amaranth. Growers are often forced to implement hand weeding and/ or cultivation (Culpepper et al. 2009b). As of the spring of 2009, GR Palmer amaranth populations have been confirmed in eight states (Alabama, Arkansas, Georgia, Missouri, North Carolina, South Carolina, Tennessee, and Mississippi) in the Southeast and mid-South regions of the United States; the total esti-mated infested area exceeds 700,000 ha and is increasing rapidly (Culpepper et al. 2009b; Heap 2010; Nichols et al. 2009; Steckel et al. 2009). Development and spread of glyphosate resistance in Palmer amaranth threatens a growers' ability to manage this pest utilizing currently available herbicide technologies (Culpepper et al. 2008; Marshall 2009; Whitaker et al. 2007). Future manage-ment strategies will rely heavily on an improved understanding of Palmer amaranth biology and ecology, particularly plant population demographics. Development of integrated management approaches using cultural, mechani-cal, and chemical controls may be the only economically effective option for controlling GR Palmer amaranth (Culpepper et al. 2008, 2009a, 2009b; Nichols et al. 2009; Steckel et al. 2009; Whitaker et al. 2007).

11.2   PALMER AMARANTH ORIGIN, IDENTIFICATION, AND BIOLOGY

Palmer amaranth, also commonly called careless weed, is native to the Sonoran Desert, which spans the Mexican states of Sonora and Baja California and parts of southern Arizona and California (Ehleringer 1983). The earliest recorded descriptions of Palmer amaranth (identified as *Amaranthus* (*Amblogyne*) *palmeri*) in the United States were made by Sereno Watson working from specimens collected by Edward Palmer in San Diego County, CA, and by Jean Louis Berlandier along the banks of the Rio Grande River (Watson 1877). Herbaria records documenting the earliest confirmed occur-rences of Palmer amaranth in the following states suggest an easterly and northerly movement of the species: Texas (1834), Arizona (1865), California (1875), New Mexico (1881), Utah (1888), Kansas (1895), Illinois (1896), Missouri (1897), Oklahoma (1926), Louisiana (1929), Mississippi (1971), Tennessee (1975), Arkansas (1976), and Colorado (1980). Collection of Palmer amaranth from some eastern seaboard states (New York [1936], Pennsylvania [1933], South Carolina [1957, adjacent to a wool combing mill in Florence], Virginia [1915, along the Potomac river in Arlington], and Maryland [1953, adjacent to a chrome ore pile in Baltimore]) prior to the first reported occurrence of the species in neighboring western states may indicate a series of separate and unique introductions (Sauer 1955).

A description of Palmer amaranth has been derived by Mosyakin and Robertson (2008). They described Palmer amaranth as a tall (often exceeding

Figure 11.1. Palmer amaranth female and male plants.

2.5 m in height), erect, summer annual, which may be frequently branched. Leaves are arranged in an alternate pattern on the stems. Leaf blades are obovate to elliptic in shape, 1.5–7 cm long and 1–3.5 cm wide, with entire margins. Petioles are often as long, or longer, than the leaf blades. Palmer amaranth male and female flowers are held on separate plants (i.e., dioecious species) in long (0.5 m and sometimes greater), narrowly elongated, linear or complex, terminal inflorescences (thryses) that are usually drooping on older plants (Fig. 11.1). Pistillate (female) flowers consist of five, often green, unevenly sized (1.7–3.8 mm), spatulate (spoon-shaped), and pointed tepals that are subtended by long (4.0–6.0 mm), rigid, narrow, and sharply pointed bracts that do not enfold the flower and have entire margins and excurrent midribs. These sharp bracts distinguish female inflorescences from male inflorescences, which are soft to the touch. There are two, or sometimes three, stigmas, and the styles are branched and spreading; the ovary is superior. Similarly, staminate (male) flowers have five, unequal (2.0–4.0 mm), spatulate, and apex long-acuminate or mucronulate tepals that are subtended by tapering bracts (4.0 mm). The inner tepals possess excurrent midribs. There are five four-locular anthers per staminate flower. Flowering occurs mainly during the summer and autumn, although, in the southernmost regions of the species' range, flowering specimens may be observed, occasionally, during the winter- and springtime months.

The fruit of Palmer amaranth is a papery, one-seeded, circumscissile, dehiscent utricle (1.5–2 mm long) (Mosyakin and Robertson 2008). Seeds are small (1–2 mm), subglobose or lenticular (round to lens shaped), smooth, shiny, and dark purple to black at maturity. Mature female Palmer amaranth plants produce prodigious amounts of seed, although emergence date and intra- and interspecific interference can significantly affect reproductive capabilities. Keeley et al. (1987) reported that mean seed production per plant ranged from 200,000 to 600,000 seeds for plants that emerged between March and June in California. Plants that emerged between July and October were smaller, produced fewer inflorescences, and yielded significantly fewer seed (0–80,000 seeds per plant) than the earlier emerging specimens (Keeley et al. 1987). Sellers et al. (2003) reported that Palmer amaranth plants developing from seed planted in late May and early June in Missouri produced greater than 250,000 seeds per plant. Palmer amaranth that emerged between mid-June and late July and grown in competition with wide-row (spaced 97 cm apart) soybean (*Glycine max*) produced 211,000 seeds m^{-2}, while those in narrow-row (spaced 19 cm apart) soybean only produced 139,000 seeds m^{-2} (Jha et al. 2008). In the same study, Palmer amaranth that emerged between late July and mid-August produced 97% fewer seeds (5600 seeds m^{-2}) than the earlier emerging plants (Jha et al. 2008). During the drought in Georgia in 2006 and 2007, GR Palmer amaranth female plants that emerged and grew with a competitive cotton variety ("DP555 BGR") for the entire season produced up to 460,000 seeds per plant (MacRae et al. 2008).

Germination of *Amaranthus* spp. seeds, specifically redroot pigweed (*Amaranthus retroflexus*) and smooth pigweed (*Amaranthus hybridus*), is controlled, in part, by the phytochrome system (Gallagher and Cardina 1997, 1998a, 1998b). This light cue requirement is likely an adaptation within these small-seeded species to signal the relative proximity to the soil surface. Palmer amaranth plants that become established in the field are also likely emerging from relatively shallow depths within the soil profile. Keeley et al. (1987) found that Palmer amaranth seedlings emerged more readily from a depth of 2.5 cm or less than from depths of 5.1 or 7.6 cm. With the exception of temperature, the roles of external environmental factors on Palmer amaranth seed germination have not been elucidated. Steckel et al. (2004) and Guo and Al-Khatib (2003) reported that seed germination increased as incubation temperature increased. Steckel et al. (2004) determined that maximum seed germination was achieved using a temperature regimen that alternated around 30°C. Guo and Al-Khatib (2003) reported that Palmer amaranth seed germination was greatest when the incubation temperature alternated between 30 and 35°C, although Palmer amaranth has been shown to emerge when soil temperatures reached 18°C in the field (Keeley et al. 1987).

It is currently unknown exactly how long Palmer amaranth seed persist once they enter the soil seedbank; Menges (1987) reported that 6 years of hand weeding and herbicide use reduced, but did not eliminate, the seed reservoir. In nontreated control plots, Palmer amaranth seedbank densities grew from 173 million seeds ha^{-1} to 1.1 billion seeds ha^{-1} between 1980 and 1985. Although weed management practices were able to reduce the seedbank size by 98%, relative to the control, approximately 18 million seed ha^{-1} remained in the soil of the treated plots at the end of the study (Menges 1987). Similarly, Norsworthy (2008) reported that the Palmer amaranth seedbank was ephemeral and decreased in density over time in response to tillage and herbicides. In an ongoing study in Georgia, Palmer amaranth seed mortality, in the absence of predation, was approximately 50% for seed buried between 1 and 10 cm deep for 18 months (Sosnoskie, personal observation).

11.3 COMPETITIVE ABILITIES OF PALMER AMARANTH AND CROP YIELD LOSS

Palmer amaranth's competitiveness is likely a function of its rapid growth rate. Horak and Loughin (2000) determined that Palmer amaranth plants grew at rates of 0.18–0.21 cm GDD^{-1} (growing degree days, base temperature of 10°C), which are 30–160% greater than the rates of height growth observed for common waterhemp (*Amaranthus rudis*) (0.11–0.16 cm GDD^{-1}), redroot pigweed (0.09–0.12 cm GDD^{-1}), and tumble pigweed (*Amaranthus albus*) (0.08–0.09 cm GDD^{-1}). The ability of Palmer amaranth to outcompete most other plants may also be due, in part, to rooting structure. Palmer amaranth plants had 3.7 times more roots that were 5 times longer than those of soybean with a similar root fresh weight (Wright et al. 1999). The disparity in the number of roots between the species was due to the smaller diameter of the Palmer amaranth roots compared with soybean. Roots of Palmer amaranth were more effective than soybean in penetrating soil layers with high bulk density, usually associated with compaction due to equipment traffic through the field (Place et al. 2008). These compacted soil layers often restrict root growth, nutrient and water uptake, and crop yield. Place et al. (2008) concluded that the ability to grow through these compacted soil layers is a competitive advantage for Palmer amaranth compared with soybean, as the roots of the weed will have greater access to water and nutrients.

Palmer amaranth interference significantly affects growth and yield of most agronomic crops, with cotton being one of the more sensitive commodities. Morgan et al. (2001) reported Palmer amaranth densities between 1 and 10 plants in 9.1 m^{-1} row of cotton (1-m row spacing at College Station, TX) reduced cotton canopy volume by 35% and 45% at 6 and 10 weeks after cotton emergence, respectively; cotton lint yields were reduced between 13% and 54% for the same densities. Rowland et al. (1999) determined that lint yield was reduced from 5% to 9% plot^{-1} (10 m long by 3.6 m wide at Perkins, OK; 10 m long by 4.1 m wide at Chickasha, OK) for each 1 kg increase in weed biomass. Research in Georgia in 2006 and 2007 noted that two GR Palmer amaranth spaced every 7 m of row (91-cm row spacing) reduced cotton yield by 23% (MacRae et al. 2007). In addition to reducing yields, Palmer amaranth interferes with cotton harvest. Morgan et al. (2001) suggested that mechanical harvesting of cotton with Palmer amaranth at densities greater than six plants 9.1 m^{-1} row was impractical because of the potential for damage to the equipment. Smith et al. (2000) reported that the frequency of work stoppages increased as Palmer amaranth densities increased because of the need to repeatedly dislodge weed stems from the harvester. Despite the fact that Palmer amaranth plant material comprised between 11% and 15% of the total trash in the harvested seed cotton, the residual weed matter was successfully removed using lint cleaners typically found in cotton gins (Smith et al. 2000). Cotton fiber quality parameters, such as length and micronaire, were generally unaffected by weed density (Morgan et al. 2001; Rowland et al. 1999; Smith et al. 2000).

Soybean, corn (*Zea mays*), grain sorghum (*Sorghum bicolor* ssp. *bicolor*), and peanut (*Arachis hypogaea*) yields are also influenced by interference from Palmer amaranth. Klingaman and Oliver (1994) reported that Palmer amaranth interference significantly affected soybean canopy structure. At 12 weeks after soybean emergence, a density of 10 Palmer amaranth plants m^{-1} of row (1-m row spacing in Fayetteville, AR) reduced soybean canopy width by 55% relative to the weed-free control. According to Klingaman and Oliver (1994), densities of 0.33, 0.66, 1, 2, 3.33, and 10 Palmer amaranth plants m^{-1} reduced soybean yields by 17%, 27%, 32%, 48%, 64%, and 68%, respectively. Bensch et al. (2003) reported that soybean yield loss increased as weed density increased when Palmer amaranth emerged with the crop; Palmer amaranth that emerged after soybean emergence did not significantly reduce crop yield. The maximum predicted soybean yield loss was 79% from season-long interference of eight Palmer amaranth plants m^{-1} of row (76-cm row spacing in Manhattan and Topeka, KS) (Bensch et al. 2003). At densities between 0.19 and 3 plants m^{-2}, Palmer amaranth caused greater soybean yield loss than the common waterhemp and redroot pigweed (Bensch et al. 2003). Massinga et al. (2001) reported that corn yield decreased as Palmer amaranth densities increased, although the degree of reduction was also affected by timing of weed emergence relative to the crop. When Palmer amaranth emerged with corn, yield losses ranged from 11% to 91% for weed densities of 0.5–8 plants m^{-1} of row (76-cm row spacing at Garden City, KS) (Massinga et al. 2001). For the same weed densities, corn yield losses ranged between 7% and 35% when Palmer amaranth emerged in four- to seven-leaf corn. Full-season interference of Palmer amaranth (density of 13.7 plants m^{-2}) reduced grain sorghum yields 38% to 63% (Moore et al. 2004). Peanut yield loss from season-long interference of one Palmer amaranth plant m^{-1} of row (91-cm row spacing at Rocky Mount, NC; 96-cm row spacing at Goldsboro, NC) was predicted to be 28% (Burke et al. 2007). Full-season interference from Palmer amaranth (density of 5.5 plants m^{-1} of row) reduced peanut yield by 68% (Burke et al. 2007). A comparison of competitive estimates across studies indicated that Palmer amaranth was more competitive with peanut than broadleaf signalgrass (*Urochloa platyphylla*), bristly starbur (*Acanthospermum hispidum*), tropic croton (*Croton glandulosus* var. *septentrionalis*), horsenettle (*Solanum carolinense*), wild poinsettia (*Euphorbia heterophylla*), and jimsonweed (*Datura stramonium*). However, Palmer amaranth was less competitive in peanut than common ragweed (*Ambrosia artemisiifolia*), fall panicum (*Panicum dichotomiflorum*), and common cocklebur (*Xanthium strumarium*) (Burke et al. 2007).

11.4   DEVELOPMENT AND SPREAD OF GLYPHOSATE RESISTANCE IN PALMER AMARANTH

In 2004, GR Palmer amaranth was discovered in a 250-ha cotton field in Macon County, Georgia (Culpepper et al. 2006). Production at this site had been a monoculture of GR cotton where glyphosate, often applied at reduced rates, was employed as the sole means of weed control for at least 7 years. By the spring of 2009, GR biotypes had been documented in 1, 12, 23, and 38 counties in Alabama, South Carolina, North Carolina, and Georgia, respectively, with a total estimated infestation exceeding 350,000 ha (Culpepper et al. 2009b; Heap 2010; Nichols et al. 2009). GR Palmer amaranth had also been documented in over 30 counties across Arkansas, Mississippi, Missouri, and Tennessee, infesting over 350,000 ha (Culpepper 2009b; Heap 2010; Nandula et al. 2009; Nichols et al. 2009; Steckel et al. 2009).

Glyphosate resistance in Palmer amaranth was confirmed using glyphosate rate response studies. In greenhouse studies, the original Georgia GR biotype had a glyphosate I_{50} (rate of glyphosate required to reduce shoot fresh weight by 50%) of 1.2 kg a.e.ha^{-1} (approximately eight times greater than that of the susceptible biotype, I_{50} = 0.15 kg ha^{-1}) (Culpepper et al. 2006). However, field studies indicated that glyphosate applied at 12 times the recommended rate failed to control that biotype (Culpepper et al. 2006). At least two GR Palmer amaranth biotypes in North Carolina had a glyphosate I_{50} between 0.18 and 0.36 kg ha^{-1}, two and four times greater than the susceptible biotype (I_{50} = 0.089 kg ha^{-1}) (York 2007) with one biotype having an I_{50} value that was at least 20 times the susceptible (Whitaker et al. 2007; York 2007). In Arkansas, a GR Palmer amaranth biotype had a glyphosate I_{50} of 2.8 kg ha^{-1} compared with 0.035 kg ha^{-1} for the susceptible biotype (Norsworthy et al. 2008).

It is likely that more than one resistance mechanism exists in GR biotypes of Palmer amaranth. While the mechanism(s) of Palmer amaranth resistance to glyphosate may not be entirely understood, glyphosate is known to inhibit an enzyme in the synthesis of aromatic amino acids (i.e., phenylalanine, tyrosine, and tryptophan) (Devine et al. 1993). When topically applied to susceptible plants, glyphosate binds to 5-enolpyruvylshikimate-3-phosphate synthase (EPSPS; EC 2.5.1.19) resulting in high levels of shikimate accumulation. The initial GR Palmer amaranth biotype from Georgia did not accumulate shikimate in the presence of glyphosate, while shikimate accumulation in glyphosate-susceptible Palmer amaranth biotypes increased linearly with glyphosate concentration (Culpepper et al. 2006). Initially, absence of shikimate accumulation in the presence of glyphosate in the Georgia biotype could be attributed to reduced absorption and translocation of the herbicide. Reduced absorption and translocation of glyphosate appears to be the mechanism of resistance in GR horseweed (*Conyza canadensis*) (Koger and Reddy 2005) and GR Italian ryegrass (*Lolium perenne* L. ssp. *multiflorum*) (Nandula et al. 2008). However, absorption and translocation of glyphosate in the Georgia GR Palmer amaranth biotype was similar to that of a glyphosate-susceptible Georgia biotype (Culpepper et al. 2006). The currently accepted mechanism of resistance in the Georgia GR Palmer amaranth biotype involves increased amplification of the EPSPS gene, a novel mechanism of herbicide resistance in weeds (Gaines et al. 2009). It is likely that more than one resistance mechanism exists in GR biotypes of Palmer amaranth. In contrast to the Georgia GR biotype, the Tennessee GR Palmer amaranth biotype accumulated shikimate in the presence of glyphosate, indicating that glyphosate inhibits this pathway and that an altered target site (or gene amplification) may not be the mechanism of resistance (Steckel et al. 2008).

Herbicide resistance in a weed population can either develop *de novo*, via genetic mutation, or be acquired through gene flow, which is accomplished by the movement of pollen and seed across the agricultural landscape (Jasieniuk et al. 1996). Palmer amaranth seeds are not adapted for wind dispersal but are probably spread by various means, including irrigation and other water flow, with the movement of birds and mammals, and through agricultural management practices such as plowing, mowing, compost and manure spreading, and harvesting (Costea et al. 2004, 2005; Menges 1987). Because Palmer amaranth is dioecious, there is concern that the resistance trait can be transferred between spatially segregated populations via wind-mediated pollen dispersal; there is evidence in the literature indicating that long-distance pollen dispersal events can and do occur (Alibert et al. 2005; Hanson et al. 2005; Massinga et al. 2003; Matus-Cadiz et al. 2004; Saeglitz et al. 2000). Palmer amaranth pollen grains are small (approximately $26\,\mu m$ in diameter) and settle slowly ($5.0\,cm\,s^{-1}$) (Sosnoskie et al. 2009b), and are therefore more likely to be transported greater distances away from the paternal plant compared with larger, sticky, and/or highly ornamented pollen grains (Ackerman 2000; Primack 1978). Additionally, the surfaces of Palmer amaranth pollen are covered with shallow pores (Borsch 1998; Franssen et al. 2001), which also serve to increase the grains' dispersal capabilities. Surface pores create a layer of turbulent air around the grain, which reduces pressure drag and increases flight time (Franssen et al. 2001). In 2006 and 2007, studies were conducted in Georgia to determine if the glyphosate resistance trait can be transferred via pollen movement from a GR Palmer amaranth source planted in the center of a 30-ha field to glyphosate-susceptible females planted between 1 and 300 m away (Sosnoskie et al. 2009c). Approximately 60% of the offspring derived from susceptible females at the 1-m distance were resistant to glyphosate; at 300 m, approximately 20% of the offspring were resistant.

Several species of bees (*Apis* spp. (*Hymenotera:Apidae*), *Bombus* spp. (*Hymenotera:Apidae*), *Melissodes thelypodii* (*Hymenoptera: Anthophoridae*)) have been observed visiting male Palmer amaranth flowers (L. M. Sosnoskie, personal observation) and collecting pollen under both field and laboratory conditions (Cane et al. 1992; Vaissière and Vinson 1994). Cane et al. (1992) indicated that the pollen of Palmer amaranth is approximately 18.4% crude protein (3.5% N) and is somewhat nutritious for bees. Although bees have not been confirmed to visit female Palmer amaranth flowers, Cane et al. (1992) suggested that bee leg and wing movements could aid in pollen dispersal by forcibly dislodging pollen grains from dehiscent anthers.

11.5   GR PALMER AMARANTH IMPACTS GEORGIA COTTON

In 2008, a survey of Georgia Cooperative Extension Agents was conducted to determine the impact of GR Palmer amaranth on Georgia cotton production. The survey included five counties that were severely infested (greater than 60% of the agronomic land infested), three counties with moderate infestations (20–60% of the agronomic land infested), and 12 counties with light or no infestations (<20% of the agronomic land infested) (Fig. 11.2).

Prior to the occurrence of GR Palmer amaranth, growers in severely infested areas were treating less than 26% of their acreage with herbicides having soil residual activity, while in 2008, at least 88% of the land was treated with two or more at-plant or preplant herbicides (Table 11.1). Growers in severely infested areas reduced conservation tillage by 35%, increased

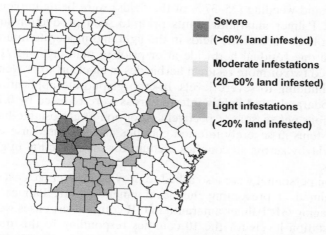

Figure 11.2. Glyphosate-resistant Palmer amaranth infestation levels in Georgia.

TABLE 11.1. Georgia Cooperative Extension Service Survey of the Impacts of Glyphosate-Resistant Palmer Amaranth in Georgia Cotton

	GR Palmer Amaranth Infestation Level					
	Severe		Moderate		Light	
Survey Questions	2004[a]	2008[a]	2004	2008	2004	2008
Acres (%) treated with a DNA[b] herbicide	25	92	75	95	70	91
Acres (%) treated with residual herbicide other than DNA	25	88	61	95	35	71
Strip-tillage production (% acres)	83	48	45	45	30	60
Herbicide incorporation through preplant tillage (% acres)	0	5	0	0	0	0
Adoption of glufosinate programs (% acres)	0	26	0	5	0	2
Cultivation (% acres)	0	20	0	25	22	12
Hand weeding (% acres)	0	45	0	35	1	37

[a]Resistance was not known to be present in 2004 but was confirmed in each county by 2008.
[b]DNA, dinitroanaline herbicide (e.g., trifluralin and pendimethalin).

herbicide incorporation using tillage by 5%, increased adoption of glufosinate-based programs by 26%, increased cultivation by 20%, and increased hand weeding by 45% in 2008 compared with years prior to having the resistant biotype (2004). Less of an impact on cotton production was noted in areas with moderate or light infestations. In these areas, conservation tillage was not negatively impacted, and growers had not adopted tillage as a method to incorporate residual herbicides. However, cultivation (12–25% of the fields) and hand weeding (35–37% of the fields) were being implemented to remove GR Palmer amaranth plants prior to seed production and pollen dispersal. At least 91% of the fields in the light and moderate infested counties received one residual herbicide at or prior to planting, and 71% of the fields received two or more residual herbicides in an effort to prevent populations from building to severe levels. In-crop glyphosate applications in the severely, moderately, and lightly infested areas were reduced by 0.7, 0.5, and 0.4 applications per season, respectively. Adoption of glufosinate-based management systems in severely infested counties and increased use of MSMA mixtures at layby across all counties account for the reduction in glyphosate applications.

Extension personnel were also asked to provide an average cost of control measures aimed at preventing the development of GR Palmer amaranth versus managing GR Palmer amaranth in GR cotton. Responses were similar across infestation levels for the 10 counties responding to this question. In 2008, preventative programs cost Georgia growers an average of $82 ha^{-1}

compared with an average of $130 ha^{-1} to manage resistance in areas with GR Palmer amaranth (excluding rebate programs). Managing Palmer amaranth was more costly because of greater residual herbicide usage. Extension personnel also noted that the $130 ha^{-1} expended to manage the resistant pest did not account for the costs of hand weeding and/or cultivation, which are often needed as herbicide programs alone may not provide adequate control.

11.6 MANAGEMENT OF GR PALMER AMARANTH

Cotton production is extremely vulnerable to GR Palmer amaranth, and managing GR Palmer amaranth in GR cotton has proven to be extremely difficult. Acceptable control of GR Palmer amaranth in cotton requires effective use of residual herbicides applied throughout the cropping season (Culpepper et al. 2008; Marshall 2009; Whitaker et al. 2007); hence, control is unpredictable in production areas without irrigation to ensure timely herbicide activation (Culpepper et al. 2008, 2009a). If GR Palmer amaranth escapes herbicides applied at or before planting in either GR or nontransgenic cotton, early-season topical options are limited to pyrithiobac (an acetolactate synthase [ALS]-inhibiting herbicide). Pyrithiobac only suppresses GR Palmer amaranth when it is less than 5 cm in height (Culpepper et al. 2008; *Staple LX Product Label* 2009). Moreover, Palmer amaranth resistant to pyrithiobac is very common throughout the Cotton Belt (Heap 2010; Wise et al. 2009), and Palmer amaranth populations with resistance to both glyphosate and pyrithiobac have been confirmed (Heap 2010; Sosnoskie et al. 2009a). Palmer amaranth with resistance to both glyphosate and ALS-inhibiting herbicides cannot be managed or even suppressed with any topical herbicide application in GR or nontransgenic cotton. Fluometuron plus MSMA can be directed to small cotton, but the height differential necessary for directed application is rarely achieved because of the rapid growth rate of Palmer amaranth (Horak and Loughin 2000). Fluometuron and MSMA can be applied overtop of cotton, but these herbicides applied in this manner often adversely affect yield and maturity of cotton (Byrd and York 1987; Guthrie and York 1989) and do not adequately control Palmer amaranth at rates suitable for topical application. Some directed herbicide options, such as diuron plus MSMA, can effectively control emerged Palmer amaranth (Culpepper et al. 2008), but these applications are restricted by cotton size at time of application, often requiring cotton to be at least 30 cm in height (*Direx Product Label* 2009). By the time cotton reaches the height required for layby directed herbicide options, Palmer amaranth is often far too large for effective control.

Soybean and corn producers have also been dependent on GR technology and the use of glyphosate (Powles 2008). In comparison to cotton, both of these crops are often more competitive with Palmer amaranth and have more options of herbicide modes of action available to help control this pest

(Marshall 2009). Additionally, both GR soybean and corn producers have effective alternative topical herbicides that can be applied to control GR Palmer amaranth if it escapes at-plant residual herbicides. In soybean, growers can effectively use herbicides such as acifluorfen, fomesafen, imazethapyr, lactofen, and thifensulfuron (Beyers et al. 2002; Hager et al. 2003; Johnson et al. 1978; Shoup and Al-Khatib 2004; Sweat et al. 1998) to control emerged *Amaranthus* species, as long as applications are made to small plants and the treated populations of *Amaranthus* are not resistant to the respective mode of actions, which is becoming more common (Heap 2010; Patzoldt et al. 2005). In North Carolina, five Palmer amaranth populations have been confirmed to be resistant to both glyphosate and thifensulfuron, an ALS-inhibiting herbicide (Whitaker 2009). In corn, producers have even more effective herbicide options to control emerged *Amaranthus* such as atrazine, dicamba, 2,4-D, mesotrione, and nicosulfuron (Armel et al. 2003; Bijanzadeh and Ghadiri 2006; Krausz and Kapusta 1998). Palmer amaranth with resistance to both the triazines and ALS herbicides has been documented (Heap 2010).

Glufosinate-resistant cotton, corn, and soybean have been commercialized and provide an alternative technology and herbicide mode of action for the control of GR Palmer amaranth. Glufosinate applied very timely in glufos-inate-resistant crops can control *Amaranthus* species (Beyers et al. 2002; Culpepper et al. 2008; Gardner et al. 2006; Marshall 2009). Although glufos-inate is typically less effective than glyphosate on non-GR Palmer amaranth, glufosinate-based systems have been more effective than glyphosate-based systems controlling GR Palmer amaranth (Culpepper et al. 2008; Marshall 2009). This technology offers growers an opportunity, provided that glufos-inate applications are timely, to manage emerged GR and ALS-resistant Palmer amaranth in cotton, corn, and soybean. The challenge is making appli-cations of glufosinate to target plants less than 10 cm in height (Coetzer et al. 2002; *Ignite 280 SL Product Label* 2009), especially considering that Palmer amaranth can grow 2.5–5.0 cm per day under ideal field conditions (A. C. Culpepper, personal observation). Despite these challenges, growers have adopted glufosinate-resistant technology to effectively manage severe infesta-tions of GR Palmer amaranth in cotton (Table 11.1).

The development of new herbicide chemistry is obviously needed but is not likely to occur in the near future. However, crops with resistance to multiple herbicide modes of action are being developed and nearing commercializa-tion. Both soybean and cotton producers may soon have an option of applying dicamba or 2,4-D over their crops to assist in the management of GR Palmer amaranth. It appears likely that these growers may even have the choice of spraying glyphosate, glufosinate, 2,4-D, dicamba, and/or traditional herbicides in their respective crops. This increased flexibility of herbicide selection will aid growers in their ability to manage GR Palmer amaranth as well as increase the diversity of herbicide selection, thereby reducing the development of additional weed resistance.

The impact of deep soil inversion (i.e., moldboard plow), cultivation, or preplant incorporation of effective soil residual herbicides for the control of GR Palmer amaranth has been explored recently. Palmer amaranth plant emergence was reduced 50% to 60% by inverting the soil when residual herbicides were not applied (Culpepper et al. 2009a). When using soil residual herbicides in the absence of an activating rainfall within 5 days of application, soil inversion improved Palmer amaranth control by 15% and cotton yield by 19% compared with the same herbicide program without soil inversion. Soil inversion did not significantly impact control or yield when the herbicide program began with an effective herbicide incorporated into a moist soil at planting. Cultivation can also be an effective tool to remove GR Palmer amaranth from row middles and increase cotton yields. In Georgia, cultivation improved GR Palmer amaranth control by 11% and yield by 10% when residual herbicides applied preemergence at planting were not immediately activated by rainfall.

The most effective and economical programs for managing GR Palmer amaranth are still those that are preventative in nature. These preventative systems often rely on the use of soil residual herbicides with multiple modes of action, including dinitroanilines (e.g., pendimethalin and trifluralin), chloroacetamides (e.g., alachlor, metolachlor, s-metolachlor), protox inhibitors (e.g., fomesafen and flumioxazin), triazines (e.g., atrazine), and substituted ureas (e.g., diuron and fluometuron). Other integrated approaches are commonly used in conjunction with the residual herbicides, including cultivation, herbicide incorporation, soil inversion, alternative herbicide-resistance technology (e.g., glufosinate-resistant crops), and crop rotation that provide greater weed management options and herbicide modes of action. Extension personnel throughout the Southeast and mid-South regions of the United States have conducted thousands of educational programs to increase growers' awareness of the problem and to encourage proactive management programs. Although resistant Palmer amaranth continues to spread at an alarming rate, most growers have become more aggressive in their management programs. It is hoped that these aggressive programs will reduce the rate of spread and begin to reduce the number of resistant seeds currently present in soil seedbanks.

REFERENCES

Ackerman, J. D. 2000. Abiotic pollen and pollination: ecological, functional, and evolutionary perspectives. *Plant Systematics and Evolution* 222:167–185.

Alibert, B., H. Sellier, and A. Souvre. 2005. A combined method to study gene flow from cultivated sugar beet to ruderal beets in the glasshouse and open field. *European Journal of Agronomy* 23:195–208.

Armel, G. R., H. P. Wilson, R. J. Richardson, and T. E. Hines. 2003. Mesotrione combinations in no-till corn (*Zea mays*). *Weed Technology* 17:111–116.

Bensch, C. N., M. J. Horak, and D. Peterson. 2003. Interference of redroot pigweed (*Amaranthus retroflexus*), Palmer amaranth (*A. palmeri*), and common waterhemp (*A. rudis*) in soybean. *Weed Science* 51:37–43.

Beyers, J. T., R. J. Smeda, and W. G. Johnson. 2002. Weed management programs in glufosinate-resistant soybean (*Glycine max*). *Weed Technology* 16:267–273.

Bijanzadeh, E. and H. Ghadiri. 2006. Effect of separate and combined treatments of herbicides on weed control and corn (*Zea mays*) yield. *Weed Technology* 20:640–645.

Borsch, T. 1998. Pollen types in the Amaranthaceae. *Grana* 37:129–142.

Burke, I. C., M. Schroeder, W. E. Thomas, and J. W. Wilcut. 2007. Palmer amaranth interference and seed production in peanut. *Weed Technology* 21:367–371.

Byrd, J. D. Jr. and A. C. York. 1987. Interaction of fluometuron and MSMA with sethoxydim and fluazifop. *Weed Science* 35:270–276.

Cane, J. H., S. L. Buchmann, and W. E. Laberge. 1992. The solitary bee *Melissodes thelypodii thelypodii* Cockerell (Hymenoptera, Anthophoridae) collects pollen from wind-pollinated *Amaranthus palmeri* Watson. *Pan-Pacific Entomologist* 68:97–99.

Coetzer, E., K. Al-Khatib, and D. E. Peterson. 2002. Glufosinate efficacy on *Amaranthus* species in glufosinate-resistant soybean (*Glycine max*). *Weed Technology* 16:326–331.

Costea, M., S. E. Weaver, and F. J. Tardif. 2004. The biology of Canadian weeds. 130. *Amaranthus retroflexus* L., *A. powellii* S. Watson and *A. hybridus* L. *Canadian Journal of Plant Science* 84:631–668.

Costea, M., S. E. Weaver, and F. J. Tardif. 2005. The biology of invasive alien plants in Canada. 3. *Amaranthus tuberculatus* (Moq.) Sauer var. *rudis* (Sauer) Costea & Tardif. *Canadian Journal of Plant Science* 85:507–522.

Culpepper, A. S., T. L. Grey, W. K. Vencill, J. M. Kichler, T. M. Webster, S. M. Brown, A. C. York, J. W. Davis, and W. W. Hanna. 2006. Glyphosate-resistant Palmer amaranth (*Amaranthus palmeri*) confirmed in Georgia. *Weed Science* 54:620–626.

Culpepper, A. S. and A. C. York. 1998. Weed management in glyphosate-tolerant cotton. *Journal of Cotton Science* 2:174–185.

Culpepper, A. S., A. C. York, and J. Kichler. 2009a. Impact of tillage on managing glyphosate-resistant Palmer amaranth in cotton. In *Proceedings of the Beltwide Cotton Conferences*, p. 1343.

Culpepper, A. S., A. C. York, and M. Marshall. 2009b. Glyphosate-resistant Palmer amaranth in the Southeast. *Abstracts of Weed Science Society of America* 364.

Culpepper, A. S., A. C. York, A. W. MacRae, and J. Kichler. 2008. Glyphosate-resistant Palmer amaranth response to weed management programs in Roundup Ready and Liberty Link Cotton. In *Proceedings of the Beltwide Cotton Conferences*, p. 1689.

Devine, M. D., S. O. Duke, and C. Fedtke. 1993. *Physiology of Herbicide Action*. Englewood Cliffs, NJ: Prentice Hall.

Direx Product Label. 2009. Wilmington, DE: Dupont Crop Protection.

Ehleringer, J. 1983. Ecophysiology of *Amaranthus palmeri*, a Sonoran Desert summer annual. *Oecologia* 57:107–112.

Franssen, A. S., D. Z. Skinner, K. Al-Khatib, and M. J. Horak. 2001. Pollen morphological differences in *Amaranthus* species and interspecific hybrids. *Weed Science* 49:732–737.

Gaines, T. A., W. Zhang, D. Wang, B. Bukun, S. T. Chisolm, D. L. Shaner, S. J. Nissen, W. L. Patzoldt, P. J. Tranel, A. S. Culpepper, T. L. Grey, T. M. Webster, W. K. Vencill, R. D. Sammons, J. Jiang, C. Preston, J. E. Leach, and P. Westra. 2010. Gene amplification confers glyphosate resistance in *Amaranthus palmeri*. *Proc. Natl. Acad. Sci. U.S.A.* 107:1029–1034.

Gallagher, R. S. and J. Cardina. 1997. Soil water thresholds for photoinduction of redroot pigweed germination. *Weed Science* 45:414–418.

Gallagher, R. S. and J. Cardina. 1998a. Phytochrome-mediated *Amaranthus* germination I: effect of seed burial and germination temperature. *Weed Science* 46:48–52.

Gallagher, R. S. and J. Cardina. 1998b. Phytochrome-mediated *Amaranthus* germination II: development of very low fluence sensitivity. *Weed Science* 46:53–58.

Gardner, A. P., A. C. York, D. L. Jordan, and D. W. Monks. 2006. Management of annual grasses and *Amaranthus* spp. in glufosinate-resistant cotton. *Journal of Cotton Science* 10:328–338.

Grichar, W. J., B. A. Besler, K. D. Brewer, and B. W. Minton. 2004. Using soil-applied herbicides in combination with glyphosate in a glyphosate-resistant cotton herbicide program. *Crop Protection* 23:1007–1010.

Guo, P. G. and K. Al-Khatib. 2003. Temperature effects on germination and growth of redroot pigweed (*Amaranthus retroflexus*), Palmer amaranth (*A. palmeri*), and common waterhemp (*A. rudis*). *Weed Science* 51:869–875.

Guthrie, D. S. and A. C. York. 1989. Cotton (*Gossypium hirsutum*) development and yield following fluometuron postemergence applied. *Weed Technology* 3:501–504.

Hager, A. G., L. M. Wax, G. A. Bollero, and E. W. Stoller. 2003. Influence of diphenylether herbicide application rate and timing on common waterhemp (*Amaranthus rudis*) control in soybean (*Glycine max*). *Weed Technology* 17:14–20.

Hanson, B. D., C. A. Mallory-Smith, W. J. Price, B. Shafii, D. C. Thill, and R. S. Zemetra. 2005. Interspecific hybridization: potential for movement of herbicide resistance from wheat to jointed goatgrass (*Aegilops cylindrica*). *Weed Technology* 19:674–682.

Heap, I. M. 2010. International Survey of Herbicide Resistant Weeds. http://www. weedscience.org/in.asp (accessed April 8, 2010).

Horak, M. J. and T. M. Loughin. 2000. Growth analysis of four *Amaranthus* species. *Weed Science* 48:347–355.

Ignite 280 SL Product Label. 2009. Research Triangle Park, NC: Bayer CropScience.

Jasieniuk, M., A. L. Brûlé-Babel, and I. N. Morrison. 1996. The evolution and genetics of herbicide resistance in weeds. *Weed Science* 44:176–193.

Jha, P., J. K. Norsworthy, W. Bridges, and M. B. Riley. 2008. Influence of glyphosate timing and row width on Palmer amaranth (*Amaranthus palmeri*) and pusley (*Richardia* spp.) demographics in glyphosate-resistant soybean. *Weed Science* 56:408–415.

Johnson, W. O., G. E. Kollman, C. Swithenbank, and R. Y. Yih. 1978. RH-6201 (Blazer): a new broad spectrum herbicide for postemergence use in soybeans. *Journal of Agricultural and Food Chemistry* 26:285–286.

Keeley, P. E., C. H. Carter, and R. J. Thullen. 1987. Influence of planting date on growth of Palmer amaranth (*Amaranthus palmeri*). *Weed Science* 35:199–204.

Klingaman, T. E. and L. R. Oliver. 1994. Palmer amaranth (*Amaranthus palmeri*) interference in soybeans (*Glycine max*). *Weed Science* 42:523–527.

Koger, C. H. and K. N. Reddy. 2005. Role of absorption and translocation in the mechanism of glyphosate resistance in horseweed (*Conyza conadensis*). *Weed Science* 53:84–89.

Krausz, R. F. and G. Kapusta. 1998. Total postemergence weed control in imidazilone-resistant corn (*Zea mays*). *Weed Technology* 12:151–156.

MacRae, A. W., A. S. Culpepper, T. M. Webster, and J. M. Kichler. 2007. The effect of glyphosate-resistant Palmer amaranth density and time of establishment on yield of cotton. In *Proceedings of Southern Weed Science Society*, p. 228.

MacRae, A. W., A. S. Culpepper, T. M. Webster, L. M. Sosnoskie, and J. M. Kichler. 2008. Glyphosate-resistant Palmer amaranth competition with Roundup Ready cotton. In *Proceedings of the Beltwide Cotton Conferences*, p. 1696.

Marshall, M. W. 2009. Complementary herbicide programs for Palmer amaranth (*Amaranthus palmeri* S. Wats.) control in glyphosate-tolerant cotton and soybeans. *Abstracts of Weed Science Society of America* 49:99.

Massinga, R. A., K. Al-Khatib, P. St. Amand, and J. F. Miller. 2003. Gene flow from imidazolinone-resistant domesticated sunflower to wild relatives. *Weed Science* 51:854–862.

Massinga, R. A., R. S. Currie, M. J. Horak, and J. Boyer. 2001. Interference of Palmer amaranth in corn. *Weed Science* 49:202–208.

Matus-Cadiz, M. A., P. Hucl, M. J. Horak, and L. K. Blomquist. 2004. Gene flow in wheat at the field scale. *Crop Science* 44:718–727.

Menges, R. M. 1987. Weed seed population dynamics during six years of weed management systems in crop rotations on irrigated soil. *Weed Science* 35:328–332.

Moore, J. W., D. S. Murray, and R. B. Westerman. 2004. Palmer amaranth (*Amaranthus palmeri*) effects on the harvest and yield of grain sorghum (*Sorghum bicolor*). *Weed Technology* 18:23–29.

Morgan, G. D., P. A. Baumann, and J. M. Chandler. 2001. Competitive impact of Palmer amaranth (*Amaranthus palmeri*) on cotton (*Gossypium hirsutum*) development and yield. *Weed Technology* 15:408–412.

Mosyakin, S. L. and K. R. Robertson. 2008. 39. Amaranthaceae, 3. *Amaranthus*, 5. *Amaranthus palmeri*. *Flora of North America* 4:405–418. http://www.efloras.org/florataxon.aspx?flora_id=1&taxon_id=101257 (accessed November 2, 2009).

Nandula, V., R. Bond, D. Poston, C. Koger, K. Reddy, and J. Bond. 2009. Glyphosate-resistant Palmer amaranth from Mississippi. *Abstracts of Weed Science Society of America* 49:71.

Nandula, V. K., K. N. Reddy, D. H. Poston, A. M. Rimando, and S. O. Duke. 2008. Glyphosate tolerance mechanism in Italian ryegrass (*Lolium multiflorum*) from Mississippi. *Weed Science* 56:344–349.

Nichols, R. L., J. Bond, A. S. Culpepper, D. Dodds, V. Nandula, C. L. Main, M. W. Marshall, T. C. Mueller, J. K. Norsworthy, A. Price, M. Patterson, R. C. Scott, K. L. Smith, L. E. Steckel, D. Stephenson, D. Wright, and A. C. York. 2009. Glyphosate-resistant Palmer amaranth (*Amarantus palmeri*) spreads in the southern United States. *Resistant Pest Management Newsletter* 18(2):8–10.

Norsworthy, J. K. 2008. Effect of tillage intensity and herbicide programs on changes in weed species density and composition in the southeastern coastal plains of the United States. *Crop Protection* 27:151–160.

Norsworthy, J. K., G. M. Griffith, R. C. Scott, K. L. Smith, and L. R. Oliver. 2008. Confirmation and control of glyphosate-resistant Palmer amaranth (*Amaranthus palmeri*) in Arkansas. *Weed Technology* 22:108–113.

Patzoldt, W. L., P. J. Tranel, and A. G. Hager. 2005. A waterhemp (*Amaranthus tuberculatus*) biotype with multiple resistance across three herbicide sites of action. *Weed Science* 53:30–36.

Place, G., D. Bowman, M. Burton, and T. Rutty. 2008. Root penetration through a high bulk density soil layer: differential response of a crop and weed species. *Plant Soil* 307:179–190.

Powles, S. B. 2008. Evolution in action: glyphosate-resistant weeds threaten world crops. *Outlooks on Pest Management (Glyphosate-Resistant Weeds)* 19:256–259.

Primack, R. B. 1978. Evolutionary aspects of wind pollination in genus *Plantago* (Plantaginaceae). *New Phytologist* 81:449–458.

Rowland, M. W., D. S. Murray, and L. M. Verhalen. 1999. Full-season Palmer amaranth (*Amaranthus palmeri*) interference with cotton (*Gossypium hirsutum*). *Weed Science* 47:305–309.

Saeglitz, C., M. Pohl, and D. Bartsch. 2000. Monitoring gene flow from transgenic sugar beet using cytoplasmic male-sterile bait plants. *Molecular Ecology* 9:2035–2040.

Sauer, J. 1955. Revision of the dioecious amaranths. *Madrono* 13:4–46.

Sellers, B. A., R. J. Smeda, W. G. Johnson, J. A. Kendig, and M. R. Ellersieck. 2003. Comparative growth of six *Amaranthus* species in Missouri. *Weed Science* 51:329–333.

Shoup, D. E. and K. Al-Khatib. 2004. Control of protoporphyrinogen oxidase inhibitor-resistant common waterhemp (*Amaranthus rudis*) in corn and soybean. *Weed Technology* 18:332–340.

Smith, D. T., R. V. Baker, and G. L. Steele. 2000. Palmer amaranth (*Amaranthus palmeri*) impacts on yield, harvesting, and ginning in dryland cotton (*Gossypium hirsutum*). *Weed Technology* 14:122–126.

Sosnoskie, L. M., J. M. Kichler, R. Wallace, and A. S. Culpepper. 2009a. Multiple resistance to glyphosate and ALS-inhibitors in Palmer amaranth in GA. In *Proceedings of Beltwide Cotton Conferences*, pp. 1351–1352.

Sosnoskie, L. M., T. M. Webster, D. Dales, G. Rains, T. L. Grey, and A. S. Culpepper. 2009b. Pollen grain size, density and settling velocity for Palmer amaranth (*Amaranthus palmeri*). *Weed Science* 57:404–409.

Sosnoskie, L. M., T. M. Webster, A. MacRae, T. L. Grey, and A. S. Culpepper. 2009c. Movement of glyphosate-resistant Palmer amaranth pollen in-field. *Abstracts of Weed Science Society of America* 49:63.

Staple LX Product Label. 2009. Wilmington, DE: DuPont Crop Protection.

Steckel, L. E., C. L. Main, A. T. Ellis, and T. C. Mueller. 2008. Palmer amaranth (*Amaranthus palmeri*) in Tennessee has low level glyphosate resistance. *Weed Technology* 22:119–123.

Steckel, L. E., K. Smith, B. Scott, D. Stephenson, T. Koger, J. Bond, D. Miller, S. Steward, and D. Dodd. 2009. Glyphosate resistant Palmer amaranth in the mid-south. *Abstracts of Weed Science Society of America* 49:365.

Steckel, L. E., C. L. Sprague, E. W. Stoller, and L. M. Wax. 2004. Temperature effects on germination of nine *Amaranthus* species. *Weed Science* 52:217–221.

Sweat, J. K., M. J. Horak, D. E. Peterson, R. W. Lloyd, and J. E. Boyer. 1998. Herbicide efficacy on four *Amaranthus* species in soybean (*Glycine max*). *Weed Technology* 12:315–321.

Vaissière, B. E. and S. B. Vinson. 1994. Pollen morphology and its effect on pollen collection by honey bees, *Apis mellifera* L. (Hymenoptera: Apidae), with special reference to upland cotton, *Gossypium hirsutum* L. (Malvaceae). *Grana* 33:128–138.

Watson, S. 1877. Descriptions of new species of plants, with revisions of certain genera. *Proceedings of the American Academy of Arts and Sciences* 12:246.

Webster, T. M. 2005. Weed survey—southern states: broadleaf crops subsection. In *Proceedings of Southern Weed Science Society*, pp. 291–306.

Wise, A. M., T. L. Grey, E. P. Prostko, W. K. Vencill, and T. M. Webster. 2009. Establishing the geographic distribution level of acetolactate synthase resistance of Palmer amaranth (*Amaranthus palmeri*) accessions in Georgia. *Weed Technology* 23:214–220.

Whitaker, J. R. 2009. Distribution, biology and management of glyphosate-resistant Palmer amaranth in North Carolina. PhD thesis. http://www.lib.ncsu.edu/theses/available/etd-03272009-143230/unrestricted/etd.pdf (accessed September 28, 2009).

Whitaker, J., A. C. York, and A. S. Culpepper. 2007. Glyphosate-resistant Palmer amaranth distribution and control in North Carolina and Georgia. In *Proceedings of the Beltwide Cotton Conferences*, pp. 1226–1227.

Wright, S. R., M. W. Jennette, H. D. Coble, and T. W. Rufty. 1999. Root morphology of young *Glycine max*, *Senna obtusifolia*, and *Amaranthus palmeri*. *Weed Science* 47:706–711.

York, A. C. 2007. Updates from states (situation, distribution, impacts, research efforts): North Carolina. In M. McClelland, ed. *2007 Managing Glyphosate-Resistant Palmer Amaranth Roundtable*. Little Rock, AR: Cotton Incorporated, p. 4. http://www.cottoninc.com/2007%2DGlyphosate%2DResistant%2DPalmer%2DAmaranth/?S=AgriculturalResearch (accessed November 2, 2009).

12

MANAGING GLYPHOSATE-RESISTANT WEEDS AND POPULATION SHIFTS IN MIDWESTERN U.S. CROPPING SYSTEMS

STEPHEN C. WELLER, MICHEAL D. K. OWEN, AND WILLIAM G. JOHNSON

12.1 INTRODUCTION

Weeds are a major limiting factor to crop production and yield maximization. Various methods to manage weeds have been developed throughout history, but the weeds have always adapted or evolved mechanisms to avoid their elimination from agriculture. Johnson et al. (2009a) described the history of weed management from the period before herbicides when management was based on manual removal and cultural practices, through the period when cultivation with tractors became common, until after World War II when herbicides became the major tool. After World War II, the introduction of the phenoxy herbicides provided the first selective herbicides for the management of troublesome dicot weeds in corn, wheat, and other Gramineae crops. Later, other herbicides were introduced that broadened weed control in both monocot and dicot crops. The use of a variety of herbicides with differing mechanisms of action became common practice and provided excellent weed control. There are now more than 315 common names of herbicides (WSSA

Glyphosate Resistance in Crops and Weeds: History, Development, and Management
Edited by Vijay K. Nandula
Copyright © 2010 John Wiley & Sons, Inc.

2007). Although, weed species shifts have occurred along with evolved herbicide resistance to many different herbicides, the techniques of crop rotation,
cultivation, and mixtures of herbicides with differing mechanisms of action can
reduce the onset of high levels of herbicide resistance and crop yield loss due
to weeds. However, growers often focus on single herbicide tactics to control
weeds, which has predictably resulted in weed shifts and evolved resistance to
these herbicides.

Historically, safe herbicide use in crops is based on metabolic selectivity
within a crop or by placement so the crop does not come into contact with a
normally phytotoxic herbicide. Selectivity meant either the crop had a physiological mechanism to metabolize the herbicide or the herbicide was physically
placed in order to avoid contact with the crop. Natural tolerance to herbicides
is common to a wide variety of herbicides. The earliest examples were demonstrated by phenoxy herbicides (e.g., 2,4-dichlorophenoxyacetic acid (2,4-D))
that are most active on dicots, so monocot crops such as wheat and corn are
safe from injury caused by the herbicide. Later, many crops were tolerant to
the triazines because of their ability to metabolize the herbicide. This has been
the mechanism for tolerance in crops to many of these later introduced herbicide groups such as grass-specific herbicides that inhibit acetyl coenzyme A
carboxylase (ACCase) and inhibitors of acetolactate synthase (ALS). Gressel
(2009) described how various weeds have evolved resistance to herbicides
based on target-site resistances and through metabolic processes both similar
to and differing from the mechanisms found in crops. He provided an excep-
tional summary of herbicide selectivity as a predictor of resistance mechanisms
including excellent examples of the evolution (defined by this chapter's authors
as changes in genotype frequencies that result from selection pressure on
genetic variation within a population of a weed species) and management of
both single-mutation target-site resistances and metabolic resistances in agronomic crop weeds. For example, in the U.S. corn belt, weeds resistant to triazine
herbicides have never become a major problem. This is the result of farmers
mixing chloracetamide herbicides with atrazine, which mitigates the rapid
evolution of triazine resistance. In most instances, atrazine resistance is most
likely to occur only where it was used alone repeatedly and at higher dosages.

There is weed resistance to 19 different herbicide modes of action or, in
other words, most of the herbicide families now used in agriculture as documented by Heap (2010). Although herbicide resistance is common, the presence of these resistant weeds has typically not resulted in loss of weed control
in most situations since farmers use mixtures of herbicides to control weeds
(Sammons et al. 2007). With the introduction of glyphosate-resistant (GR)
crops in 1996, there have been major changes in crop production and weed
management practices (Carpenter and Gianessi 1999). These crop production
changes present new challenges for weed management such as the probability
of weeds evolving resistance to glyphosate and, in some cases, resistance to
multiple herbicides and the inevitable shift in weed species present in crop
fields.

This chapter will discuss the current situation regarding GR cropping systems in the midwestern United States. Emphasis is placed on how implementation of these GR systems has affected patterns of herbicide use, the evolution of herbicide-resistant weeds, grower opinions in regard to herbicide resistance management strategies, and perspectives on weed management systems that must be developed to maintain long-term utility of GR cropping systems.

12.2   BACKGROUND ON GR CROPPING SYSTEMS

Genetically engineered (GE) GR crops were first commercially introduced in 1996 and have been the most rapidly and globally accepted new agronomic crop trait in the history of agriculture (Cerdeira and Duke 2006). Although data vary on the area of GE crops, there were a reported 124 million ha of GE crops grown in 25 countries by more than 12 million farmers in 2008 (James 2008). Since 1996, more than a billion cumulative acres of GE crops have been planted in the United States (Marvier et al. 2008; Sankula 2006). For 2007, the National Agriculture Statistics Service (NASS) of the United States Department of Agriculture (USDA) reported that 70% of upland cotton, 52% of corn, and 91% of soybean ha were planted with GE GR cultivars (USDA-NASS 2007), whereas the Biotechnology Industry Organization (BIO) reported slightly higher percentages for corn (71%) and cotton (87%) (BIO 2008). For example, in Indiana in 2008, 95% of soybeans and 70% of corn grown was GR. This is typical of most of the midwestern corn belt states.

U.S. farmers account for approximately 50% of the worldwide hectares of GE GR crops grown (James 2008). Glyphosate use for weed control increased eightfold between 1995 and 2005 (USDA-NASS 2008). The rapid adoption of GE GR crops occurred because glyphosate is highly effective against almost all economically important weeds; weed management programs were greatly simplified; crop tolerance was exceptionally good; and its use facilitated widespread adoption of reduced- and no-tillage systems that conserve both soil and energy resources. No-till systems in the United States increased from 15 million ha to over 25 million ha from 1994 to 2004 (NCRMS 2004), largely due to GR crops (Young 2006). This has resulted in a decrease of multiple herbicide mixture applications to cropland, an increase in glyphosate use, and the practice of multiple applications of glyphosate (USDA-NAAS 2008; Young 2006).

The relative benefits and risks of the widespread adoption of GE GR crops on the agroecosystem and for society has been a contentious topic of debate (Ermakova 2007; John 2007; Liefert 2007; Marshall 2007). Benefits to farmers include highly reduced effort and time needed to implement a weed management system that significantly increases potential crop production. Other glyphosate use benefits include its low mammalian and environmental toxicity (WSSA 2007), low cost compared with other herbicides (Gianessi et al. 2002),

and the broad-spectrum control of many previously difficult to control perennial and annual weed species (Johnson et al. 2000). Risks resulting from the
use of glyphosate include effects on ecosystems such as possible risks of
decreased species biodiversity, weed spectrum shifts, the tendency of some
growers to rely solely on glyphosate for weed management, which reduces
benefits from the time-proven practices of integrated weed management
(IWM), and the likelihood that weeds will evolve resistance to glyphosate
(Ammann 2005; Heard et al. 2003; Watkinson et al. 2000). The evolved weed
resistance issue and societal concerns associated with GE crops, including food
safety, GE pollen movement to wild species, volunteer GE GR crops, and
other issues, have been the focus of many scientific and press publications
(Arntzen et al. 2003; Freckleton et al. 2003; Gura 2001; Hails and Kinderlerer
2003; Ledford 2007; Madsen and Sandoe 2005; Sandermann 2006). Concerns
over the benefits and risks of GE crops are valid and deserve further discussion. GE GR technology has been adopted by farmers with a high level of
satisfaction in most cases, implying great benefit (Johnson et al. 2009b).
However, the concerns of the advocate groups and lay public about these
technologies are nevertheless important and must be seriously addressed with
due scientific consideration.

As will be discussed further, there has been an increase in weeds difficult
to control with glyphosate and the evolution of weeds resistant to glyphosate.
In the United States, there are currently 10 GR species, and worldwide, there
are 18 species (see Chapter 2) (Heap 2009). Seven of the 10 GR weed biotypes identified in the United States have all evolved in conjunction with
either GE GR soybean or cotton (the others evolved in orchards or noncrop
areas), again suggesting that there may be undefined and nonuniform ecological risks associated with the widespread adoption of GE crops.

Another issue with the GR technology is that as overall crop production
prices decreased (Martinez-Ghersa et al. 2003), major chemical companies
have lowered the priority on herbicide discovery programs (Johnson et al.
2009a). Nondevelopment of new herbicides could be detrimental in future
management of herbicide-resistant weeds that evolve in GR cropping systems
(Rüegg et al. 2007). A primary concern for the viability of the GE GR cropping system is whether weed shifts, the evolution of GR weeds, or GE GR
volunteer crops will become a pervasive problem. These are issues that university researchers, government agency officials, and private sector life sciences companies have and will continue to consider. The widespread adoption
of GE GR crops has, to a degree, already changed the abundance and types
of weed species found in agronomic fields, and this impact will certainly
increase in the future (Johnson et al. 2009a; Owen 2008).

The full implications of these inevitable changes are currently unclear.
There is now an agreement that the evolution of GR weed biotypes was inevitable, although, there is some disagreement as to the ultimate degree and
nature of the impact of glyphosate resistance on agricultural practices
(Sandermann 2006; Shaner 2000). A further concern, as management practices
are developed in GR crops, is whether there will be increased potential for

weeds to evolve multiple herbicide resistance. It is important to recognize that the impact of GE GR technology on weed communities is not directly attributable to the use of a GE GR crop, but rather an indirect effect of the management of the GR crop (Cattaneo et al. 2006; Owen 2008) (e.g., how and which herbicide is applied), which is different from other GE crops (i.e., cultivars that include GE *Bt*). In order to address all these possible outcomes related to weed management, we need to briefly review the history of weed responses to various crop production practices. This knowledge and experience is important in the ultimate design of effective weed management strategies used to maintain acceptable weed control in GR crops.

12.3 AGRONOMIC PRACTICE INFLUENCES ON WEED POPULATIONS/SHIFTS

All agriculture production practices impart selection pressure on weed populations that can result in shifts of species composition within a field (Owen 2008). The weeds present can be more difficult to manage, or more easily controlled, depending on the practices used. The three practices most common and selective prior to GR crop introduction were tillage, herbicide(s) used, and crop rotation. Within the GR system, crop type/rotation plays less of a role since glyphosate is the primary herbicide applied in all systems.

12.3.1 Tillage and Crop Rotation

In agriculture prior to GR crops, herbicide use and tillage systems had significant impacts on weed population dynamics, but as Owen pointed out, "they cannot easily be separated from the specific effect of the crop rotation on the weed community" (Owen 2008). The effectiveness of these practices was dependent on the integration of tactics. Relying on one tactic over others results in weeds adapting to the practice and thus increasing problems for weed management. Tillage is widely used for preparing fields prior to planting, eliminating existing weeds, and in-season weed control. Tillage has an effect on weed emergence, weed control, weed seed production, and distribution of weed in the soil profile (Buhler 1995; Grundy et al. 2003). Many studies have shown the effect of tillage and crop rotation on weed communities and their control (Grundy et al. 2003; Mulugeta et al. 2001; Swanton et al. 2006). Comparisons of tillage type (e.g., moldboard plow, chisel plow, and no-till) and crop rotation (e.g., continuous corn, corn–soybean, corn–soybean–winter wheat) on weed species diversity and population density and resultant seedbanks showed that tillage had a greater effect than crop rotation (Swanton et al. 2006). As tillage intensity decreased from moldboard to no-till, weed population density declined but weed species diversity increased with no effect on crop yield between tillage systems, presuming equal management levels. The increased weed species diversity in no-till was due to species that were not summer annuals and therefore were less competitive in the crop (Swanton et al. 2006). No-tillage systems do favor species able to germinate at or near the soil surface including shallow-germinating summer annuals such as common lambsquarters (*Chenopodium album* L.), common waterhemp (*Amaranthus rudis* L.), and wind-dispersed species such as horseweed (*Conyza canadensis* L. Cronq.) (Felix and Owen 1999; Owen 2008). As Johnson et al. (2009a) pointed out, in a no-till GR system, increased weed species diversity increases the possibility of selecting species that have greater tolerance to the herbicides used.

Other studies have shown how cropping systems play a role in weed community composition. Liebman and Dyke (1993) suggested that differing crop rotations create an inhospitable and unstable ecosystem that would minimize drastic weed shifts. Cardina et al. (2002) showed that a no-till corn–corn–corn rotation had 45% higher soil weed seedbank population density of yellow wood sorrel (*Oxalis stricta* L.) and common chickweed (*Stellaria media* L.) than a corn–oats (*Avena sativa* L.)–alfalfa (*Medicago sativa* L.) rotation, and this was attributed to the lowered competitive effect of the corn rotation on these species. Other researchers have shown the general effect that the more diverse the rotation, the more diverse the community of weed species (Anderson and Beck 2007; Anderson et al. 2007; Cardina et al. 2002; Heggenstaller and Liebman 2005; Murphy et al. 2006; Samarajeewa et al. 2005; Teasdale et al. 2005). Research with cover crops has shown similar changes in community diversity with some weed species increasing and other decreasing (Chikoye et al. 2005; Ngouajio and Mennan 2005; Samarajeewa et al. 2005; Teasdale et al. 2005).

This weed species diversity facilitates the possibility of weed shifts based on practices used and is especially relevant in GR crops. A review by Owen (2008) of weed species shifts in GR crops should be referred to for more detail on specific factors affecting such shifts and factors that must be considered when managing weeds in GR systems. Owen (2008) suggested that the selection pressure by glyphosate will result in weed species shifts attributable to natural tolerance within species or the evolution of resistance within a specific weed population.

In terms of tillage, growers recognized that GR crops provided an effective and consistent method of controlling weeds while requiring less tillage. This resulted in an increase in no-tillage and reduced tillage in these systems, a reduction in soil losses, and increased savings to the grower. However, many of the effects of tillage, herbicide use patterns, and/or crop rotation described above may not hold in GR crops as changes in tillage and herbicide use patterns have resulted in changes in weed spectrums in crop fields. Specifically, some small-seeded annual broad-leaved weeds and others have become more common (Owen 1997) and as described below, weeds such as horseweed, Palmer amaranth (*Amaranthus palmeri* S. Wats.), common waterhemp (*Amaranthus tuberculatus* (Moq.) J.D. Sauer), and giant ragweed (*Ambrosia trifida* L.) have become more prevalent (Owen 2008) and, in some cases, are now resistant to glyphosate.

12.3.2 Herbicide Effects

Herbicide use patterns play a key role in weed shifts and the rate at which they occur. This has been shown quite dramatically by some of the earliest herbicides introduced into agriculture, the phenoxies and the triazines. Shortly after 2,4-D was introduced in the 1940s, Harper (1956) mentioned the possibility of it causing shifts in weed species present in agronomic fields where it was used. Lee (1948) observed that 2,4-D was not the answer to total weed control in cornfields as cultivation was needed for managing the grassy weeds that 2,4-D did not control. In the late 1940s and into the 1950s, weed communities where 2,4-D was routinely used were dominated by grasses, proving 2,4-D's influence on weed species shifts. In terms of resistance, even though 2,4-D has been widely used in cropping systems since the 1940s, resistance has been slow to develop. There have been many reports of variable response of certain species to 2,4-D, but prior to 1981, there were only four species in the world listed as resistant. Since 1981, additional 12 species are now listed as resistant (Heap 2010).

Triazines, first introduced in the 1950s, provided a wider use pattern in crops such as corn and sorghum. Selectivity was generally due to the crop metabolizing the herbicide. These herbicides provided control of both dicots and monocots and became the most widely used herbicides in Poaceae crops, woody crops, and noncropland until the 1990s. The triazines are still widely used today. However, this widespread use has led to the first documented resistance in 1970 of common groundsel (*Senecio vulgaris* L.) in an orchard where simazine had been repeatedly used for many years (Ryan 1970). To date, there are over 68 species of weeds resistant to triazines in the world (Heap 2010). Atrazine, the most widely used triazine (and like many other triazines), has both soil and foliar activity against weeds, and a long soil residual life that is important in selecting for resistant weeds. In fact, resistant weeds evolved due to an altered site of action, and in two cases, velvetleaf (*Abutilon theophrasti* Medik.) (Anderson and Gronwald 1991; Patzoldt et al. 2003) and fall panicum (Thompson et al. 1971), the ability to metabolize atrazine.

The shift in weed species composition and evolution of resistance has continued to be commonly observed with most herbicides introduced since the phenoxies and triazines. The ALS-inhibiting herbicide group, first introduced in 1982, is an excellent example of a group whose widespread and repeated use resulted in weed shifts. These herbicides were so effective that they were quickly and widely adopted by farmers. Although still commonly used in a wide variety of crops and noncrop situations, their use not only has caused weed shifts due to their selective nature but has resulted, again due to their longevity when soil applied, to high selection pressure resulting in evolution of resistance. Resistance at the site of action due to selection for ALS *R* alleles (Tranel and Wright 2002) has occurred within 5 years of the initial use of chlorsulfuron in prickly lettuce (*Lactuca serriola* L.) (Mallory-Smith et al. 1990). There are now 108 ALS-resistant weed species worldwide (Heap 2010).

Glyphosate use patterns can impact weed species diversity and influence species that are most common. Owen (2008) pointed out that "when weed population density and diversity is low, the effect of a single weed control tactic on those species will be greater." The number of glyphosate applications within a cropping season also has an effect on weed diversity (Scursoni et al. 2006). A single glyphosate application resulted in greater diversity than any treatment including an untreated control, while diversity decreased with two applications (Scursoni et al. 2006), and decreased further if multiple glyphosate applications were combined with interrow cultivation when compared with soil-applied residual herbicides (Wicks et al. 2001). The authors suggested that the potential of and level of weed shifts were no greater in GR crops than those associated with other herbicides and conventional crops production systems (Wicks et al. 2001).

The tremendous increase in the use of glyphosate in GR cropping systems and the ability of the weed populations to adapt to the tillage system has and will continue to have a great impact on the ecosystem and the weed species therein. Weed population shifts should be described in this context to include both populations that evolve resistance to glyphosate or have natural tolerance that developed as a result of the selection pressure imposed by the cropping system, including tillage and herbicide type(s) used, and patterns of their use (Owen 2008). An important point to remember is that regardless of the system used (GR, conventional, etc.), any type of high selection pressure will typically result in weed populations within the ecosystem with adaptive traits that overcome the management tactic used with an increase in their population density (Owen 2008)

12.4   PRESENCE OF RESISTANT WEEDS

The first GR weed in row crops identified in the United States was horseweed reported in Delaware in 2000 (VanGessel 2001). Horseweed has always been a troublesome weed in no-till production; however, the appearance of GR biotypes occurred quickly once GE GR soybeans were widely cultivated. This resistance rapidly spread across the eastern and midwestern United States. The rapid spread was related to high seed production, rapid seed dispersal, adaptation to conservation tillage, and the fact that glyphosate resistance was due to a single semidominant gene (Zelaya et al. 2004, 2007). Recently, other GR weed populations have been reported including common ragweed (*Ambrosia artemisiifolia* L.) (Heap 2010) in Missouri; Palmer amaranth in Georgia, Arkansas, Tennessee, Mississippi, and North Carolina (Heap 2009); tall waterhemp (*A. tuberculatus* (Moq.) J.D. Sauer) in Minnesota, Missouri, Iowa, Illinois, and Kansas (Heap 2010); and giant ragweed in Minnesota, Indiana, Iowa, and Ohio, and kochia (*Kochia scoparia* L.) in Kansas (Heap 2010). Frequent glyphosate failures in common lambsquarters control (Gibson et al. 2005; Johnson et al. 2004) have also been reported.

All the GR weeds are major economic problems in agronomic crops in the corn-, soybean-, and cotton-growing regions of the United States. GR horseweed has widespread distribution throughout the United States. It is usually accepted that the recurrent use of glyphosate will increase selection pressure for the evolution of additional GR weed biotypes. In terms of weeds with multiple resistance to glyphosate and other herbicides, the list is relatively short for the United States. According to Heap (2010), there are only three weed biotypes in the midwestern United States that exhibit multiple resistance to glyphosate and other herbicides. These include common waterhemp in Missouri that has resistance to ALS inhibitors, protoporphyrinogen oxidase (PPO) inhibitors, and glyphosate, and common waterhemp in Illinois that has resistance to ALS inhibitors and glyphosate. In both cases, these biotypes have only been found at one site in the state. In Ohio, there is a horseweed biotype that has resistance to ALS inhibitors and glyphosate, and it has been found at five sites in the state. These low numbers of multiple resistance occurrences within the Midwest U.S. region suggest that, as of now, the distribution of weeds with multiple resistances to glyphosate and other herbicides is limited, although all of the common waterhemp in Iowa is ALS resistant, suggesting that any biotypes resistant to any other herbicide will have multiple resistances. Since multiple resistance has been identified in several locations and in several important weed species, weed managers must keep this in mind when developing weed management approaches in GR crops to avoid further episodes of multiple resistance.

In addition to resistance, several species seem to be naturally adapted to glyphosate that include common lambsquarters, velvetleaf, Asiatic dayflower (*Commelina communis* L.), tropical spiderwort (*Commelina benghalensis* L.), *Dicliptera chinensis* (Jussieu), evening primrose (*Oenothera biennis* L.), wild parsnip (*Pastinaca sativa* L.), pokeweed (*Phytolacca americana* L.), and field horsetail (*Equisetum arvense* L.) (Owen 2008). Thus, glyphosate effectiveness in GE GR crop systems is at serious risk unless programs are developed to effectively educate farmers to choose weed management tactics that prevent or delay glyphosate resistance evolution.

12.5   GROWER OPINIONS

A survey of farmers' weed management practices and views on GR weeds and tactics used to prevent or manage GR weed populations in GE GR crops was conducted in the fall of 2005. The survey was conducted with over 1200 farmers in Indiana, Illinois, Iowa, Nebraska, North Carolina, and Mississippi. It was designed to assess the level of concern among farmers about GR weeds and their perceptions of tactics that they believe would help to manage or delay the selection and spread of GR weeds. The full summary of this survey was reported in *Weed Technology* (Givens et al. 2009; Johnson et al. 2009b; Kruger et al. 2009; Shaw et al. 2009).

Briefly, the survey showed that only 30% of farmers thought GR weeds were a serious issue. Few farmers thought field tillage and/or using a non-GR crop in rotation with GR crops would be an effective strategy. Most farmers did not recognize the role that recurrent herbicide use plays in the evolution of resistance. A substantial number of farmers underestimated the potential for GR weed populations to evolve in an agroecosystem dominated by glyphosate as the weed control tactic and were largely unaware of the potential risks to the sustainability of the GR cropping systems in regard to weed resistance.

Because farmers are the ultimate decision makers for the use and management of GE GR crops, it is important to understand their attitudes and perceptions about the likelihood of selecting for weed resistance to glyphosate. Once farmer attitudes are understood, they need to be coupled with science-based knowledge that guides the development of farmer educational programs. These programs must increase awareness and knowledge of GR weeds, how to minimize their appearance, and how to manage glyphosate resistance when it appears in weed populations. The programs must provide knowledge that allows farmers to clearly consider other concomitant risks associated with GE GR crops. These include maintaining the long-term sustainability of this technology and other effects on the agroecosystem that will be impacted by their management decisions. A greater educational emphasis on appropriate IWM in GE GR crops will prevent farmers from choosing a weed management approach that may lead to a catastrophic loss of presently available chemical weed control tools.

12.6   MANAGEMENT APPROACHES

Johnson et al. (2009a) described three herbicide use patterns that were largely abandoned due in part to the adoption of GR cropping systems. The three included the use of tank mixtures of herbicides, the use of alternative herbicides in rotation with glyphosate, and the use of residual herbicides before or at planting. The abandonment of these important weed management approaches in favor of applying only glyphosate was due, in great part, to the effectiveness in most situations of the sole use of glyphosate for weed control in GR systems, the initial marketing of the technology, and a general attitude that resistance to glyphosate was not a major concern.

In the grower survey mentioned above (Givens et al. 2009; Johnson et al. 2009b; Kruger et al. 2009; Shaw et al. 2009), a majority of farmers thought that following the glyphosate label rate recommendation was the most effective strategy for reducing or preventing GR weeds, whereas very few thought that tillage and not using a GE GR crop would be effective strategies. Apparently, most respondents do not understand the role that repeated use of glyphosate alone can play in the evolution of glyphosate resistance (Gressel 1995a, 1995b). To date, there are only two other peer-reviewed reports of farmer perceptions about the impact of GR weeds (Johnson and Gibson 2006; Llewellyn et al. 2002). Llewellyn et al. (2002) surveyed farmers in Western Australia regarding herbicide resistance in *Lolium rigidum* (Gaudin). This survey concluded that farmers believed new herbicides would be ultimately introduced in time to control herbicide-resistant weeds, and farmers generally chose not to adopt alternative weed management tactics in systems that were effectively managing weeds (Llewellyn et al. 2002). The attitudes among U.S. farmers appear to be unfortunately similar (Givens et al. 2009; Johnson and Gibson 2006; Johnson et al. 2009b; Kruger et al. 2009; Shaw et al. 2009).

The problem with this attitude is that no new herbicides with novel mechanisms of action are in the latter stages of development, and no herbicides with new mechanisms of action have been released since 1990, although, as discussed below, companies are in the process of developing crop plants with resistance to several widely used agronomically important herbicides. Since development time of a new pesticide is at least 11 years and the current cost estimate is greater than $190 million, it is unlikely that herbicides with new modes of action will become available to farmers in the next 5 years or longer (Fernandez-Cornejo et al. 1998). In addition, USDA Ag Census data (USDA-NASS 2003) indicated that over 80% of the total number of farms has 199 ha or less, suggesting that there is a large number of part-time farmers. These surveyed farmers were less aware that glyphosate resistance in weeds has happened than were the larger, full-time farmers. Thus, a substantial percentage of farmers will continue to underestimate the potential for GR weed populations to evolve in a landscape dominated by frequent and repeated use of glyphosate. The first step in a proactive program is to add information to herbicide labels that educate farmers on techniques of herbicide use and stewardship practices that will avoid or reduce the incidence of evolved GR weeds.

The second step is related to the fact that printed farm press publications appear to be the most important source of information to farmers concerning herbicide resistance. Survey data showed that farmers' primary sources of information on glyphosate resistance were farm press publications, followed by agriculture chemical dealers, universities/cooperative extension services; the Internet was mentioned by less than 1% of the respondents. Information in farm press and retailer publications originates largely from land-grant university-based research results and extension information. This information is often supplemented with results from life sciences companies. However, the presentation of this information is not consistent, which leads to some confusion on the farmers' part as to what exactly is the most appropriate IWM approach(es) to use in GE GR crops. This situation suggests that all organizations (farmer groups, universities, life sciences companies, and government agencies) must work together closely to properly provide a consistent message describing the best objective and scientifically based recommendations on herbicide resistance management and other priorities that minimize the impact of GE crops on the agroecosystem. Farmers will then have the educational base and confidence to choose appropriate approaches to IWM in GE GR crops.

Sammons et al. (2007) suggested that the best method of herbicide resistance management is to have weed-free fields. This is true from a theoretical resistance management perspective but is not environmentally or economically practical; thus, other management tools must be used. Most current GR weeds have evolved a relatively low level of resistance to glyphosate (Sammons et al. 2007). This is sometimes referred to as creeping resistance (Gressel 2009; Grignac 1978; Heap 1988). Creeping resistance can be due to many different genes conferring a small modicum of resistance, and as Gressel (2009) describes, there may be a gradual but rapid creep or shift in the mean response of the weed population to an applied herbicide. A population of weeds with this creeping resistance could be eliminated by using higher rates of the herbicide, but if they survive and sexually recombine, the increment of resistance can increase and the entire population would shift to a higher level of resistance. Many genes may be involved in creeping resistance and would involve polygenic inheritance. With multiple genes involved, they may interact in the resistance mechanism such that one gene allows creep by partially suppressing the rate of herbicide uptake, faster herbicide catabolism, herbicide translocation, sequestration, target-site modification, increases in protein level (as shown in 5-enolpyruvylshikimate-3-phosphate synthase [EPSPS] mRNA in *Lolium* from Australia with low-level glyphosate resistance [Feng et al. 1999]), or combinations. There is evidence that two such creeping genes are responsible for resistance in two *Conyza* species (Dinelli et al. 2006, 2008). The reader is referred to Gressel (2009) for a more complete and thorough description of creeping resistance and thoughts on understanding the evolution of herbicide resistance.

It has been argued that low-level glyphosate resistance can be overcome by adjusting the rate of glyphosate applied. This approach would require farmers to adjust the glyphosate rate to target the most difficult-to-control weeds in their field in hopes of delaying or preventing the evolution of GR weeds (Sammons et al. 2007). Growers can address this type of resistance by using higher rates, and if not higher rates, by alternating between lower and higher rates (Gardner et al. 1998). Alternating rates would kill the creeping resistance members with the high rates and then the lower rates would allow susceptible members of the population to survive, and they could then dilute the resistant populations through sexual crosses (Gressel 2009). There is some research to support this alteration of rate approach (Gardner et al. 1998) but no scientific consensus that using only high rate approach is valid; in fact, there is documentation that increasing the rate of glyphosate may expedite the evolution of GR weeds where the resistance is controlled by a single partially dominant nuclear gene (Zelaya et al. 2004).

Even though a herbicide rate adjustment approach is easiest and may work to lessen the probabilities of herbicide resistance evolution in some weeds, the most sustainable and effective approach to GR weed management should include several tactics such as applying tank mixtures of herbicides with different mechanisms of action, tillage, crop rotation, and other IWM approaches (Sammons et al. 2007). As Sammons et al. (2007) pointed out, herbicide resistance in a few weed species to various herbicide types has not made herbicide use impractical or uneconomical (Sammons et al. 2007). Whereas this may be true, the evolution of herbicide resistance, particularly glyphosate resistance, could deplete management options for many problematic weeds and force growers to use more herbicides within a given crop to control a variety of weeds resistant to more than one herbicide. The tank-mix approach using residual herbicides appears to be favored by many farmers. Care must be used in choosing the specific tank-mix herbicides to avoid selecting for resistance of weeds to other herbicides and causing antagonistic interactions between herbicides that result in reduced weed control.

Another popular commercial approach that is now being considered to address weed resistance to herbicides is to switch to crops engineered with resistance to another herbicide or stacked resistance to more than one herbicide. Considerable research to discover genes responsible for conferring resistance to an array of herbicides and then include these genes in crop cultivars by genetic engineering is ongoing (Behrens et al. 2007; Castle et al. 2004). GE crops with resistance to dicamba (Behrens et al. 2007), glyphosate (Castle et al. 2004), glufosinate (Service 2007), 2,4-D (Wright et al. 2005), and ALS inhibitors are either commercially available or under development. The concept is that the use of GE crops resistant to multiple herbicides may delay, prevent, or allow better control of the evolution of herbicide resistance in weeds. However, this approach must also be carefully implemented and managed to avoid weed species evolving with multiple and/or cross-resistance to herbicides that are widely used in the United States (Heap 2010). How resistance to multiple herbicides occurs and even the specific mechanism(s) of cross-resistance remain unknown. Furthermore, there has been no assessment of the actual risk of multiple herbicide-resistant GE crops to agroecosystems.

12.7 ACTION PLANS FOR WEED MANAGEMENT IN GR CROPS IN THE U.S. MIDWEST CORN BELT

Weed scientists in the midwestern United States and throughout the country are concerned about the possibility of future problems of weed resistant to glyphosate causing management difficulties in GR cropping systems. An example of this is from the Iowa State University weed scientists who offer the following advice for effective weed management and mitigation of herbicide resistance in weeds:

- Know which weeds you have in your fields and which ones are the most problematic to manage. Understand weed-emergence patterns, and use this information to plan weed management tactics.

- Early-season weed management is critical and should be the first and foremost recommendation for any grower in any crop; begin with the application of a residual herbicide applied before or immediately after planting.
- Timeliness of postemergence applications is critical to achieving effective weed management and protecting crop yield potential.
- Use an early preplant or preemergence herbicide followed by a postemergence herbicide in both Roundup Ready® Corn 2 and Roundup Ready® Soybeans to more effectively, consistently, and profitably manage weeds.
- Use full, labeled rates of glyphosate. Using lower rates can result in variable control.

In addition to the information shown from Owen, and in fact, provided by all university-based weed scientists within the U.S. Midwest, a national glyphosate stewardship forum (NGSF) has been organized. The two goals of the NGSF are to inform key representatives from commodities, industry, and government agencies on the current status of GR weeds and to facilitate a planning process to improve glyphosate stewardship (Boerboom and Owen 2009). The NGSF has had two meetings, and the group has suggested the following processes to ensure that glyphosate use stewardship succeeds:

- Provide uniform labeling statements on glyphosate products based on five core practices, which will be provided by the Herbicide Resistance Action Committee (HRAC). Include the Weed Science Society of America (WSSA) group number on glyphosate labels. The labeling process should be facilitated by the Environmental Protection Agency (EPA).
- Seek to have seed dealers deliver a uniform message on the risks of GR weeds and core management practices by working with the American Seed Trade Association (ASTA).
- Seek increased education of growers through state pesticide safety education programs (i.e., pesticide applicator training programs) by contacting the American Association of Pesticide Safety Educators (AAPSE).
- Request the USDA Economic Research Service to analyze the cost of glyphosate resistance to production agriculture.
- Education on glyphosate stewardship should continue in multiple venues including extension, certified crop advisors (within the Certified Crop Advisors (CCA) exams), communications from the Natural Resources Conservation Service (NRCS), popular press articles, and newsletters.

12.8 FINAL THOUGHTS

The weed science community must develop widely accepted glyphosate resistance management strategies based on proven science and consistently educate growers on these strategies. We actually know little about the specific mechanisms of resistance to glyphosate in weeds. More research into these mechanisms is needed before successful scientifically based management practices can be fully developed. It is also imperative that a more in-depth and broad-based assessment of the societal and ecological benefits and risks of GE crops and specifically, GE GR crops be conducted. We have a long journey ahead in achieving economically sustainable, environmentally acceptable chemical management of weeds.

The sustainability of managing glyphosate resistance in weeds is now being tested in millions of hectares of cropland around the world, although in a nonscientific manner. We suggest that the solution to the sustainability of herbicide weed management, in general, and specifically, GR weed management must involve more than finding new herbicides and developing new herbicide-resistant crops. A truly effective and economically sustainable strategy will require a systems approach to weed management based on the integration of multiple crop improvement and farm management tools that have been developed over the last 60 years driven by science-based knowledge. The ecological perspective of weed management must not be ignored, and weed management systems that focus on the agroecosystem and not specifically on individual species must be developed (Mortensen et al. 2000). These strategies must be packaged into educational modules that offer attractive choices to farmers that result in consistent and effective weed control while reducing selection pressure for herbicide resistance development in weeds.

REFERENCES

Ammann, K. 2005. Effects of biotechnology on biodiversity: herbicide-tolerant and insect-resistant GM crops. *Trends in Biotechnology* 23:388–394.

Anderson, M. and J. Gronwald. 1991. Atrazine resistance in a velvetleaf (*Abutilon theophrasti*) biotype due to enhanced glutathione S-transferase activity. *Plant Physiology* 96:104–109.

Anderson, R. L. and D. L. Beck. 2007. Characterizing weed communities among various rotations in central South Dakota. *Weed Technology* 21:76–79.

Anderson, R. L., C. E. Stymiest, B. A. Swan, and J. R. Rickertsen. 2007. Weed community response to crop rotations in western South Dakota. *Weed Technology* 21:131–135.

Arntzen, C. J., A. Coghlan, B. Johnson, J. Peacock, and M. Rodemeyer. 2003. GM crops: science, politics and communication. *Nature Reviews Genetics* 4:839.

Behrens M. R., N. Mutlu, S. Chakraborty, R. Dumitru, W. Z. Jiang, B. J. LaVallee, P. L. Herman, T. E. Clemente, and D. P. Weeks. 2007. Dicamba resistance: enlarging and preserving biotechnology-based weed management strategies. *Science* 316:1185–1188.

Biotechnology Industry Organization (BIO). 2008. Agricultural biotechnology continues to increase crop yield and farmer income worldwide while supporting the environment. http://www.bio.org/news/pressreleases/newsitem.asp?id=2008_0213_01 (accessed March 23, 2010).

Boerboom, C. and M. Owen. 2009. *National Glyphosate Stewardship Forum II*. East Lansing, MI: The North Central IPM Center.

Buhler, D. D. 1995. Influence of tillage systems on weed population dynamics and management in corn and soybean in the central USA. *Crop Science* 35:1247–1258.

Cardina, J., C. P. Herms, and D. J. Doohan. 2002. Crop rotation and tillage system effects on weed seedbanks. *Weed Science* 50:448–460.

Carpenter, J. and L. Gianessi. 1999. Herbicide tolerant soybeans: why growers are adopting Roundup Ready varieties. *AgBioForum* 2:65–72.

Castle, L. A., D. L. Siehl, R. Gorton, P. A. Pattern, Y. H. Chen, S. Bertain, J.-J. Cho, N. Duck, J. Wong, D. Liu, and M. W. Lassner. 2004. Discovery and directed evolution of a glyphosate tolerance gene. *Science* 304:1151–1154.

Cattaneo M. G., C. Yafuso, C. Schmidt, C. Y. Huang, M. Rahman, C. Olson, C. Ellers-Kirk, B. J. Orr, S. E. Marsh, L. Antilla, P. Dutilleu, and Y. Carriere. 2006. Farm-scale evaluation of the impacts of transgenic cotton on biodiversity, pesticide use, and yield. *Proceedings of the National Academy of Sciences of the United States of America* 103:7571–7576.

Cerdeira, A. L. and S. A. Duke. 2006. The current status and environmental impacts of glyphosate-resistant crops: a review. *Pest Management Science* 35:1633–1658.

Chikoye, D., U. E. Udensi, and S. Ogunyemi. 2005. Integrated management of cogongrass (*Imperata cylindrical* L. Rauesch.) in corn using tillage, glyphosate, row spacing, cultivar, and cover cropping. *Agronomy Journal* 97:1164–1171.

Dinelli G., I. Marotti, A. Bonetti, P. Catizone, J. M. Urbano, and J. Barnes. 2008. Physiological and molecular bases of glyphosate resistance in *Conyza bonariensis* biotypes from Spain. *Weed Research* 48:257–265.

Dinelli G., I. Marotti, A. Bonetti, M. Minelli, P. Catizone, and J. Barnes. 2006. Physiological and molecular insight on the mechanisms of resistance to glyphosate in *Conyza canadensis* (L.) Cronq. biotypes. *Pesticide Biochemistry and Physiology* 86:30–41.

Ermakova, I. V. 2007. GM soybeans—revisiting a controversial format. *Nature Biotechnology* 25:1351–1354.

Felix, J. and M. D. K. Owen. 1999. Weed population dynamics in land removed from the conservation reserve program. *Weed Science* 47:511–517.

Feng, P. C. C., J. E. Pratley, and J. A. Bohn. 1999. Resistance to glyphosate in *Lolium rigidum*. II. Uptake, translocation, and metabolism. *Weed Science* 47:412–415.

Fernandez-Cornejo, J., S. Jans, and M. Smith. 1998. Pesticide economic issues: a review article. *Reviews of Agricultural Economics* 20:462–488.

Freckleton, R. P., W. J. Sutherland, and A. R. Watkinson. 2003. Deciding the future of GM crops in Europe. *Science* 302:994–996.

Gardner, S. N., J. Gressel, and M. Mangel. 1998. A revolving dose strategy to delay the evolution of both quantitative vs. major monogene resistances to pesticides and drugs. *International Journal of Pest Management* 44:161–180.

Gianessi, L. P., C. S. Silvers, S. Sankula, and J. E. Carpenter. 2002. Case study 26-herbicide tolerant soybean. *Plant Biotechnology: Current and Potential Impact for Improving Pest Management in U.S. Agriculture: An Analysis of 40 Case Studies*. Washington, DC: National Center for Food and Agricultural Policy, June 2002. http://www.ncfap.org/documents/SoybeanHT.pdf (accessed March 31, 2010).

Gibson, K. D., W. G. Johnson, and D. Hillger. 2005. Farmer perceptions of problematic corn and soybean weeds in Indiana. *Weed Technology* 19:1065–1070.

Givens, W. A., D. R. Shaw, G. R. Kruger, W. G. Johnson, S. C. Weller, B. G. Young, R. G. Wilson, M. D. K. Owen, and D. Jordan. 2009. Survey of tillage trends following the adoption of glyphosate-resistant crops. *Weed Technology* 23:150–155.

Gressel, J. 1995a. Creeping resistances: the outcome of using marginally effective or reduced rates of herbicide. *Proceedings of Brighton Crop Protection Conference, Weeds* 2:587.

Gressel, J. 1995b. Catch 22—mutually exclusive strategies for delaying/preventing poly-genically vs. monogenically inherited resistances. In N. N. Ragsdale, P. C. Kearney, and J. R. Plimmer, eds. *Options 2000—Eighth International Congress of Pesticide Chemistry*. Washington, DC: American Chemical Society, pp. 330–349.

Gressel, J. 2009. Evolving understanding of the evolution of herbicide resistance. *Pest Management Science* 65:1164–1173.

Grignac, P. 1978. The evolution of resistance to herbicides in weedy species. *Agro-Ecosystems* 4:377–385.

Grundy A. C., A. Mead, and S. Burston. 2003. Modeling the emergence response of weed seeds to burial depth interactions with seed density, weight, and shape. *Journal of Applied Ecology* 40:757–770.

Gura, T. 2001. The battlefields of Britain. *Nature* 412:760–763.

Hails, R. and J. Kinderlerer. 2003. The GM public debate: context and communication strategies. *Nature Reviews Genetics* 4:819–825.

Harper, J. L. 1956. The evolution of weeds in relation to resistance to herbicides. In *Proceedings of the 3rd Brighton Weed Control Conference*. Hampshire, UK: BCPC Publications, pp. 179–188.

Heap, I. 1998. *Resistance to herbicides in annual ryegrass (Lolium rigidum)*. PhD thesis, University Waite Agriculture Institute.

Heap, I. 2010. International Survey of Herbicide Resistant Weeds. http://www.weedscience.org (accessed April 8, 2010).

Heard, M. S., C. Hawes, G. T. Champion, S. J. Clark, L. G. Firbank, A. J. Haughton, A. M. Parish, J. N. Perry, P. Rothery, R. J. Scott, M. P. Skellern, G. R. Squire, and M. I. Hill. 2003. Weeds in fields with contrasting conventional and genetically modified herbicide-tolerant crops. I. Effects on abundance and diversity. *Philosophical Transactions of the Royal Society B: Biological Sciences* 358:1819–1832.

Heggenstaller, A. H. and M. Liebman. 2005. Demography of *Abutilon theophrasti* and *Setaria faberi* in three crop rotation systems. *Weed Research* 46:138–151.

James, C. 2008. *Global Status of Commercialized Biotech/GM Crops: 2008*. Ithaca, NY: International Service for the Acquisition of Agri-Biotech Applications.

John, B. 2007. GM soybeans-revisiting a controversial format. *Nature Biotechnology* 25:1354–1355.

Johnson, B., J. Barnes, K. Gibson, and S. Weller. 2004. Late season weed escapes in Indiana soybean fields. *Crop Management* doi:10.1094/CM-2004-0923-01-BR.

Johnson, W. G., P. R. Bradley, S. E. Hart, M. L. Buesinger, and R. E. Massey. 2000. Efficacy and economics of weed management in glyphosate-resistant corn (*Zea mays*). *Weed Technology* 14:57–65.

Johnson, W. G., V. M. Davis, G. R. Kruger, and S. C. Weller. 2009a. Influence of glyphosate-resistant cropping systems on weed species shifts and glyphosate-resistant weed populations. *European Journal of Agronomy* 31:162–172.

Johnson, W. G. and K. D. Gibson. 2006. Glyphosate-resistant weeds and resistance management strategies: an Indiana grower perspective. *Weed Technology* 20:768–772.

Johnson, W. G., M. D. K. Owen, G. R. Kruger, B. G. Young, D. R. Shaw, R. G. Wilson, J. W. Wilcut, D. L. Jordan, and S. C. Weller. 2009b. US farmer awareness of glyphosate-resistant weeds and resistance management strategies. *Weed Technology* 23:308–312.

Kruger G. R., W. G. Johnson, S. C. Weller, M. D. K. Owen, D. R. Shaw, J. W. Wilcut, D. L. Jordan, R. G. Wilson, M. L. Bernards, and B. G. Young. 2009. US grower views on problematic weeds and changes in weed pressure in glyphosate-resistant corn, cotton, and soybean cropping systems. *Weed Technology* 23:162–166.

Ledford, H. 2007. Out of bounds. *Nature* 445:132–133.

Lee, O. C. 1948. The effect of 2,4-D as a selective herbicide in growing corn and sorghums. *Proceedings of the North Central Weed Science Society* 5:18–21.

Liebman, M. and E. Dyke. 1993. Crop rotation and intercropping strategies for weed management. *Ecological Applications* 3:92–122.

Liefert, C. 2007. GM soybeans-revisiting a controversial format. *Nature Biotechnology* 25:1355.

Llewellyn, R. S., R. K. Lindner, D. J. Pannel, and S. B. Powles. 2002. Resistance and the herbicide resource: perceptions of Western Australian grain growers. *Crop Protection* 21:1067–1075.

Madsen, K. H. and P. Sandoe. 2005. Ethical reflections on herbicide-resistant crops. *Pest Management Science* 61:318–325.

Mallory-Smith, C. A., D. C. Thill, and M. J. Dial. 1990. Identification of sulfonylurea herbicide-resistant prickly lettuce (*Lactuca serriola*). *Weed Technology* 4:163–168.

Marshall, A. 2007. GM soybeans and health safety—a controversy revisited. *Nature Biotechnology* 25:981–987.

Martinez-Ghersa, M. A., C. A. Worster, and S. R. Radosevich. 2003. Concerns a weed scientist might have about herbicide-tolerant crops: a revolution. *Weed Technology* 17:202–210.

Marvier M., Y. Carriere, N. Ellstrand, P. Gepts, P. Kareiva, E. Rosi-Marshall, B. E. Tabashnik, and L. L. Wolfenbarger. 2008. Harvesting data from genetic engineered crops. *Science* 320:452–453.

Mortensen D. A., L. Bastiaans, and M. Sattin. 2000. The role of ecology in the development of weed management systems: an outlook. *Weed Research* 40:49–62.

Mulugeta, D., D. E. Stoltenberg, and C. M. Boerboom. 2001. Weed species-area relationships as influenced by tillage. *Weed Science* 49:217–223.

Murphy, S. D., D. R. Clements, S. Belaoussoff, P. G. Kevan, and C. J. Swanton. 2006. Promotion of weed species diversity and reductin of weed seedbanks with conservation tillage and crop rotation. *Weed Science* 54:69–77.

National Crop Residue Management Survey (NCRMS). 2004. Conservation Technology Information Center (CTIC). http://www.ctic.purdue.edu/ (accessed November 18, 2009).

Ngouajio, M. and H. Mennan. 2005. Weed populations and pickling cucumber (*Cucumis sativus*) yield under summer and winter cover crop systems. *Crop Protection* 24:521–526.

Owen, M. D. K. 1997. Risks and benefits of weed management technologies. In R. De Prado, J. Jorrin, and L. Garcia-Torres, eds. *Weed and Crop Resistance to Herbicides.* London: Kluwer Academic Publishers, pp. 291–297.

Owen, M. D. K. 2008. Weed specieis shifts in glyphosate-resistant crops. *Pest Management Science* 64:377–387.

Patzoldt, W. L., B. S. Dixon, and P. J. Tranel. 2003. Triazine resistance in *Amaranthus tuberculatus* (Moq) Sauer that is not site-of-action mediated. *Pest Management Science* 59:1134–1142.

Rüegg, W. T., M. Quadranti, and A. Zoschke. 2007. Herbicide research and development: challenges and opportunities. *Weed Research* 47:271–275.

Ryan, G. F. 1970. Resistance of common groundsel to simazine and atrazine. *Weed Science* 18:614–616.

Samarajeewa, K. B. D. P., T. Horiuchi, and S. Oba. 2005. Effects of Chinese milk vetch (*Astragalus sinicus* L.) as a cover crop on weed control, growth, and yield of wheat under different tillage systems. *Plant Production Science* 8:79–85.

Sammons, R. D., D. C. Herring, N. Dinicola, H. Glick, and G. A. Elmore. 2007. Sustainability and stewardship of glyphosate and glyphosate-resistant crops. *Weed Technology* 21:347–354.

Sandermann, H. 2006. Plant biotechnology: ecological case studies on herbicide resistance. *Trends in Plant Science* 11:324.

Sankula, S. 2006. *Quantification of the Impacts on U.S. Agriculture of Biotechnology-Derived Crops Planted in 2005.* Washington, DC: National Center for Food and Agriculture Policy. http://www.ncfap.org/documents/2007biotech_report/Quantification_of_the_Impacts_on_US_Agriculture_of_Biotechnology.pdf (accessed November 18, 2009).

Scursoni J. A., F. Forcella, J. Gunsolus, M. Owen, R. Oliver, R. Smeda, and R. Vidrine. 2006. Weed diversity and soybean yield with glyphosate management along a north-south transect in the United States. *Weed Science* 54:713–719.

Service, R. F. 2007. AGBIOTECH: a growing threat down on the farm. *Science* 316:1114–1117.

Shaner, D. L. 2000. The impact of glyphosate-tolerant crops on the use of other herbicides and on resistance management. *Pest Management Science* 56:320–326.

Shaw, D. R., W. A. Givens, L. A. Farno, P. D. Gerard, J. Dordan, W. G. Johnson, S. C. Weller, B. G. Young, R. G. Wilson, and M. D. K. Owen. 2009. Using a grower survey to assess the benefits and challenges of glyphosate-resistant cropping systems for weed management in US corn, cotton, and soybean. *Weed Technology* 23:134–149.

Swanton, C. J., B. D. Booth, K. Chandler, D. R. Clements, and A. Shrestha. 2006. Management in a modified no-tillage corn-soybean-wheat rotation influences weed population and community dynamics. *Weed Science* 54:47–58.

Teasdale, J. R., P. Parthan, and R. T. Collins. 2005. Synergism between cover crop residue and herbicide activity on emergence and early growth of weeds. *Weed Science* 53:521–527.

Thompson, L. Jr., J. M. Houghton, F. W. Slife, and H. S. Butler. 1971. Metabolism of atrazine by fall panicum and large crabgrass. *Weed Science* 19:409–412.

Tranel, P. J. and T. R. Wright. 2002. Resistance of weeds to ALS-inhibiting herbicides: what have we learned? *Weed Science* 50:700–712.

U.S. National Agricultural Statistics Service, Indiana Census of Agriculture (USDA-NASS). 2003. 2002 Census Publications. http://www.agcensus.usda.gov/Publications/2002/index.asp (accessed April 13, 2010).

U.S. National Agricultural Statistics Service, Indiana Census of Agriculture (USDA-NASS). 2007. 2007 Census Publications. http://www.agcensus.usda.gov/Publications/2007/index.asp (accessed April 13, 2010).

U.S. National Agricultural Statistics Service, Indiana Census of Agriculture (USDA-NASS). 2008. 2008–2009 Indiana Agriculture Statistics. http://www.nass.usda.gov/Statistics_by_State/Indiana/Publications/Annual_Statistical_Bulletin/0809/09index.asp (accessed April 13, 2010).

VanGessel, M. J. 2001. Glyphosate resistant horseweed from Delaware. *Weed Science* 49:703–705.

Watkinson, A. R., R. P. Freckleton, R. A. Robinson, and W. J. Sutherland. 2000. Predictions of biodiversity response to genetically modified herbicide-tolerant crops. *Science* 289:1554.

Weed Science Society of America (WSSA). 2007. *Herbicide Handbook*, 9th ed. Lawrence, KS: Weed Science Society of America, pp. 243–246.

Wicks, G. A., P. W. Stahlman, J. M. Tichota, and T. M. Price. 2001. Weed shifts in no-till glyphosate-tolerant crops in semiarid areas of the great plains. *Proceedings of the North Central Weed Science Society* 56:231.

Wright, T. R., J. M. Lira, D. J. Merlo, N. Hopkins, inventors; Dow Agrosciences, assignee. Novel herbicide resistance genes. 2005. World Intellectual Property Organization Patent WO/2005/107437, filed November 17, 2005.

Young, B. C. 2006. Changes in herbicide use pattern and production practices resulting from glyphosate-resistant crops. *Weed Technology* 20:301–307.

Zelaya, I. A., M. D. K. Owen, and M. J. VanGessel. 2004. Inheritance of evolved glyphosate resistance in *Conyza canadensis* (L.) Cronq. *Theoretical and Applied Genetics* 110:58–70.

Zeleya I. A., M. D. K. Owen, and M. J. VanGessel. 2007. Transfer of glyphosate resistance: evidence of hybridization in *Conyza* (Asteraceae). *American Journal of Botany* 94:660–673.

13

GLYPHOSATE-RESISTANT RIGID RYEGRASS IN AUSTRALIA

CHRISTOPHER PRESTON

13.1 INTRODUCTION

Rigid ryegrass (*Lolium rigidum* Gaudin) is a widespread weed of cropped agricultural systems in Australia (Kloot 1983). This species infests millions of hectares of agricultural land in southern Australia, and its range is continuing to expand to the north. Rigid ryegrass was originally introduced to Australia in the nineteenth century (Mullett 1919). However, it proved to be well suited to the Mediterranean climate of southern Australia as well as an excellent stock feed growing in the winter season. For these reasons, it was widely planted across much of southern Australia (Monaghan 1980). With the decline of the Australian wool industry since the 1970s, land was moved from sheep pasture to grain production. As a result of this change, rigid ryegrass became one of the most abundant weeds of grain cropping in this area (Gill 1996a).

Across the agricultural regions of southern Australia, rigid ryegrass typically germinates following opening rains in autumn, often as a small number of major cohorts (Chauhan et al. 2006a; McGowan 1970). It is an annual species that grows over the wet winter months, flowering in spring, setting seed, and dying in early summer (Gill 1996b). The seed has a period of after-ripening dormancy that largely prevents seed germinating during occasional summer rain periods (Steadman et al. 2003). Rigid ryegrass is an obligate outcrossing, wind-pollinated species, and these traits maintain high

Glyphosate Resistance in Crops and Weeds: History, Development, and Management
Edited by Vijay K. Nandula
Copyright © 2010 John Wiley & Sons, Inc.

levels of genetic diversity in populations (Powles and Matthews 1992). These features allow rigid ryegrass to persist and spread despite the changing agricultural environment in southern Australia.

Prior to the 1980s, rigid ryegrass was controlled by repeated tillage operations prior to crop seeding in southern Australia. The advent of selective grass herbicides, starting with trifluralin in the 1970s, led to a reduction in the amount of tillage conducted and increasing reliance on herbicides for weed control in cropping operations (Pratley and Rowell 1987). Now, much of the grain production area in Australia is sown using no-till techniques (D'Emden and Llewellyn 2006). These use narrow-point or disk openers to plant the crop seed in a direct seeding operation, with no prior tillage events. The move to reliance on herbicides only for control of rigid ryegrass inevitably resulted in the evolution of herbicide resistance in this species (Holtum and Powles 1991). Resistance to diclofop-methyl was first documented in 1982 (Heap and Knight 1982), and resistance to other herbicides occurred soon afterward (Heap and Knight 1986). By the late 1990s, herbicide-resistant rigid ryegrass was widespread across much of the cropped area of Western Australia and South Australia (Llewellyn and Powles 2001; P. Boutsalis and C. Preston, unpublished data). Resistance has been documented in this species to most of the herbicides registered for its control in Australia.

The reduction of tillage and the loss of herbicides to resistance led to greater reliance on glyphosate for control of rigid ryegrass (Preston et al. 2009). The intense selection pressure applied to an abundant and widespread weed by one herbicide year after year inevitably resulted in the evolution of glyphosate resistance in this species (Powles and Preston 2006). Glyphosate resistance in rigid ryegrass was first documented in 1996 in Australia, and it was the first weed to evolve glyphosate resistance anywhere in the world (Powles et al. 1998; Pratley et al. 1999a). This chapter will examine the current understanding of glyphosate resistance in rigid ryegrass in Australia and discuss the strategies employed to manage glyphosate resistance in this species.

13.2 EVOLUTION OF GLYPHOSATE RESISTANCE IN RIGID RYEGRASS

The first report of glyphosate resistance in rigid ryegrass was from a no-till grain cropping operation in Victoria in 1996 (Pratley et al. 1999a, 1999b). This was rapidly followed by resistance occurring in an apple orchard in New South Wales (NSW) (Powles et al. 1998). Following these two reports, there were further reports of resistance occurring in a variety of situations. Currently, there are 87 sites with confirmed glyphosate-resistant rigid ryegrass in Australia (Table 13.1). These sites occur in several land management uses. The situation contributing the most number of sites is a winter chemical fallow system in northern NSW. Other systems with considerable number of glyphosate resistance sites are farm fence lines and vineyards.

TABLE 13.1. Occurrence of Glyphosate-Resistant Rigid Ryegrass Populations in Australia

Situation		Number of Sites
Broadacre cropping	Winter chemical fallow	25
	No-till winter grains	13
Horticulture	Tree crops	4
	Vine crops	14
Other	Driveway	1
	Fence line and crop firebreak	22
	Irrigation channel	6
	Airstrip	1
	Railway	1

The largest numbers of glyphosate-resistant rigid ryegrass weed populations occur in chemical fallows in northern NSW (Table 13.1). This area has a rainfall pattern that is more evenly distributed between summer and winter, unlike much of the rest of the cropping region in southern Australia that has a winter-dominated rainfall pattern. Either summer or winter crops can be planted in this region; however, moisture needs to be stored in the soil during the off-season to grow the crop (Osten et al. 2007). Weeds use stored moisture, so a chemical fallow involving repeated applications of glyphosate is employed to control weeds. That fallow may occur either in winter or summer (or both) depending on the crop to be grown. Typically, no other weed management is employed during the fallow (Storrie and Cook 2002). Growers who have opted for mainly summer crop production have used glyphosate extensively several times each winter for fallow weed control. After a decade or more of this practice, glyphosate-resistant rigid ryegrass has evolved on many farms (Storrie and Cook 2002).

Another situation where large numbers of glyphosate-resistant rigid ryegrass populations occur are vineyards and orchards (Table 13.1). In vineyards in Australia, it is common practice to have a permanent sward or a cover crop in the mid-row, but the area under the vines is kept weed-free. Increasingly, herbicides are used for under-vine weed management and the preferred herbicide in winter is glyphosate (Wakelin and Preston 2008). A similar situation occurs in orchards, where the area under trees will be kept bare with glyphosate applications during winter. After 15 or more years of glyphosate used two to three times each winter, glyphosate-resistant rigid ryegrass evolved in both systems (Powles et al. 1998; Wakelin and Preston 2006a).

The long period when sheep were an important part of agriculture across southern Australia means that many crop fields are surrounded by fences originally constructed to keep stock in. Crop producers in this area prefer to keep the areas around their fields free of weeds and other vegetation. There are several reasons why this is done. It acts as a firebreak around crops, limiting the potential losses as a result of crop fires; reduces the potential invasion of weeds into the cropped area; and reduces habitat for crop pests such as snails. Management strategies vary, but many growers use glyphosate as the main or only weed control method along fences and crop edges. After 15 or more years of once annual glyphosate application during winter, glyphosate-resistant rigid ryegrass evolved from this use (Wakelin et al. 2004).

One of the most common uses of glyphosate in southern Australian agriculture is for weed control prior to seeding the crop. Following the opening rains in autumn, weeds will germinate in crop fields. Farmers often control these weeds prior to crop seeding. Due to the widespread adoption of no-till crop seeding practices across most of the cropped area, herbicides are relied on for weed control prior to crop seeding. The most common seeding equipment uses a narrow-point tine, although seeding disks are employed (D'Emden and Llewellyn 2006). Both types of seeding equipment provide low levels of soil disturbance, meaning early-season weed control is totally dependent on herbicides (Chauhan et al. 2006b). Much of this area also has a high frequency of resistance to postemergent herbicides in rigid ryegrass. Field surveys show that more than half of all cropped fields in South Australia and Western Australia contain rigid ryegrass with resistance to acetyl coenzyme A carboxylase (ACCase)-inhibiting and acetolactate synthase (ALS)-inhibiting herbicides (Owen et al. 2007; Boutsalis and Preston, unpublished data). These two factors mean that the presowing and preemergent herbicides provide most of the early weed control. As a result, it was no surprise that the first reported case of glyphosate resistance in rigid ryegrass occurred in a no-till grain production field in Victoria following 15 years of glyphosate use (Pratley et al. 1999a).

All the situations where glyphosate-resistant rigid ryegrass has evolved in Australia can be characterized as having intensive use of glyphosate for many years and no other effective weed control practiced. In these situations alternative management practices are not favored or resistance to other herbicides is present in rigid ryegrass, resulting in the exclusive use of glyphosate for weed control (Neve et al. 2004; Stanton and Broster 2004). However, despite the relatively large area of no-till winter grain production in southern Australia where glyphosate is relied on for presowing knockdown weed control, a relatively small number of glyphosate-resistant weed populations have evolved in this system.

13.3 RESISTANCE MECHANISMS IN RIGID RYEGRASS IN AUSTRALIA

Understanding resistance mechanisms in rigid ryegrass in Australia has been important largely because management of glyphosate-resistant rigid ryegrass in Australia has continued to involve the use of glyphosate, such as glyphosate mixtures. Much of the work conducted on management strategies for glyphosate-resistant rigid ryegrass has occurred on a small number of populations. This may not be a problem if all populations respond similarly to management. However, if resistant populations have different resistance mechanisms and as a result differ in their response to management, then management strategies suggested by research on a single population may not be generally applicable to other populations.

Early work to understand the mechanism of glyphosate resistance in rigid ryegrass identified a reduction in translocation of the herbicide as a possible mechanism of resistance (Lorraine-Colwill et al. 2003). Glyphosate is a very mobile herbicide, and this property plays a significant role in the efficacy of this herbicide (Baylis 2000). In susceptible plants, glyphosate moves rapidly throughout the plant and accumulates in sink tissues, such as roots, meristems, buds, and fruits (Franz et al. 1997). Studies with glyphosate-resistant rigid ryegrass showed a different pattern of herbicide distribution within the plant. Instead of widespread translocation throughout the plant, glyphosate preferentially accumulated in the tips of the treated leaves in resistant plants (Lorraine-Colwill et al. 2003). There was significantly less glyphosate accumulation in the stem meristem compared with susceptible plants (Wakelin et al. 2004). Therefore, glyphosate-resistant rigid ryegrass plants survive glyphosate application because insufficient glyphosate accumulates at the growing point to kill the plant. Surveys of resistance mechanisms in rigid ryegrass in Australia found that this mechanism of resistance was by far the most common mechanism detected. More than 85% of all populations tested contain this resistance mechanism (Preston et al. 2009). Typically, this mechanism provides between 5- and 12-fold resistance to glyphosate.

With time, it became clear that other mechanisms of glyphosate resistance were present in resistant populations of rigid ryegrass. Sequencing of the 5-enolpyruvylshikimate-3-phosphate synthase (EPSPS) gene in resistant and susceptible populations identified one population with a Thr substitution for Pro 106 in EPSPS (Wakelin and Preston 2006a). This substitution is known to produce a glyphosate-resistant EPSPS and to endow plants with resistance to glyphosate (Healy-Fried et al. 2007). The population containing the mutant EPSPS had threefold resistance to glyphosate. A survey of 40 glyphosate-resistant rigid ryegrass populations identified two other populations containing a target-site mutation (Preston et al. 2009). Less than 10% of glyphosate-resistant rigid ryegrass populations in Australia contain the target-site mutation. The relatively low level of resistance endowed by mutations in EPSPS would seem to act against its selection in the field. In addition to the above two mechanisms, there appears to be at least one other mechanism of glyphosate resistance in rigid ryegrass that has not yet been identified (Baerson et al. 2002; Preston and Wakelin 2008). This is possibly over-expression of EPSPS (Baerson et al. 2002; Harrison et al. 2004), although that has not been confirmed.

Both of the glyphosate resistance mechanisms so far identified in rigid ryegrass in Australia provide only modest levels of resistance to the herbicide.

This means that some level of control can be achieved through the application of glyphosate on these populations. Therefore, producers continue to use glyphosate, often in mixtures, to control glyphosate-resistant populations (Storrie and Cook 2002; Wakelin and Preston 2008). However, as rigid ryegrass is an obligate outcrossing species, it is possible to accumulate resistance mechanisms within individual plants, provided the genes for both mechanisms are present in the population. To date, multiple mechanisms of resistance to glyphosate have evolved in at least two populations of rigid ryegrass in Australia (Preston et al. 2009). Both populations contain a target-site modification as well as the decreased translocation mechanism (Preston et al. 2009). Populations containing both mechanisms of resistance have much greater resistance to glyphosate than populations containing the individual mechanisms. This means that glyphosate at normal use rates provides virtually no control of these populations.

There appears to be a relationship between selection history and the mechanism of glyphosate resistance selected in rigid ryegrass in Australia. The reduced translocation mechanism of resistance has been selected with a variety of glyphosate uses in Australia. This mechanism has occurred in situations as varied as no-till winter grain production, where glyphosate is used once at 360–450 g a.e. ha^{-1} each year, to orchards and vineyards, where glyphosate is used at 720–1100 g ha^{-1} multiple times each year (Preston and Wakelin 2008). It is clear that this mechanism is preferentially selected because of the relatively higher level of resistance it provides. In contrast, to date, all the examples of target-site resistance to glyphosate in rigid ryegrass have evolved in vineyards where high rates of glyphosate are used several times per year (Preston et al. 2009). This is despite the lower level of resistance this mechanism provides. It is not clear why this mechanism is only appearing in situations of greater glyphosate use. The populations with multiple mechanisms of resistance to glyphosate have also occurred in vineyards. Vineyard managers have had less experience with herbicide-resistant weeds and have often responded to poor control by increasing the glyphosate use rate (Wakelin and Preston 2008). It is likely that continual selection with glyphosate will ultimately result in other rigid ryegrass populations with multiple mechanisms of resistance to glyphosate. These populations will have very high levels of resistance to glyphosate and will not be controlled by this herbicide either alone at higher rates or in mixtures with other herbicides, unless the mixing partner provides effective control on its own.

13.4 INHERITANCE OF GLYPHOSATE RESISTANCE IN RIGID RYEGRASS

In the period immediately after the evolution of glyphosate resistance in rigid ryegrass in Australia, there were suggestions that the relatively low use rates of this herbicide in Australia had selected for polygenic resistance

(Gressel 1999). Furthermore, it has been suggested that a rotating dose strategy of high and low doses could be used to mitigate resistance selection for polygenic resistance (Gardner et al. 1998). Therefore, it became important to determine the mode of inheritance of glyphosate in rigid ryegrass populations to determine whether this was the best strategy of management.

Crosses between resistant and susceptible individuals demonstrated that the glyphosate resistance genes were located on the nuclear genome and were inherited in a dominant or partially dominant fashion (Lorraine-Colwill et al. 2001). Determining the number of genes contributing to glyphosate resistance in rigid ryegrass has not been simple. The relatively modest levels of resistance mean that the dose–response curves of both parental types and the F_1 cross overlap. This means that it is not possible to use a single glyphosate rate to unambiguously identify individuals as homozygous or heterozygous for either resistance or susceptibility. One approach taken to resolve this problem was to create backcross populations and determine the response to glyphosate of every individual in the population. This was done by allowing plants to grow until well tillered, breaking the plants into small pieces and then applying a series of glyphosate doses across the pieces. In this way, the response of an individual plant to a number of glyphosate doses could be determined, and plants could be classified on the basis of whether they responded like the susceptible parent or not (Wakelin and Preston 2006b).

Using this approach, Lorraine-Colwill et al. (2001) was able to determine that in one population of rigid ryegrass, resistance was the result of a single gene with intermediate dominance. Wakelin and Preston (2006b) examined six more populations of rigid ryegrass and found that in all six, glyphosate resistance was encoded on the nuclear genome and was intermediate or fully dominant. In five of the six populations, resistance was inherited as a single gene. In the sixth population, the pattern of inheritance was not clear. This population contained an overabundance of resistant individuals in the back-cross generation, suggesting additional genes contributed to resistance. These seven populations were all resistant due to reduced glyphosate translocation. The reduced translocation mechanism is one where the potential exists for multiple genes to contribute to a resistance mechanism. However, in the work conducted to date, a single gene contributes to this resistance mechanism in most populations studied. The inheritance of glyphosate resistance has been studied in one population where a target-site mutation was the mechanism of resistance. In this population, resistance was also inherited as a single domi-nant allele (Preston et al. 2009).

In Australia, it is clear that glyphosate selection patterns are overwhelm-ingly selecting for single gene resistance rather than for polygenic resistance. This clearly has implications on how various resistance mitigation strategies might work. As resistance is mostly dominant, selection for resistance will occur readily regardless of the rate of glyphosate used. Rather it will be the intensity of glyphosate use that will be the dominant factor in selection (Powles and Preston 2006; Preston et al. 2009). However, as there are two known mechanisms for resistance, and the possibility of other resistance mechanisms existing, populations with multiple mechanisms of resistance should be expected (Preston et al. 2009). In these populations, resistance will be inherited by more than one gene, with at least one gene for each mechanism.

13.5   FITNESS OF GLYPHOSATE-RESISTANT RIGID RYEGRASS POPULATIONS IN AUSTRALIA

Despite the extensive use of glyphosate for the control of weed populations prior to seeding grain crops in Australia, there are relatively few glyphosate-resistant populations that have evolved following this usage pattern. Many of the populations that have appeared on grain farms have occurred on fence lines, on crop margins, in irrigation channels, and in winter fallows (Table 13.1); situations that have much lower levels of competition. The low level of competition in these situations allows any survivors of glyphosate treatment to set large amounts of seed unhindered. However, seed production of weeds should be lower in situations where the crop provides competition for light and other resources. The pattern of locations where glyphosate-resistant rigid ryegrass occurs suggested that lack of competition could be a key factor in the evolution of resistance.

Investigations have confirmed apparent fitness penalties in some glyphosate-resistant populations of rigid ryegrass. In one study, individuals resistant and susceptible to glyphosate were isolated from a single population and compared (Pedersen et al. 2007). This study found no differences in growth rates when comparing resistant and susceptible plants. However, resistant plants in the population produced fewer, although larger, seeds compared with susceptible plants.

A second approach to examining fitness penalties in glyphosate-resistant rigid ryegrass used segregating populations created from crosses between resistant and susceptible populations. These experiments attempted to distribute background genetics among the population to avoid the need to create near-isogenic lines. Populations were planted into wheat crops and no herbicides were used. At the end of each season, seed was collected from the surviving plants and pooled. At the end of the experiment, seed from each year was grown out, treated with a single rate of herbicide and survival compared with the original segregating population. In one trial, a single population was planted at three different sites. At each site, the frequency of glyphosate-resistant individuals in the population declined over 3 years (Preston et al. 2009). However, there were differences in the rate of decline between the different environments. In a second trial, segregating populations of four glyphosate-resistant populations were planted into a single site. For each population, there was a decline in frequency of glyphosate-resistant individuals over 2 years (Wakelin and Preston 2006c).

All of the fitness studies so far have been conducted with populations containing the translocation mechanism of resistance. The evidence to date suggests that this resistance mechanism carries a substantial fitness penalty (Preston et al. 2009). To date, no work has been conducted with populations containing the target-site mechanism or with populations containing both mechanisms. It is not known whether these populations would have a similar, a smaller, or a larger fitness penalty. The existence of a fitness penalty in glyphosate-resistant rigid ryegrass could explain why selection for resistance to glyphosate has been so difficult. A fitness penalty would keep resistance alleles at very low frequencies in populations in the absence of selection (Jasieniuk et al. 1996). It would also delay the advent of resistance, as resistant individuals would contribute fewer seed to the next generation. Lastly, a fitness penalty offers opportunities for management of glyphosate-resistant populations.

13.6 MANAGEMENT OF GLYPHOSATE-RESISTANT RIGID RYEGRASS IN AUSTRALIA

In managing the risks of glyphosate-resistant weeds in Australia, a whole of industry approach has been taken. In order to coordinate information flow, the Grains Research and Development Corporation and the Cooperative Research Centre for Australian Weed Management set up the Australian Glyphosate Sustainability Working Group. The group comprises weed researchers, representatives from key agricultural industries, and agrochemical companies. Its role is to identify key extension messages for glyphosate users to reduce the risk of glyphosate resistance. Users are more likely to take action to manage resistance if they receive a consistent message, rather than mixed messages from different sectors (Preston 2009).

The evolution of glyphosate-resistant rigid ryegrass in winter fallows presents a major problem for managers. Glyphosate has been the preferred option for weed control in this system because of its ease of use, wide spectrum of weeds controlled, and lack of soil residual effects restricting cropping options (Storrie and Cook 2002). Following research showing that clethodim was an effective control for glyphosate-resistant rigid ryegrass in winter fallows, a mixture of glyphosate and clethodim became the preferred control option. Some growers used weed-detecting spray equipment to apply clethodim to the clumps of surviving rigid ryegrass plants several weeks after glyphosate application to reduce herbicide costs. Following 6 years of this management practice, populations of rigid ryegrass with resistance to both glyphosate and clethodim have evolved in these winter fallow systems (A. Storrie, pers. comm.). These multiple-resistant populations of rigid ryegrass create additional management problems. Other possible management options in this system are occasional cultivation, paraquat following glyphosate application, or the addition of residual herbicides (Walker et al. 2004), but these are yet to be widely employed.

Glyphosate has been the preferred herbicide for winter weed control in vineyards when vines are dormant because of its relative safety to humans, wide weed spectrum, and lack of soil activity. In summer, glufosinate or paraquat are more likely to be used as they pose less of a risk if they drift onto nondormant vines. Vine growers are reluctant to move away from glyphosate during the winter period, despite the evolution of resistance to the herbicide, because glyphosate still controls other weeds. Therefore, the main strategies used have included increasing the rate of glyphosate or mixing other herbicides with glyphosate. The most successful strategy for managing glyphosate-resistant rigid ryegrass in vineyards includes the use of residual herbicide chemistry in autumn when vines are dormant, which reduces rigid ryegrass emergence and problems through the winter period. Changing the timing of spring applications of glufosinate or paraquat to the period when rigid ryegrass is flowering helps reduce rigid ryegrass seed set and will ultimately reduce populations (Wakelin and Preston 2008).

Glyphosate remains the most popular product for presowing weed control in southern Australia because of its ease of use, low cost, and efficacy on a wide range of weed species. The evolution of glyphosate resistance in no-till farming systems has seen an increase in the use of paraquat + diquat for weed control prior to seeding, a strategy called the "double knock" (Borger and Hashem 2007; Neve et al. 2003). The double knock is widely employed as a strategy to limit the evolution of glyphosate resistance. Other tactics that have proven successful at managing glyphosate-resistant rigid ryegrass include crop competition and weed seed set control. It is well understood in Australia that some crop species, particularly cereals, are more competitive against grass weeds (Lemerle et al. 1995). In addition, increased competition can be achieved by increasing cereal crop seeding rates (Lemerle et al. 2004). Increased competition helps exploit fitness penalties in rigid ryegrass carrying glyphosate resistance alleles (Wakelin and Preston 2008). In Australia, application of paraquat at low rates late in the growing season is widely used in crops such as field peas, lupins, and chickpeas to reduce seed set of rigid ryegrass (Peck and McDonald 2001). This practice, called crop topping, can reduce rigid ryegrass seedbanks by up to 70% (Matthews et al. 1996). Crop topping or cutting infested areas for hay prior to rigid ryegrass seed set can be an effective means of limiting the buildup of glyphosate-resistant rigid ryegrass populations (Wakelin and Preston 2008).

Strategies employed for the management of glyphosate-resistant rigid ryegrass in Australia depend very much on the system in which resistance occurs. In some situations, there are a limited number of control options available, greatly constraining management. In several cases, simply changing to a different herbicide is not an available option. The least successful control strategies have continued to rely mostly or entirely on glyphosate for control. The most successful strategies include a number of control strategies and employ tactics to control seed production of resistant individuals (Preston et al. 2009). In these strategies, glyphosate may still be used, but is relied on less.

13.7 CONCLUSIONS AND FUTURE PROSPECTS

Glyphosate resistance in rigid ryegrass has evolved in Australia in situations where glyphosate has been used intensively for weed control and few other effective weed management tactics are used (Table 13.1). Glyphosate-resistant rigid ryegrass has evolved most commonly in winter chemical fallows, vineyards, fence lines, and other areas where no crop is present. Despite the extensive use of glyphosate in no-till grain cropping systems, there have been relatively few glyphosate-resistant populations reported from this system.

The understanding that multiple mechanisms of glyphosate resistance occur in rigid ryegrass has changed ideas about the management of glyphosate-resistant rigid ryegrass in Australia. Continuing to use glyphosate for management of these resistant populations will lead to accumulation of resistance mechanisms, resulting in higher levels of glyphosate resistance in populations. However, there is no apparent relationship between glyphosate rate or use pattern and the resistance mechanism selected. The reduced translocation mechanism provides greater resistance to glyphosate than target-site resistance mechanism, and is therefore selected more often. However, all populations with target-site resistance have been found in vineyards where higher rates of herbicide and greater intensity of use occur.

The understanding of mechanisms, inheritance, and fitness of glyphosate resistance in rigid ryegrass has been used to inform management of this weed in Australia. It is now recognized that crop competition can play an important part in managing glyphosate-resistant weeds, in addition to changing herbicides or other strategies (Wakelin and Preston 2008). The reduced fitness of rigid ryegrass populations carrying glyphosate resistance alleles also means that alternative management strategies are likely to be more effective on glyphosate-resistant individuals than susceptible individuals (Preston et al. 2009). In addition, the activities of the Australian Glyphosate Sustainability Working Group have ensured that messages concerning the management of glyphosate-resistant weeds are consistent and based on the best understanding available.

The continued widespread use of glyphosate for weed management in Australia will inevitably lead to the appearance of more populations of glyphosate-resistant rigid ryegrass. In addition, glyphosate resistance is likely to be selected in other weed species. In the past 2 years, glyphosate resistance has been detected in two additional weed species: jungle rice (*Echinochloa colona* (L.) Link.) and panic liverseed grass (*Urochloa panicoides* P. Beauv.) (Heap 2010; Preston 2009). Glyphosate-resistant populations of both of these grass weeds have occurred in summer fallows where glyphosate was the only weed control employed over the summer period.

An additional complication for the management of glyphosate-resistant rigid ryegrass in southern Australian grain production systems is the introduction of glyphosate-resistant canola. This crop was grown on a commercial scale for the first time in NSW and Victoria in 2008 and in Western Australia in 2009. The recommended use pattern of glyphosate with this crop is for two applications of glyphosate in crop (Monsanto Australia Ltd. 2009). Adopting glyphosate-resistant canola in an environment of high existing glyphosate use will inevitably lead to more selection pressure for glyphosate-resistant weeds (Neve et al. 2003; Preston and Rieger 2000; Preston et al. 1999). In response to this increased risk, glyphosate-resistant canola has a resistance management plan in Australia that includes growers conducting a risk assessment of their fields and taking action to manage risks of glyphosate-resistant weeds evolving (Monsanto Australia Ltd. 2009). This management plan includes the recommendation that growers do not use glyphosate in the year after growing glyphosate-resistant canola.

REFERENCES

Baerson, S. R., D. J. Rodriguez, N. A. Biest, M. Tran, J. You, R. W. Kreuger, J. E. Pratley, and K. J. Gruys. 2002. Investigating the mechanism of glyphosate resistance in rigid ryegrass (*Lolium rigidum*). *Weed Science* 50:721–730.

Baylis, A. D. 2000. Why glyphosate is a global herbicide: strengths, weaknesses and prospects. *Pest Management Science* 56:299–308.

Borger, C. and A. Hashem. 2007. Evaluating double knockdown technique: sequence, application interval and annual ryegrass growth stage. *Australian Journal of Agricultural Research* 58:265–271.

Chauhan, B. S., G. S. Gill, and C. Preston. 2006a. Influence of environmental factors on seed germination and seedling emergence of rigid ryegrass (*Lolium rigidum*). *Weed Science* 54:1004–1012.

Chauhan, B. S., G. S. Gill, and C. Preston. 2006b. Influence of tillage systems on vertical distribution, seedling recruitment and persistence of rigid ryegrass (*Lolium rigidum*) seed bank. *Weed Science* 54:669–676.

D'Emden, F. H. and R. S. Llewellyn. 2006. No-till adoption decisions in southern Australian cropping and the role of weed management. *Australian Journal of Agricultural Research* 46:563–569.

Franz, J. E., M. K. Mao, and J. A. Sikorski. 1997. *Glyphosate: A Unique Global Pesticide.* Washington, DC: American Chemical Society.

Gardner, S. N., J. Gressel, and M. Mangel. 1998. A revolving dose strategy to delay the evolution of both quantitative vs major monogene resistances to pesticides and drugs. *International Journal of Pest Management* 44:161–180.

Gill, G. S. 1996a. Why annual ryegrass is a problem in Australian agriculture. *Plant Protection Quarterly* 11:193–195.

Gill, G. S. 1996b. Ecology of annual ryegrass. *Plant Protection Quarterly* 11:195–198.

Gressel, J. 1999. Needed: new paradigms for weed control. In A. C. Bishop, M. Boersma, and C. D. Barnes, eds. *Proceedings of the 12th Australian Weeds Conference.* Devonport, Tas.: Tasmanian Weed Society, pp. 462–486.

Harrison, D. K., W. Liu, S. W. Adkins, P. M. Gresshoff, and R. W. Williams. 2004. Glyphosate-resistant annual ryegrass (*Lolium rigidum* Gaudin) from Tamworth, Australia, has upregulated EPSPS expression. In B. M. Sindel and S. B. Johnson, eds.,

Proceedings of the 14th Australian Weeds Conference. Sydney, NSW: Weed Society of New South Wales, pp. 418–420.

Healy-Fried, M. L., T. Funke, M. A. Priestman, H. Han, and E. Schönbrunn. 2007. Structural basis of glyphosate tolerance resulting from mutations of Pro[101] in *Escherichia coli* 5-enolpyruvylshikimate-3-phosphate synthase. *Journal of Biological Chemistry* 282:32949–32955.

Heap, I. 2010. The International Survey of Herbicide Resistant Weeds. http://www. weedscience.com (accessed April 8, 2010).

Heap, J. and R. Knight. 1982. A population of ryegrass tolerant to the herbicide diclofop-methyl. *Journal of the Australian Institute of Agricultural Science* 48:156–157.

Heap, I. and R. Knight. 1986. The occurrence of herbicide cross-resistance in a population of annual ryegrass, *Lolium rigidum*, resistant to diclofop-methyl. *Australian Journal of Agricultural Research* 37:149–156.

Holtum, J. A. M. and S. B. Powles. 1991. Annual ryegrass: an abundance of resistance, a plethora of mechanisms. In *Proceedings of the Brighton Crop Protection Conference—Weeds*, Farnham, Surrey, UK, pp. 1071–1078.

Jasieniuk, M., A. L. Brûlé-Babel, and I. N. Morrison. 1996. The evolution and genetics of herbicide resistance in weeds. *Weed Science* 44:176–193.

Kloot, P.M. 1983. The genus *Lolium* in Australia. *Australian Journal of Botany* 31:421–435.

Lemerle, D., R. D. Cousens, G. S. Gill, S. J. Peltzer, M. Moerkerk, C. E. Murphy, D. Collins, and B. R. Cullis. 2004. Reliability of higher seeding rates of wheat for increased competitiveness with weeds in low rainfall environments. *Journal of Agricultural Science* 142:395–409.

Lemerle, D., B. Verbeek, and N. E. Coombes. 1995. Losses in grain yield of winter crops from *Lolium rigidum* competition depend on crop species, cultivar and season. *Weed Research* 35:503–509.

Llewellyn, R. S. and S. B. Powles. 2001. High levels of herbicide resistance in rigid ryegrass (*Lolium rigidum*) in the wheat belt of Western Australia. *Weed Technology* 15:242–248.

Lorraine-Colwill, D. F., S. B. Powles, T. R. Hawkes, P. H. Hollinshead, S. A. J. Warner, and C. Preston. 2003. Investigations into the mechanism of glyphosate resistance in *Lolium rigidum*. *Pesticide Biochemistry and Physiology* 74:62–72.

Lorraine-Colwill, D. F., S. B. Powles, T. R. Hawkes, and C. Preston. 2001. Inheritance of evolved glyphosate resistance in *Lolium rigidum*. *Theoretical and Applied Genetics* 102:545–550.

Matthews, J. M., R. Llewellyn, S. Powles, and T. Reeves. 1996. Integrated weed management for the control of herbicide resistant annual ryegrass. *Proceedings of the 8th Australian Agronomy Conference*. http://www.regional.org.au/au/asa/1996/contributed/417matthews.htm (accessed May 29, 2009).

McGowan, A. A. 1970. Comparative germination patterns of annual grasses in northeastern Victoria. *Australian Journal of Experimental Agriculture and Animal Husbandry* 10:401–404.

Monaghan, N. M. 1980. The biology and control of *Lolium rigidum* as a weed of wheat. *Weed Research* 20:117–121.

Monsanto Australia Ltd. 2009. Roundup Ready® Canola 2009 Crop Management Plan. http://www.nuseed.com.au/Assets/43/1/rrc_cmp08.pdf (accessed May 29, 2009).

Mullett, H. A. 1919. *Lolium subulatum*, vis. "Wimmera" ryegrass. *Journal of the Department of Agriculture, Victoria* 17:266–278.

Neve, P., A. J. Diggle, F. P. Smith, and S. B. Powles. 2003. Simulating evolution of glyphosate resistance in *Lolium rigidum*. II: past, present and future glyphosate use in Australian cropping. *Weed Research* 43:418–427.

Neve, P., J. Sadler, and S. B. Powles. 2004. Multiple herbicide resistance in a glyphosate-resistant rigid ryegrass (*Lolium rigidum*) population. *Weed Science* 52:920–928.

Osten, V. A., S. R. Walker, A. Storrie, M. Widderick, P. Moylan, G. R. Robinson, and K. Galea. 2007. Survey of weed flora and management relative to cropping practices in the north-eastern grain region of Australia. *Australian Journal of Experimental Agriculture* 47:57–70.

Owen, M. J., M. J. Walsh, R. S. Llewellyn, and S. B. Powles. 2007. Widespread occurrence of multiple herbicide resistance in Western Australian annual ryegrass (*Lolium rigidum*) populations. *Australian Journal of Experimental Agriculture* 58:711–718.

Peck, D. M. and G. K. McDonald. 2001. Survey of pea production practices in South Australia. *Proceedings of the 10th Australian Agronomy Conference*. http://www.regional.org.au/au/asa/2001/2/a/peck1.htm (accessed May 29, 2009).

Pedersen, B. P., P. Neve, C. Andreasen, and S. B. Powles. 2007. Ecological fitness of a glyphosate-resistant *Lolium rigidum* population: growth and seed production along a competition gradient. *Basic and Applied Ecology* 8:258–268.

Powles, S. B., D. F. Lorraine-Colwill, J. J. Dellow, and C. Preston. 1998. Evolved resistance to glyphosate in rigid ryegrass (*Lolium rigidum*) in Australia. *Weed Science* 46:604–607.

Powles, S. B. and J. M. Matthews. 1992. Multiple herbicide resistance in annual ryegrass (*Lolium rigidum*): a driving force for the adoption of integrated weed management. In I. Denholm, A. L. Devonshire, and D. W. Hollomon, eds. *Resistance '91: Achievements and Developments in Combating Pesticide Resistance*. London: Elsevier Applied Science, pp. 75–87.

Powles, S. B. and C. Preston. 2006. Evolved glyphosate resistance in plants: biochemical and genetic basis of resistance. *Weed Technology* 20:282–289.

Pratley, J. E. and D. L. Rowell. 1987. Evolution of Australian farming systems. In P. S. Cornish and J. E. Pratley, eds. *Tillage—New Directions in Australian Agriculture*. Melbourne, Vic.: Inkata Press, pp. 2–23.

Pratley, J., N. Urwin, R. Stanton, P. Baines, J. Broster, K. Cullis, D. Schafer, J. Bohn, and R. Krueger. 1999a. Resistance to glyphosate in *Lolium rigidum*. I. Bioevaluation. *Weed Science* 47:405–411.

Pratley, J., R. Stanton, N. Urwin, P. Baines, D. Hudson, and G. Dill. 1999b. Resistance of Annual ryegrass (*Lolium rigidum*) biotypes to glyphosate. In A. C. Bishop, M. Boersma, and C. D. Barnes, eds. *Proceedings of the 12th Australian Weeds Conference*. Devonport, Tas.: Tasmanian Weed Society, pp. 223–225.

Preston, C. 2009. Australian Glyphosate Resistance Register. Australian Glyphosate Sustainability Working Group. http://www.glyphosateresistance.org.au (accessed March 30, 2010).

Preston, C. and M. A. Rieger. 2000. Risks of herbicide resistant weeds from use of herbicide tolerant crops. *Plant Protection Quarterly* 15:77–79.

Preston, C., R. T. Roush, and S. B. Powles. 1999. Herbicide resistance in weeds of southern Australia: why are we the worst in the world? In A. C. Bishop, M. Boersma, and C. D. Barnes, eds., *Proceedings of the 12th Australian Weeds Conference.* Davenport, TAS: Tasmanian Weed Society, pp. 454–459.

Preston, C. and A. M. Wakelin. 2008. Resistance to glyphosate from altered herbicide translocation patterns. *Pest Management Science* 64:372–376.

Preston, C., A. M. Wakelin, F. C. Dolman, Y. Bostamam, and P. Boutsalis. 2009. A decade of glyphosate-resistant *Lolium* around the world: mechanisms, genes, fitness and agronomic management. *Weed Science* 57:435–441.

Stanton, R. and J. Broster. 2004. Multiple resistance to glyphosate in annual ryegrass (*Lolium rigidum* Gaudin) populations. In B. M. Sindel and S. B. Johnson, eds. *Proceedings of the 14th Australian Weeds Conference.* Sydney, NSW: Weed Society of New South Wales, pp. 442–444.

Steadman, K. J., G. P. Bignell, and A. J. Ellery. 2003. Field assessment of thermal after-ripening time for dormancy release prediction in *Lolium rigidum* seeds. *Weed Research* 43:458–465.

Storrie, A. and T. Cook. 2002. Glyphosate resistance in northern New South Wales—a growing concern. In H. Spafford-Jacob, J. Dodd, and J. H. Moore, eds. *Proceedings of the 13th Australian Weeds Conference.* Perth, WA: Plant Protection Society of Western Australia, Inc., pp. 601–603.

Wakelin, A. M., D. F. Lorraine-Colwill, and C. Preston. 2004. Glyphosate resistance in four different populations of *Lolium rigidum* is associated with reduced translocation of glyphosate to meristematic zones. *Weed Research* 44:453–459.

Wakelin, A. M. and C. Preston. 2006a. A target-site mutation is present in a glyphosate-resistant *Lolium rigidum* population. *Weed Research* 46:432–440.

Wakelin, A. M. and C. Preston. 2006b. Inheritance of glyphosate resistance in several populations of rigid ryegrass (*Lolium rigidum*) from Australia. *Weed Science* 54:212–219.

Wakelin, A. M. and C. Preston. 2006c. The cost of glyphosate resistance: is there a fitness penalty associated with glyphosate resistance in annual ryegrass? In C. Preston, J. H. Watts, and N. D. Crossman, eds. *Proceedings of the 15th Australian Weeds Conference.* Adelaide, SA: Weed Management Society of South Australia, Inc., pp. 515–518.

Wakelin, A. M. and C. Preston. 2008. Impact of management on glyphosate-resistant *Lolium rigidum* populations on farm. In R. D. van Klinken, V. A. Osten, F. D. Panetta, and J. C. Scanlon, eds. *Proceedings of the 16th Australian Weeds Conference.* Brisbane, Qld.: Queensland Weeds Society, pp. 80–82.

Walker, S., M. Widderick, A. Storrie, and V. Osten. 2004. Preventing glyphosate resistance in weeds of the northern grain region. In B. M. Sindel and S. B. Johnson, eds. *Proceedings of the 14th Australian Weeds Conference.* Sydney, NSW: Weed Society of New South Wales, pp. 428–431.

14

GLYPHOSATE RESISTANCE IN LATIN AMERICA

BERNAL E. VALVERDE

14.1 INTRODUCTION

The Latin American pesticide market continues growing at a higher rate than those of other regions, with herbicides prevailing as the most important component. Thus, Brazil became the largest pesticide market in the world in 2008, reaching a value of US$6900–7000 million. This represents a 30% increase over the previous year (ABIQUIM 2009; Farm Chemicals International 2009). In the same year, the Argentinean market reached US$1777 million, of which the herbicide segment comprised 71% (US$1265 million). Glyphosate alone has a market share of US$914 million (Kleffmann & Partner SRL 2009).

Glyphosate is used in several crops to control weeds before planting or "selectively" by direct spraying avoiding contact to the crop. This herbicide has become a major component of conservation agriculture (minimum and no-tillage systems) that are widespread in several countries (Christoffoleti et al. 2008b). Additionally, a large area of transgenic Roundup Ready® (RR; Monsanto Company, St. Louis, MO) crops are planted in Latin America, on which glyphosate is used selectively to control annual and perennial weeds. In Argentina, where soybean farmers are allowed to save seed and do not have to pay royalty fees (Green 2009), the entire soybean area is planted with RR soybeans (James 2008). An estimated 117 million L of the 182 million L of formulated glyphosate used in Argentina are applied in chemical fallows or as preplant treatments, particularly in no-till agriculture that covers 17 million

Glyphosate Resistance in Crops and Weeds: History, Development, and Management
Edited by Vijay K. Nandula
Copyright © 2010 John Wiley & Sons, Inc.

ha. About 65 million L are applied to glyphosate-resistant (GR) crops (J. Delucchi, pers. comm.).

Latin America contributes nine species to the growing list of confirmed GR weeds. Most of them have appeared in crops and situations that are considered prone to glyphosate resistance evolution: no-till agriculture, RR crops, and plantation crops with a long history of glyphosate use. In Brazil alone, an estimated more than 2 million ha are infested with herbicide-resistant (HR) weeds, resulting in crop losses amounting to nearly US$80 million per year (Vidal et al. 2006). About 100,000 ha of the 400,000 ha of wheat planted in Chile have HR grass weeds at varying infestation levels (N. Espinoza, pers. comm.).

Information about herbicide resistance in Latin America is scarce, except for countries where major cases have emerged and there is a strong research effort to document and understand the resistance problem. A previous review on herbicide resistance in grass weeds in Latin America has been recently published (Valverde 2007). Similar to that review, this chapter compiles updated information from all possible sources, aiming to provide a thorough account of GR weeds in the region. It includes information published in refereed scientific journals as well as regional and local journals, papers, and reports in Spanish, Portuguese, and English of more limited distribution, conference abstracts and extended summaries, and undergraduate and graduate theses. Information provided by local scientists and agronomists for the previous review that is relevant to this chapter, as well as new unpublished material, is incorporated as personal communications, including from my own studies and experience. The website on herbicide resistance maintained by Dr. Ian Heap (http://www.weedscience.org) is by far the best source of global updated information on this topic. Some of the cases discussed here have not yet been included in that database. Thus, there may be some incongruence in relation to the number and species evolving resistance in this part of the world. Presentation of resistance cases is organized by species.

14.2   BROAD-LEAVED WEED SPECIES RESISTANT TO GLYPHOSATE

14.2.1   *Conyza bonariensis* and *Conyza canadensis*

Conyza bonariensis, native to Latin America, is a common weed in Central America and several South American countries, including Argentina, Uruguay, Paraguay, and Brazil. In Brazil, where it has evolved resistance to glyphosate, it is more common in the south, southeast, and midwest regions. Seeds germinate in autumn and winter with canopy closure in summer, making it both a winter and summer weed. It is an important weed in wheat, soybeans, and maize (Vargas et al. 2007b). In Colombia, it is a common weed in perennial crops, pastures, coffee plantations, and nonagricultural areas. *Conyza canadensis*, native to North America but cosmopolitan (Holm et al. 1997), is also found in South America, including southern Brazil, where there are GR populations. In fact, both species are frequently found growing together in Brazil, where they thrive in no-till or minimum-till conditions (Lazaroto et al. 2008).

14.2.1.1 Resistance Evolution
Resistance to glyphosate in these species has evolved in Brazil in citrus orchards and GR soybeans; *C. bonariensis* additionally has evolved resistance in coffee plantations in Colombia.

Two putative resistant populations of each species collected in orange orchards in the Matão and Cajobi municipalities, Sao Paulo state, were compared with their respective susceptible controls. Biotypes from Matão (Araraquara region) originated from a farm with at least 12 years under citrus plantations; the current one being only 2 years old. During this period, glyphosate has been regularly applied to control weeds. The second site (Cajobi municipality, Catanduva region) also has a history of glyphosate use of at least 10 years (Moreira 2008). Plants were treated with increasing doses of glyphosate at the five-leaf growth stage and evaluated 28 days after treatment (DAT) for control and dry biomass. Resistance indices (RI) or ratios calculated based on visual estimates of control were always higher than those estimated from dry bioamass reduction. Although not discussed by the authors, upon inspection of the parameters for the regression equations from which the ED_{50} (effective herbicide dose to control 50% of plant growth) values were derived, it is apparent that some of the models did not fit properly and that untreated plants had different sizes at the time of evaluation. However, it is clear that putative resistant plants had a diminished response to glyphosate (Moreira et al. 2007b).

In a similar study, biotypes of both species were collected in Rio Grande do Sul (RGS) state from areas with a 20-year history of no-till agriculture, of which, almost a decade was under soybean monoculture and thus exposed to glyphosate for 35–40 times during this period. The response of the plants to glyphosate applied at the 8- to 10-leaf growth stage was compared to that of respective susceptible biotypes from Porto Alegre (Table 14.1). Based on visual estimates of control, the biotypes of both species with a history of exposure to glyphosate were confirmed resistant to this herbicide (Lamego and Vidal 2008). GR *Conyza* spp. have also been reported in soybeans in the state of Parana (Fornarolli et al. 2008).

Seed of putative *C. bonariensis* resistant biotypes was collected from three coffee farms in Palestina (biotypes Las Américas and El Rodeo) and Chinchiná (biotype La Suiza), Department of Caldas, Colombia. These farms had a long history (>10 years) of glyphosate use at least four times per year, where it was no longer being controlled satisfactorily with the herbicide. A susceptible reference population was collected at a certified organic coffee farm in Los Santos (biotype El Roble), Department of Santander, where glyphosate had not been used for at least 10 years. The response of the populations to increasing doses

TABLE 14.1. Resistance Indices (RI) to Glyphosate of *Conyza bonariensis* and *Conyza canadensis* Populations from South America

Biotype	Collection Site	RI Based on Visual Assessment of Control	RI Based on Fresh or Dry Biomass Determination	Comments	Reference
			C. bonariensis		
Matão	SP	14.8	5.0	Control biotype from Piracicaba, SP	Moreira et al. (2007b)
Cajobi	SP	10.4	1.5	Control biotype from Piracicaba, SP	Moreira et al. (2007b)
Santa Rosa	RGS	2.3	—	Control biotype from Porto Alegre, RGS	Lamego and Vidal (2008)
Cruz Alta	RGS	ND	ND	Some plants survived glyphosate at 5769 g ha^{-1}	Vargas et al. (2007b)
Unnamed	RGS	—	6.2	Identified as *Conyza* sp.	Magro et al. (2008)
			C. canadensis		
Matão	SP	10.8	6.2	Control biotype from Piracicaba, SP	Moreira et al. (2007b)
Cajobi	SP	7.1	6.6	Control biotype from Piracicaba, SP	Moreira et al. (2007b)
Victor Graeff	RGS	2.4	—	Control biotype from Porto Alegre, RGS	Lamego and Vidal (2008)

ND, not determined; RGS, Rio Grande do Sul; SP, Sao Paulo.

of glyphosate was tested both in Petri dish assays and whole plant (20-cm tall) bioassays. Because the selected dose ranges were the same for the alleged GR and glyphosate-susceptible (GS) biotypes, and the lowest dose was already quite effective, it was not possible to obtain clear dose–response curves and calculate a reliable RI. The whole plant bioassays, however, clearly showed a distinct response to glyphosate between the three putative GR biotypes and El Roble (GS). Glyphosate at the lowest dose applied ($360\,g\,a.e.\,ha^{-1}$) reduced fresh weight of the GR biotypes by 37% and that of the GS biotype by 93%. To obtain the same growth reduction in the GR biotypes, it was necessary to apply $2160\,g\,a.e.\,ha^{-1}$ (Menza Franco and Salazar Gutiérrez 2007).

14.2.1.2 Mechanism of Resistance Limited research has been conducted on the possible mechanism of resistance. Ferreira et al. (2008a) compared the glyphosate translocation patterns in GR and GS plants from RGS. Ten hours after application (HAA) of glyphosate to the middle adaxial section of the third node leaf of 45-day-old plants, radiolabeled glyphosate accumulated preferentially at the apex and treated section of the leaf in the resistant biotype. Conversely, the susceptible biotype accumulated most of the herbicide at the base and midsection of the leaf. About 90% of the applied glyphosate remained in the treated leaf 72 HAA in the GR biotype, whereas only 69% remained in the GS biotype. Differential partitioning of glyphosate was also observed. More glyphosate translocated to other leaves, stem, and roots in the GS biotype than in the GR biotype. Comparison of a GR and a GS biotype revealed that they did not differ in the content of epicuticular waxes (Paula et al. 2008a).

GR biotypes of both *Conyza* species from Matão, Sao Paulo state, had decreased productivity (biomass) compared with GS biotypes when grown as individuals under greenhouse conditions (Moreira 2008). The implication of this difference in relation to fitness has not been determined experimentally.

14.2.1.3 Control Practices Weed control in soybeans begins with the desiccation of plants that have been growing during the fallow period. In conventional soybeans, it was always recommended to plant in a clean but undisturbed field. This should also be practiced in GR soybeans instead of relying on glyphosate to control all weeds after no-till planting (Gazziero et al. 2008). Late application of glyphosate to weeds already at advanced growth stages carries two disadvantages: first, competitive damage to the crop has probably occurred by the time of treatment, and second, a late application could be equivalent to subdosing that has important implications for the selection of resistant biotypes.

As is the case with other resistant weeds, other chemicals take the place vacated by those that have become ineffective because of resistance. In most instances, however, since glyphosate continues to be effective on other weeds and profitable, farmers continue using it either alone or in mixture with other herbicides to deal with the resistant species. Alternative herbicide mixtures containing glyphosate and a partner herbicide with a different mode of action [2,4-dichlorophenoxyacetic acid (2,4-D), metsulfuron-methyl, bromacil, atrazine, diuron, or metribuzin] controlled resistant biotypes (Moreira 2008; Moreira et al. 2007b). Three-way mixtures of glyphosate plus 2,4-D and diclosulam or flumetsulam, and glyphosate plus bromacil and diuron applied at the 10-leaf growth stage were also effective (Adegas et al. 2008; Melo et al. 2008; Paula et al. 2008b). The partner herbicides, 2,4-D and metsulfuron-methyl, have also effectively controlled GR *C. bonariensis* under controlled greenhouse conditions. When using metsulfuron-methyl, it is important to allow at least 2 months before planting soybeans (Gazziero et al. 2008). Other herbicides that have proven to be effective in field or greenhouse tests are paraquat, chlorimuron-ethyl, glufosinate, atrazine, mixtures of paraquat with diuron, monosodium methylarsonate (MSMA), or flumioxazin, glufosinate with MSMA or metsulfuron, and the three-way mixture of glufosinate, bromacil, and diuron (Fornarolli et al. 2008; Melo et al. 2008; Moreira et al. 2007a; Rizzardi et al. 2008; Vargas et al. 2007a, 2007b). Some herbicide mixtures that include glyphosate are not effective, including those in which the partner herbicide is carfentrazone, flumioxazin, cloransulam, or chlorimuron (Adegas et al. 2008). The cost of using alternative herbicides in sequential application to glyphosate in citrus production in Brazil was estimated to be BRL290 (equivalent to US$160) for three applications per year (Christoffoleti et al. 2008a). Coffee farmers in the affected area in Colombia are using both hand weeding and increased glyphosate doses to control the resistant weed (Menza Franco and Salazar Gutiérrez 2006a). Similar to the reports in Brazil, 2,4-D, glufosinate, and glyphosate in mixture with 2,4-D provided adequate control of GR *C. bonariensis* in field experiments at two commercial farms. Glyphosate alone only provided 15% control of the GR biotypes (Menza Franco and Salazar Gutiérrez 2007).

Conyza spp. are characteristic of soybean fallows, and their control in GR-soybeans is improved by integrating nonchemical control tactics such as planting oats and other species as cover crops, supplemented with chemical control at early growth stages (Gazziero et al. 2008; Rizzardi et al. 2007). When rotation is possible, there are more chemical options to control these weeds.

14.2.2 *Parthenium hysterophorus* in Colombia

In 2002, there were increasing claims of lack of control of *Parthenium hysterophorus* with glyphosate in the fruit production areas of the Valle del Cauca Department in Colombia (Gómez and Fuentes 2008). Increasing the dose of the herbicide in an attempt to control the weed probably exacerbated the emerging problem.

At one of the major farms in the region (Agropecuaria El Nilo), the entire cultivated 730 ha are infested with GR *P. hysterophorus*. Reflecting the conditions in the area, resistance was noticed at this farm as early as 2002 after intensive use of the herbicide, which is still being applied about 10 times per year in fruit-tree orchards. The extent of the infestation in Valle del Cauca is unknown, but in the ASORUT irrigation district that covers 10,500 ha in the north of the department, only the sugarcane cropping area (about 3500 ha) is not affected by GR *P. hysterophorus*. The irrigation district comprised three municipalities from which it derives its name: Roldanillo, La Union, and Toro. GR biotypes, apparently, are also spreading in the central and southern parts of Valle del Cauca. Resistance evolved simultaneously at several locations in the area. Where GR *P. hysterophorus* infestations are high at Agropecuaria El Nilo, control is now dependent on glufosinate that is also used for spot treatments. Mechanical control with line trimmers is also used (S. A. Silva Gonzalez, pers. comm.).

Resistance was confirmed by comparing, in Petri dish and whole plant bioassays, a putative GR (designated as Rioja) and a GS biotype (designated as Isla) both collected in the northern part of the department from farms growing citrus, guava, papaya, and other fruits. Glyphosate resistance had evolved after 15 years or more of using the herbicide (Rosario 2005; Rosario and Fuentes 2005). Rioja was four- to sixfold more resistant to glyphosate than Isla based on ED_{50} values and seedlings treated at the three- and six-leaf growth stage responded similarly to the herbicide (Alonso 2005). The resistant biotype was able to withstand up to 3.6 kg ha^{-1}. ^{14}C-Glyphosate uptake and total translocation was similar in both biotypes, but the herbicide had a tendency to rapidly accumulate more at the tip of the treated leaf in the GR plants (Rosario 2005). No other weeds have been reported as GR in this area, but there is concern with *Cyperus rotundus* that is showing a diminished response to glyphosate. However, no comparative bioassays have been carried out with this species (S. A. Silva Gonzalez, pers. comm.). *P. hysterophorus* has also evolved resistance to acetolactate synthase (ALS)-inhibiting herbicides in Brazil (Gazziero et al. 2006).

14.2.3   *Euphorbia heterophylla* (Multiple Resistance)

Euphorbia heterophylla is an important annual weed of soybeans in Brazil. An estimated 20 million ha are treated with herbicides to control it (Vidal et al. 2007). *E. heterophylla* biotypes resistant to ALS herbicides have been reported in soybeans in several states, including Parana, RGS, Sao Paulo, Santa Catarina, and Mato Grosso do Sul (Gelmini et al. 2005; Trezzi et al. 2005). From a screening test for ALS and Protox resistance, two biotypes from the southwest region of Parana designated as biotype 4 and biotype 23 were selected for detailed characterization of their response to herbicides of the two modes of action. Biotypes 4 and 23 were collected at Vitorino and Pato Branco municipalities, respectively, from farms with confirmed ALS resistance for more than 4 years and with use of Protox herbicides alone or in mixture with ALS herbicides as an alternative control method (Trezzi et al. 2005, 2006). Both were confirmed coresistant to the two modes of action.

Since the adoption of GR soybean varieties and given the generalized failure of ALS herbicides, farmers are relying on glyphosate to control

E. heterophylla. Thus, in RGS, it is controlled exclusively with glyphosate that has resulted in glyphosate resistance evolution (Vidal et al. 2007). Resistant plants were selected in no-till soybean monoculture production areas where glyphosate had been used at least twice a year for 15 years plus additional one or two applications per year for in-crop weed control in GR soybeans during the last 5 years. Total glyphosate use in a single field ranged from 35 to 40 times, even up to 50 times, during the entire period. In GR soybeans, glyphosate is applied at 360–540 $g\,ha^{-1}$, although the recommended label dose for controlling *E. heterophylla* is 720–1080 $g\,ha^{-1}$ (Vidal et al. 2007). GR biotypes have also been reported in Parana state (Rizzardi et al. 2007).

The glyphosate resistance level in the limited number of biotypes tested so far is low. Biotypes BE and GC, collected in areas with a long history of glyphosate use as a preplant burndown treatment and "selective" herbicide in GR soybeans in RGS, were confirmed GR. Biotype BE had an RI of 3.1 when compared with the susceptible biotype AE from the same state. At the maximum dose of 450 $g\,ha^{-1}$ tested in greenhouse bioassays, glyphosate injured the GR biotype GC 42%, compared with 70% damage to the susceptible SP biotype from Sao Paulo (Vidal et al. 2007). These biotypes are also resistant to ALS herbicides. Biotypes resistant to glyphosate and either ALS or Protox herbicides are found predominantly in the southwest region of Paraná and in the northwest region of Santa Catarina states. So far, no populations resistant to the three modes of action have been found, and those exhibiting coresistance to two modes of action do not seem to be spreading, probably because farmers are following the advice to use alternative herbicide mixtures and residual herbicides (R. Vidal, pers. comm.).

The disparity between the recommended and actual glyphosate dose in the field has been used by herbicide companies as a criterion to deny resistance (Vidal et al. 2007). Thus, it was recently reported that inadequate control of *E. heterophylla* with postemergence applications in soybeans at several locations was not due to resistance since glyphosate at 1200 $g\,ha^{-1}$ (recommended dose) and above satisfactorily controlled the weed. Instead farmers are blamed for using the herbicide at lower-than-recommended doses and applying it post emergence in fields where preplanting control was absent, thus trying to control plants that are already at late growth stages (Marochi et al. 2008).

14.3   GRASS WEED SPECIES RESISTANT TO GLYPHOSATE

Several grass weeds have evolved resistance to herbicides in Latin America, including glyphosate (Valverde 2007).

14.3.1   *Sorghum halepense* in Argentina

Without doubt glyphosate-resistance in *Sorghum halepense* (SORHA) in Argentina is the most important case of glyphosate resistance in Latin America and perhaps one of the most relevant in the world. The case of GR SORHA in Argentina was first informally disclosed at the Iberoamerican Workshop on Herbicide Resistance and Transgenic Crops held in Uruguay in December 2005. Previously, farmers had complained about the lack of efficacy of glyphosate to manufacturers and distributors without much success. In the following year, farmers' awareness about the problem increased and they became more belligerent. Lack of proper attention forced them to go the local press to gain attention. More details about the disclosure of this case of resistance can be found elsewhere (Valverde and Gressel 2006). However, the problem has now drawn so much attention that even politicians have become involved. Cordoba province Congressman Alberto Cantero introduced a bill in September 2007 aimed at eradicating the GR weed, stating that the spread of GR SORHA could increase agricultural production costs by 500–3000 million Argentine pesos (US$160–950 million) per year. Combating the strain, he added, will require the use of 25 million L of herbicides other that glyphosate each year (Romig 2007). The bill has been modified to declare the prevention, control, and management of HR weeds of public interest (Sembrando Satelital 2008) but still awaits approval.

According to farmer accounts, poor performance of glyphosate, used to control SORHA before beans (*Phaseolus vulgaris*) planting, was first noticed as early as in the late 1990s. If this was the beginning of the problem, it remained unnoticed as the weed was effectively controlled in the crop with postemergence selective graminicides. In 2003, however, lack of control became evident to farmers in soybeans. When the author, accompanied by Dr. Jonathan Gressel, first visited the affected area [north of Argentina (NOA)] as part of a consultancy agreement with the Argentinean National Agri-food Health and Quality Service (SENASA by its name in Spanish) in 2006, the typical situation was the presence of clumps of uncontrolled SORHA in a limited number of fields and farms (Valverde and Gressel 2006). The NOA region, comprising 16.7 million ha of agricultural land, is characterized by its large farms. About 3% of the farm units represent 64% of this agricultural land, almost equally divided between farms of 2500–10,000 ha and above 10,000 ha (SAGPyA 2009). Thus, at "Establecimiento Los Angeles," a large farm in Departamento General José de San Martín, Salta province, along Highway 34 near Tartagal, which was one of the first three sites where resistance was initially documented in 2003, an estimated 800 ha of the total 5000 ha were already infested and survived a spot application with a 2:10 v/v dilution of a formulated glyphosate product. Nearby, at Coronel Cornejo, severely infested fields were already observed.

The soybean production system in Argentina was prone to the evolution of resistant weeds. Only GR soybean varieties are planted under a conservation tillage system in which monoculture predominates over extensive areas. In addition to repeated in-crop application of glyphosate, this herbicide has also been used systematically in fallow periods, particularly after cost reductions, that made it a profitable treatment.

Herbicide regimes that imposed the selection pressure for resistance typically included three seasons planted with beans subjected to a preplant application of glyphosate under a no- or minimum-tillage scheme and selective, in-crop application of systemic graminicides; three seasons with no-till cotton and 6 years with no-till RR soybeans. Glyphosate application in soybeans averaged three sprays of $2.5 \, \text{L ha}^{-1}$ ($900 \, \text{g ha}^{-1}$) each for an estimated total glyphosate use of $27 \, \text{kg ha}^{-1}$ during the 12-year period.

Once the initial resistance problem was realized, the farmer's association [Asociación de Productores de Granos del Norte (PROGRANO)] conducted a survey among their members to determine the extent of the GR SORHA infestation in the NOA. Thus, in 2006, the dispersion area was estimated as 91,700 ha and the actual area infested as 17,000 ha. At this time, the dispersion occurred for 100 km along Highway 34 in the north to south direction, almost covering the entire soybean area of the San Martin Department (81,300 ha) and part of the Oran Department (10,400 ha) (J. C. Rodriguez, pers. comm.). Concomitantly, a single resistant population was documented in 2005 at a soybean farm in the Cruz Alta Department, Tucuman province (I. Olea, pers. comm.). There was no apparent relationship between this case and those of Salta, suggesting that resistance evolved locally (Fig 14.1). Indeed, the closeness among 37 GR and GS individuals from the provinces of Salta, Tucumán, Córdoba, Santiago del Estero, Santa Fe, and Buenos Aires was evaluated with 10 microsatellites representing 20 loci of the SORHA tetraploid genome. According to the obtained dendrogram, plants from Salta and Tucumán formed separate groups, indicating that GR plants found in Tucumán evolved separately from those found in Salta. No separate groups were formed by GR and GS individuals (Fernández et al. 2009).

Spreading of GR SORHA continued, and by 2007 two new locations in the Anta Department (Las Lajitas and Piquete Cabado) were added to the list of infested sites. Although this did not represent a substantial increase in the area affected, it reflected a significant increase in the distance of the spread. A new focus was found at the eastern part of Tucumán (Puesto del Medio, Burruyacu Department) close to the border with the Santiago del Estero province (Olea 2007). Some 400 km away from this location, two separate sites with GR SORHA were found at Bandera. Three additional provinces contributed locations where resistant biotypes were present. In Córdoba, where the weed is widely distributed (Leguizamón and Canullo 2008), GR SORHA was found at two sites at Piquillín and one site at Monte Buey; Santa Fe also had two sites at Las Rosas and El Trébol, and Corrientes had one site close to its capital (Fig 14.1).

Currently, the distribution of GR SORHA also includes sites at Charata and surroundings [Chaco province in the northeast of Argentina (NEA)], and at Idiazábal in Córdoba (Olea 2007). In Tucuman, the infestation has increased to eight sites (I. Olea, pers. comm.). Most recently, GR SORHA was confirmed in Colón, Buenos Aires province (D. Tuesca, pers. comm.), bringing to eight

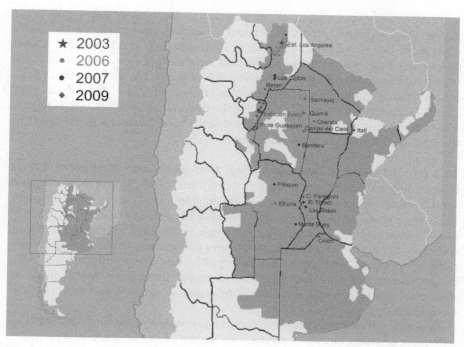

Figure 14.1. Distribution of sites where populations of *Sorghum halepense* resistant to glyphosate have been found in soybean production areas (in light gray or green) in Argentina. Prepared based on information provided by local researchers and farmers (J. C. Rodríguez, D. Tuesca, and I. Olea) and published information (Binimelis et al. 2009; Valverde and Gressel 2006).

the number of provinces where resistant populations have been found. These eight provinces represent almost the entire soybean area in Argentina.

With the exception of El Trébol, where the farmer tilled the field after detecting an initial infestation with GR SORHA, spreading it throughout the entire field, Coronel Cornejo and its surroundings is the only area where totally infested fields are found. In the rest of the country, the infestation is still at an early stage (J. C. Rodriguez, pers. comm.) but spreading.

It is likely that the situation is worse than described here according to reliable sources (Olea and Sabaté 2008; E. Hopp, pers. comm.). Based on group and individual interviews, as well as on participatory observation in the provinces of Salta, Tucumán, Santiago del Estero, Entre Rios, and Buenos Aires, Binimelis et al. (2009) concluded that farmers are still reluctant to report resistance cases in their farms because of the uncertainty of the consequences this may bring about.

There is only speculation about how resistance has spread within fields and at long distances across the country. Farmers are convinced that farm equipment has played a major role in disseminating vegetative and sexual propagules both within fields and across neighboring farms, and especially, over long distances. About 90% of the fields in the NOA are harvested under contract using combines that travel around the country (J. C. Rodriguez, pers. comm.). This represents a major risk for the dissemination of the GR SORHA. The contribution of the so-called bolsa blanca seed (farmer's saved seed) to the spread of resistance is also unknown.

14.3.1.1 Bioassay Corroboration of Resistance and Resistance Levels

Responding to complaints from farmers in late 2003, Monsanto conducted some testing both in Argentina and in the United States in 2004. Initially, there were some conflicting results, probably because the levels of resistance of the tested populations were not very high, of the limited number of plants tested, and of poor experimental procedures (differences in plant size between putative resistant and reference susceptible biotypes, inappropriate dose ranges, and lack of statistical analyses). However, there were clear indications that some biotypes had a diminished response to glyphosate (Valverde and Gressel 2006).

Resistance by proper comparison of the glyphosate dose–response curve of a putative resistant biotype with that of a known susceptible was first provided by De la Vega et al. (2006), although this report has been ignored by others (Vila-Aiub et al. 2007, 2008). A biotype from the General José de San Martín Department (Salta province), collected as rhizomes, had an RI of 2.8 when compared with a susceptible one from Tucumán Province based on the fresh biomass reduction determined 21 days after application of glyphosate.

Vila-Aiub et al. (2007) collected SORHA plant material (both rhizomes and seed) at three locations near Tartagal from fields with a history of glyphosate use and recent failure of the herbicide to control it and at a roadside where the herbicide had not been used. Additionally, two other susceptible populations from distant locations were included in the study. Plants were treated with increased doses of glyphosate (0, 500, 1000, 2000, and 4000 g ha^{-1}) when shoots arising from rhizomes reached the four- to six-leaf stage and 40-cm height or when seedlings grown from seed reached the four-leaf stage and the same height. Response to glyphosate was assessed as plant survival at 30 DAT. Additionally, plants obtained from seed collected in the greenhouse from plants derived from rhizomes that had survived a repeated application of 3128 g ha^{-1} each were sprayed with 680 g ha^{-1} at two growth stages (single 100- to 120-cm-high shoots with incipient rhizomes and tillers, and 30- to 50-cm-tall plants with no rhizome or tillers formed). Plant survival and aboveground biomass were determined at 28 DAT.

LD$_{50}$ (lethal dose of herbicide required to cause 50% mortality) values of 2182 and 1263 g ha^{-1} were estimated for two populations originating from rhizomes collected in fields with a 6-year history of glyphosate use. For the corresponding two reference (susceptible) populations, an LD_{50} could not be calculated because plants were killed by glyphosate at the lowest dose included in the experiment. Older, taller putative resistant plants originating from seeds (either collected directly in the field or from greenhouse-grown plants) with a few tillers and rhizome required more glyphosate to be controlled than small plants with no tillers or rhizomes and exhibited 3.5- to 10.5-fold resistance levels compared with the known susceptible types.

Dose of glyphosate required to inhibit SORHA growth increased as plants became more developed, from the seedling stage (plants about 8-cm tall) to young plants (about 45 cm in height without tillers or rhizomes) and to adult plants (75 cm in height with incipient rhizomes and tillers at early flowering stage). The resistance differential increased with plant age, from 1.6–2.1 at the seedling stage to 2.0–4.0 for young plants and about 5.5 for adult plants (Vila-Aiub et al. 2008). These RIs are similar to those found when comparing a putative GR SORHA biotype from Salta province with susceptible biotypes from Costa Rica, Nicaragua, and Venezuela (B. E. Valverde, unpublished data).

Glyphosate resistance also has been confirmed in biotypes from El Trébol (RI = 11 compared with a susceptible biotype from Zavalla), both from Santa Fe province. In field trials at two locations in this province (Las Rosas and Etruria), glyphosate at 720, 1440, and 2880 g ha^{-1} provided only 6–15%, 20–35%, and 42–60% control of SORHA, respectively. Both populations, however, were highly susceptible to haloxyfop and clethodim (Papa et al. 2008).

14.3.1.2   *Mechanism of Resistance*   Despite recommendations to initiate a strong research program and effort to implement coordinated activities to study the problem and design appropriate management strategies (Valverde and Gressel 2006), it is still unclear how resistance evolved and what type of mechanism(s) confer resistance. Resistance does not seem to have been selected by repeated applications of glyphosate at low doses, yet the levels of resistance are not very high. Both symptom development, resprouting ability of treated plants, and low resistance levels suggest that the resistance mechanism(s) probably involves limited translocation of the herbicide within the resistant plants. Some preliminary studies indeed suggest that differential glyphosate absorption and translocation may be responsible for glyphosate resistance in Argentinean GR SORHA. Plants of a resistant biotype absorbed half of the glyphosate through the leaf compared with susceptible plants. Resistant plants also translocated less glyphosate to the culms and roots (Vila-Aiub et al. 2008). A biotype from Salta accumulated more shikimate than a susceptible biotype after treatment with glyphosate. However, from the experiment it could not be determined if this was due to decreased translocation of the herbicide to the active site, an altered 5-enolpyruvylshikimate-3-phosphate synthase (EPSPS), or increased EPSPS activity in the resistant biotype (De la Vega et al. 2007). Sequencing of the EPSPS in a limited number of individuals revealed that no mutations at positions known to confer glyphosate resistance have occurred (Fernández et al. 2009; E. Hopp, pers. comm.).

14.3.1.3   Coping with Resistance   Initial response of farmers realizing that SORHA was not controlled by glyphosate was to re-treat the uncontrolled clumps with the same herbicide at high doses and later with graminicides using knapsack sprayers. They also asked glyphosate distributors and agrochemical-company technical representatives for guidance but found no support. Once the problem was publicly disclosed, SENASA authorities called for international advice, and an extensive report about the situation was prepared (Valverde and Gressel 2006) and discussed as part of a herbicide workshop held in 2006. In the workshop, national priorities and coordinated policies for ascertaining the mechanism(s) of resistance and the mode(s) of spread, determining alternative procedures, and mixtures to delay resistance were outlined. Also, a unique effort brought together SENASA, the Instituto Nacional de Tecnología Agropecuaria (INTA, the agricultural research and extension service), and the Comisión Nacional Asesora de Biotecnología Agropecuaria (CONABIA, the National Advisory Committee on Agricultural Biosafety that regulates transgenic plants) with farm groups and industry for immediate action to contain and prevent the spread of the resistant weed (Valverde et al. 2007). In August 2007, the Comisión Nacional Asesora de Plagas Resistentes (CONAPRE, the National Advisory Board on Resistant Pests) was officially created to coordinate all parties involved in the interest of sharing information, advancing research and training activities, discussing the establishment of national policies, and helping farmers prevent and control resistant populations.

As part of the consultancy report and workshop, several timely recommendations were presented to the Argentinean authorities, farmer organizations, and researchers to help them cope with GR SORHA. Some of them were rapidly adopted, but others still await implementation. We strongly encouraged installing a low-cost/highly effective monitoring system based on the excellent eyes of the growers, reporting to marketers, with automated reporting to the phytosanitary authorities. Under this system, companies marketing glyphosate would be required to have a SENASA-approved rapid-response strategy and trained rapid response teams to deal with resistant outbreaks. SENASA acted promptly and established a monitoring system based on their experience with a program called Sistema Nacional de Vigilancia y Monitoreo de Plagas (SINAVIMO, http://www.sinavimo.gov.ar/) to monitor soybean rust and other plant protection problems. The Web-based system (http://www.sinavimo.gov.ar/?q=node/777) allows farmers and farm advisors to report new cases of resistance and to provide details about the current status of the problem. Unfortunately, updated summaries are not available online and no information is provided about how authorities follow up each report. This program was to be supplemented by an aerial or satellite monitoring system to follow the extent and spread of resistance, and as a guidance tool to affected farmers. To date, this has not been accomplished because of lack of resources.

Regulatory action was also suggested. We proposed that herbicide labels should contain information on the need for early discovery of resistance, with an explanation of how resistance may appear to the farmer, and whom to contact if resistance is suspected. SENASA was asked to consider requiring that glyphosate be sold as a premix or as a combi package with mixing spout to prevent separation as part of a national strategy to deal with this resistant weed. Possible action to establish official guidelines for cleaning farm equipment, particularly combines, with verification points along main highways used to transport them across the soybean and cereal growing regions, was also discussed as well as monitoring the contribution of soybean saved seed (locally called "bolsa blanca"—white bag) to dispersal at long distances.

Agronomically, we recommended devising and promoting rotation schemes to avoid or delay resistance. These should include rotating soybeans with transgenics bearing other herbicide resistances, rotations with other herbicides applied preplant, rotating RR soybeans with a conventional variety every 3–4 years so that alternative herbicides could delay the evolution of GR populations, and requiring that RR-soybeans be rotated only with non-RR crops.

The need to integrate other control tactics was also stressed; particularly in relation to preventing seed set in putative resistant clumps and the spreading of resistant seed locally and at long distances. Herbicides with alternative modes of action could play an important role in preventing plants from setting seed. Affected farmers are using acetyl coenzyme A carboxylase (ACCase) grass killers and ALS herbicides to control resistant clumps for preplant and in-crop postemergence control of GR SORHA. These two herbicide families, however, are resistance prone, and thus should be carefully managed to avoid selection of new resistant and multiple-resistant biotypes. The less resistance-prone arsenical herbicide, MSMA, is also being used alone and in mixture with glyphosate to control this weed. There could be opportunities to use other herbicides, including the rare Protox herbicides, that kill this weed. Nonselective herbicides could also be used with rope wick applicators set above the soybean canopy or as spot treatments to eliminate resistant clumps, provided that their use is authorized by the regulatory authorities. Farmers have previous experience with this type of equipment, and some still keep them in their shops. There may be also opportunity for precision monitoring with automated digital weed detection and Global positioning system (GPS)-controlled patch spraying that have been successfully tested elsewhere.

To be successful, a resistance management program requires the support of relevant research and strong dissemination of results and increased farmer awareness. Among relevant topics to address in a coordinated national research program, we suggested ascertaining by DNA fingerprinting of GR and GS biotypes throughout Argentina whether there have been multiple evolutionary events in the appearance of GR SORHA. In this respect, an effort was initiated at INTA-Castelar and preliminary results indicate that GR SORHA collected at several locations is very diverse without an apparent founder effect with seed and not rhizomes being most likely responsible for its dissemination. Genetic affinity is stronger among plants collected from the same geographical region, and those from Salta and Tucuman differ in their genetic background, suggesting that resistance foci in Tucuman arose independently from those in Salta (S. Passalacqua, pers. comm.). It is also important to determine if resistance is conferred by a single gene or by additive gene effects, as well as the possible modes of resistance. Incipient cross-resistances of GR SORHA to other graminicides or other alternative herbicides should be detected as early as possible to prevent their evolution at multiple locations and their dissemination throughout the fields and at long distances. Because of its modes of reproduction, we also suggested instating biosafety quarantine restrictions for the resistant biotype(s) to ensure that research with the biotype(s) would not be a cause of spread. To help researchers obtain and better characterize biological material, we also encouraged the establishment of a national seed and clonal S. halepense repository and database.

To complement the proposed activities, it was defined as imperative to increase awareness and understanding of the problem. Groups such as SENASA, CONAPRE, farmer's organizations, and the Obispo Colombres Experiment Station in Tucuman have been very active in organizing training workshops and preparing written dissemination material with valuable information about prevention and management of GR SORHA (http://www.eeaoc. org.ar/pastoruso.htm#SARG).

In practice, farmers are implementing some preventative measures to delay the evolution and dissemination of GR SORHA, but rely mostly on a limited number of alternative herbicides alone or in combination with glyphosate to control resistant clumps. Before planting soybeans, resistant clumps are treated with herbicides at the end of the fallow period, aiming to control them with systemic herbicides or at least to prevent production of additional seed and deplete plant resources. In the NOA, SORHA continues growing vegetatively in autumn and winter and completes more generations that in other parts of the country. The most widely used treatments include systemic graminicides (ACCase inhibitors) and ALS-inhibiting imidazolinones (imazapyr and imazapic) and the sulfonylurea nicosulfuron, which can prevent sprouting. An opportunity to use nicosulfuron selectively would be the introduction of sulfonylurea-tolerant soybean varieties. Nicosulfuron controlled GR SORHA at doses of 30–42 g ai ha^{-1} under experimental field conditions in Argentina, but was slightly phytotoxic to the sulfonylurea-tolerant variety NA8087RG, causing a yield decrease of about 15% (Sabaté et al. 2007). In heavily infested fields, the local advisors recommend adding MSMA to the conventional mixture of glyphosate and 2,4-D for the chemical fallow to burn down the foliage and facilitate planting soybeans (Olea et al. 2008). In the NOA, rotations are very limited; some fields are planted with safflower or, less frequently to wheat as a cover crop. In other regions, possible rotation crops include safflower, maize, and sorghum. Non-tillage agriculture helps contain resistant clumps because the rhizomes are not divided by soil-preparing equipment and seed is not incorporated into the soil profile, leaving only that lying on the surface for the establishment of new plants. In desperation, however, some farmers in Salta are plowing severely infested fields to desiccate rhizomes with its long rain-free period.

Once the soybeans crop is planted, acetochlor and S-metolachlor can be used for preemergence control of GR SORHA seedlings. Glyphosate continues to be used as part of the GR soybean production system since it controls the remnant susceptible SORHA and other weeds. In postemergence soybean growth stages (V3–V5), locally recommended herbicides include mixtures of glyphosate and graminicides ("fops" and "dims," mostly haloxyfop) and glyphosate plus imazathapyr, for which, there is a registered formulated mixture, but taking into consideration that some varieties may be susceptible to the last treatment. Late-emerging plants will require an additional application of graminicide and in the NOA, perhaps a third one according to current recommendations. These regimes will inevitably contribute to the selection of ACCase-resistant biotypes. Spot treatment of resistant clumps with knapsack sprayers is profitable even in such large operations where the infestation with GR SORHA is not extreme. The estimated per hectare cost of this operation is only US$3.0 (J. C. Rodriguez, pers. comm.). Herbicides used for this purpose include selective graminicides that allow at least partial control of large plants and prevent seed filling when applied before panicle emergence and the non-selective ALS inhibitors. Farmers understand that, where there are resistant clumps, no soybean will be harvested because of the intense competition imposed by the GR SORHA, so it is acceptable to apply such extreme measures to eliminate those clumps. Finally, farmers are putting more attention to using clean combines to avoid infesting clean fields and spreading seed from contained clumps to the rest of the field. At harvest, infested fields are left for last and where dense clumps are present, the crop is not harvested with the combine to prevent dissemination. Those clumps are later treated with herbicides.

Substituting glyphosate for a systemic grass killer such as haloxyfop plus crop oil, applied twice, once in the spring before planting and later in postemergence during the cropping cycle, increases the herbicide cost (without cost of application) by US$31.20 per hectare per year (Papa et al. 2008). Under heavy infestations, the total cost of applying herbicides to control GR SORHA ranges from US$45–65 per hectare (Olea et al. 2008). In Argentina, over 50% of the cultivated land is leased predominantly under annual contracts that impose a high pressure to quickly obtain maximum revenue. Yield losses and incremental control costs associated with resistance evolution have induced changes in the lease regime (both in the price and length of the contract) as a consequence of the depreciation of the value of affected lands. Glyphosate allowed farmers to manage more land and increase overall productivity and profitability. This advantage could be lost, and increased control costs leave middle-sized farms in a precarious situation (Binimelis et al. 2009).

14.3.2   *Lolium* spp.

Lolium multiflorum was the first species to become resistant to glyphosate in Latin America (Kogan and Pérez 2002; Pérez and Kogan 2001, 2003). Initial GR populations were found in grape orchards that were sprayed an average of three times per year with glyphosate ($1.08-1.44\,kg\,ha^{-1}$) in combination with 2-methyl-4-chlorophenoxyacetic acid (MCPA) ($1.5\,kg\,a.e.\,ha^{-1}$) for 8–10 years. Populations from San Bernardo and Olivar located in Region VI (central area) of Chile were two- and fourfold, respectively, more resistant to glyphosate (Pérez and Kogan 2003) than the corresponding susceptible population. Another resistant population (San Fernando) was later found in the same region (Pérez-Jones et al. 2007; M. Kogan pers. comm.). A few years later, GR *L. multiflorum* also appeared in Region IX (southern Chile). A GR biotype designated as Vilcún was documented in chemical fallows after 12 years of glyphosate use prior to planting wheat or oats (Espinoza et al. 2005).

    L. multiflorum is also an important weed in Brazil, where it also became the first GR species. This grass is well adapted to the southern region of Brazil (RGS, Santa Catarina, and Paraná), where it is used in no-till agriculture as a winter cover crop, as a cover crop or mulch in orchards, and as a forage crop. In the first two situations, plants are desiccated with herbicides, mostly glyphosate (Galli et al. 2005). *L. multiflorum* is weedy in several crops including soybean, wheat, and corn. It is also the most common grass weed in apple orchards where three to four glyphosate applications (at $720-1080\,g\,ha^{-1}$) are made yearly (Vargas et al. 2004, 2005). GR populations have been found at several locations in RGS in annual crops and orchards (Roman et al. 2004); some plants can withstand doses of glyphosate as high as $11.5\,kg\,ha^{-1}$, and when initially damaged by the herbicide at high doses, they later are able to recover and develop new tillers (Vargas et al. 2004, 2006). The first case was identified at Vacaria; later, resistant biotypes were also found at several municipalities, including Lagoa Vermelha, Capão Bonito, Sananduva, Ciríaco, Tapejara, Bento Gonçalves, Caxias, Flores da Cunha, Marau, Passo Fundo, Carazinho, Ernestina, Tio Hugo, Tapera, Espumoso, Ibirubá, and Tupanciretã. GR populations of this species also have been found in São Joaquim, Santa Catarina (Vargas and Gazziero 2008). All 15 cases of suspected GR *L. multiflorum* sent by farmers to Embrapa Trigo for testing during the 2006/2007 soybean cropping cycle were confirmed resistant (Vargas and Gazziero 2008). At least in one case, contaminated crop seed appears to be involved in introducing a resistant biotype to a farm from a location where GR had been previously confirmed (Galli et al. 2005).

    More recently, GR *L. multiflorum* was found in Argentina (Vigna et al. 2008b), and there are strong indications of its appearance in Uruguay (Formoso et al. 2008). GR populations of *Lolium rigidum* in Argentina have been found in the south of the province of Buenos Aires (Pampean region), specifically in the neighboring partidos (municipalities) of General Dorrego, General Pringles, and Bahia Blanca. In this area, as well as the southwest of Buenos

Aires, *L. multiflorum* had become established in conventional and no-till agriculture and in pastures with some farmers planting locally selected varieties as pastures (Vigna et al. 2008b). Wheat is the most important winter crop in the south of Buenos Aires province (Scursoni and Gigón 2007), and *L. multiflorum* is one of the most important weeds in this monoculture system, including in the fallow periods (Vigna et al. 2008c). Seeds usually germinate at the end of the summer or beginning of autumn coinciding with the fallow period (López et al. 2008). In the southeast of the province, where the average farm size is 75 ha, *L. multiflorum* was the second most frequent grass weed after *Avena fatua* in a 2004/2005 survey (Scursoni and Gigón 2007); but its presence had remained the same since the 1980s. The same was found in the south and southwest wheat region, where 183 commercial fields were surveyed in the 2006/2007 and 2007/2008 cropping seasons. Fields in the sampled area were predominantly under conventional agriculture (71%), with only 29% under no-till production. *L. multiflorum* infestation was substantially lower in no-till fields (24%) than in conventional farming (48%), but in both cases remained second after *A. fatua* (Gigón et al. 2008). In these parts of the province, it is also abundant in the fallow periods.

Wheat farmers traditionally have used glyphosate at low doses (390 g ha^{-1}) compared with the recommended commercial dose of 600–1200 g ha^{-1}. (Vigna et al. 2008c). Only a handful of GR-biotypes have been partially characterized:

- A1, collected from a fallow field at Coronel Falcon (Bahia Blanca), confirmed to be resistant when tested along with cultivars (Vigna et al. 2008a; M. Vigna, pers. comm.).
- A2, collected by a farmer at Coronel Pringles in 2005.
- A3, from a wheat field (2006) in Coronel Dorrego. At this site, glyphosate had been used from 1999 to 2005 two to three times per year at 395 g ha^{-1}, frequently in mixture with 2,4-D. Once the farmer suspected of resistance beginning in 2006, application of the herbicide increased to five times per year at 482 g ha^{-1}.
- A4a, also from a wheat field at Coronel Dorrego, initially collected by a farmer in 2005 and for experimental purposes again in 2006 (M. Vigna, pers. comm.).
- A4b, collected by researchers in 2006 at a margin of field A4a.

Two susceptible biotypes have been used as reference materials: B1, collected in 2006 at a roadside along National Highway 3 in Coronel Dorrego, at a location very close to field A4 that probably has never been treated with glyphosate; and B2, collected in 2005 from a field not previously exposed to glyphosate at the Jacinto Arauz Department (La Pampa province).

Field, greenhouse, and Petri dish bioassays were conducted by Vigna et al. (2008c) with populations collected in those areas where farmers had complained about glyphosate failure to control *L. multiflorum*. In the greenhouse, A3 and A4a withstood glyphosate at doses above $360\,g\,ha^{-1}$ that were lethal to the GS biotypes; A3 was slightly more resistant than A4. In the field, the A3 biotype was able to withstand glyphosate at doses well above the recommended one for *L. multiflorum* control. This biotype was readily controlled with haloxyfop-R-methyl. In petri dishes, the GR biotypes had resistance indices between 2.0 and 4.0, with the biotype A4b (field margin) being less resistant that A4a (Vigna et al. 2008c). The GR biotypes from the fourth location have the ability to resprout after their foliage has been necrotized by the herbicide (Vigna et al. 2008a). A reference (C1) locally selected landrace used as a forage pasture was confirmed susceptible to glyphosate. Similarly, 17 *L. multiflorum* and four *Lolium perenne* cultivars were susceptible to the herbicide (Vigna et al. 2008a).

There is also suspicion that *L. perenne* has also evolved resistance to glyphosate at least at a location (La Fe farm) in General Dorrego in Argentina (Anónimo 2009).

14.3.2.1   Mechanism of Resistance and Fitness   The mechanisms of resistance to glyphosate in the Latin American *L. multiflorum* populations have not been totally elucidated yet. Studies comparing GR ("Olivar") and GS biotypes from Region VI of Chile determined that differential absorption, translocation, or allocation of glyphosate within the plant was not the primary resistance mechanism (Pérez et al. 2004). Likewise, glyphosate leaf uptake and translocation was similar in the San Fernando GR biotype and a GS biotype. Amplification and sequencing of the EPSPS gene determined that the San Fernando biotype has a proline 106 to serine amino acid substitution (Pérez-Jones et al. 2007). Conversely, the Vilcum biotype from the Region IX of Chile has an EPSPS enzyme susceptible to glyphosate, and in this case, limited spray retention and altered herbicide translocation appear to contribute to resistance (Michitte et al. 2007).

Field and greenhouse studies with a GR biotype collected in an area of soybean–wheat rotation in Lagoa Vermelha (RGS) indicated that higher glyphosate doses are required for adequate control as the GR plants become older and some of the affected plants are able to resprout (Christoffoleti et al. 2005). Glyphosate ED_{50} values for a GR biotype found in an apple orchard and a reference susceptible biotype were 4835 and $290\,g\,ha^{-1}$, respectively (Vargas et al. 2005). Although ACCase inhibitors had never been used in the field to control the GR biotype, it was slightly less affected by diclofop, fenoxaprop, and haloxyfop when compared with the susceptible biotype, especially at low doses. Fluazifop affected both the GR and GS biotypes in the same manner (Vargas et al. 2005).

More glyphosate remained in the treated leaf of a GR *L. multiflorum* biotype from Brazil than in the GS one. In the GS biotype, more herbicide (40%) translocated to the roots compared with the GR biotype that accumulated 10% of the treated amount. Both biotypes exuded glyphosate from the roots at the same rate (Ferreira et al. 2006, 2008b). GR and GS biotypes did not differ in their content or composition of their cuticles (Guimarães et al. 2009). A GR biotype from Tepejara differed in the pattern of its trichomes that were arranged in pairs, while those of a GS biotype were distributed individually. Both biotypes absorbed glyphosate by the leaves similarly, reaching about 65% of the applied amount by 72 h, but no distinct patterns in translocation could be associated with resistance. Almost twice more shikimate accumulated in the GS biotype than in the Tepejara biotype at 48 hours after treatment (HAT). The proportion increased to 3.5 times at 72 HAT (Ribeiro 2008).

Preliminary studies indicate that GR biotypes from Argentina do not have a resistant EPSPS enzyme, since no mutations were found at positions 101 and 106 that have previously been shown to confer resistance to the herbicide (Diez de Ulzurrun 2008).

Studies conducted with Brazilian biotypes indicate that glyphosate resistance is conferred by a nuclear gene with incomplete dominance (Vargas et al. 2007c).

In analyzing the productivity of a Brazilian GR biotype, Vargas et al. (2005) found that it accumulated more aboveground biomass, but less tillers, inflorescences, and seed, and matured later than the susceptible biotype. The studies, however, were not conducted under competitive conditions to indicate fitness differences. A susceptible biotype was more competitive to wheat than a resistant biotype, indicating that the resistant biotype appeared to be less fit than the resistant one (Ferreira et al. 2008c). In a pot study comparing the competitive ability of a single GR or GS plant of *L. multiflorum* growing in association with up to four plants of the opposite biotype, the GR biotype also appeared to be less fit based on the accumulation of aboveground biomass; although it was less efficient in water use (Concenço et al. 2007).

14.3.2.2 Control of Resistant Populations

In general, herbicides with alternative modes of action effectively control the GR *L. multiflorum* biotypes (Christoffoleti et al. 2005; Vargas et al. 2004, 2005), with the most widely used being the ACCase graminicides (clethodim, tepraloxydim, sethoxydim, cletodim + fenoxaprop, and haloxyfop). These herbicides are sometimes applied tank mixed with glyphosate itself (Christoffoleti et al. 2005; Gazziero et al. 2008; Vargas et al. 2005). Amitrol and paraquat alone or in mixture with diuron are also used as alternative treatments (Christoffoleti et al. 2005; Roman et al. 2004; M. Kogan, pers. comm.).

In a field study at Coronel Dorrego, Argentina, where glyphosate applied at 2000 g ha^{-1} did not provide adequate control, both clethodim alone (192 g ai ha^{-1}) or in mixture with glyphosate (96 and 500 g ha^{-1}, respectively), and haloxyfop alone (87.5 g a.e. ha^{-1}) or in mixture with glyphosate (62.5 and 500 g ha^{-1}, respectively) adequately controlled GR *L. multiflorum* (López et al. 2008). Paraquat is also effective in controlling the GR-biotypes (Vigna et al. 2008b).

14.3.2.3 Multiple Resistance There are some indications that a few GR biotypes of *L. multiflorum* are also resistant to herbicides with other modes of action. The Vilcún biotype, for example, was also cross-resistant to both iodosulfuron and flucarbazone but susceptible to the ACCase inhibitors, diclofop and clethodim (Espinoza et al. 2005). Another biotype from Chile confirmed resistant to glyphosate (RI = 6.8) was also resistant to clethodim (RI = 10), but not to diclofop. It also exhibited a low level (RI = 2.0) of resistance to the ALS-inhibiting herbicide, iodosulfuron (Espinoza et al. 2003).

14.4 OTHER GR GRASS WEEDS

14.4.1 *Digitaria insularis* in Paraguay and Brazil

Digitaria insularis is a perennial, rhizomatous grass weed than can be effectively controlled with glyphosate when originating from seed. Developed plants that have formed rhizomes are more difficult to control. In no-till agriculture and coffee production systems, it may become prevalent because the weed is not controlled at the commonly used glyphosate doses (Machado et al. 2006). Its small seeds spread easily and find safe sites in undisturbed soils in no-till soybeans (Gazziero et al. 2001).

Populations of *D. insularis* resistant to glyphosate have appeared in soybeans in Paraguay and Brazil (Heap 2010), but they have not been well characterized yet. As a reference, an ED_{50} of $128\,g\,ha^{-1}$ was calculated for a susceptible biotype treated at the four-leaf growth stage under greenhouse conditions (Lacerda and Victoria Filho 2004). In areas infested with GR *D. insularis*, local advisors recommend adding an ACCase-inhibiting herbicide to glyphosate to control it (Gazziero et al. 2008).

14.4.2 *Eleusine indica* in Colombia, Bolivia, and Costa Rica

In Colombia, GR *Eleusine indica* was reported at the same locations in the Caldas Department, where GR *C. bonariensis* was found (Menza Franco and Salazar Gutiérrez 2006b). Bioassay studies identical in methodology to those explained for *C. bonariensis* demonstrated a diminished response of three biotypes of *E. indica* to glyphosate. Resistance levels are low, and in this particular study, reliable comparison of entire dose–response curves was not possible. Both fluazifop-butyl and glufosinate controlled resistant populations in field experiments at two commercial farms, while glyphosate provided less than 60% control (Menza Franco and Salazar Gutiérrez 2007).

Resistant populations were also reported in Santa Cruz, Bolivia (Franco 2008; Valverde 2007), but the resistant biotypes are not fully characterized.

In Costa Rica, a GR population was found at a pejibaye (for hearts of palm) plantation on the Caribbean side. According to preliminary bioassays, its resistance level is low, but under field conditions glyphosate at recommended doses failed to control this population (B. E. Valverde, unpublished data). Previously, *E. indica* had evolved resistance to imazapyr at a single location in the Central Valley of Costa Rica (Valverde et al. 1993).

14.4.3 *Paspalum paniculatum* in Costa Rica

At the same location where GR *E. indica* was found in Costa Rica, populations of *Paspalum paniculatum* had become the most troublesome weed because glyphosate failed to control it. Seed and seedlings surviving after a commercial application were collected. Both, seedlings originating from seed and from subdividing collected seedlings, were able to withstand glyphosate at higher doses than a putative susceptible biotype from the same region and another biotype (from a road side) unlikely to have been exposed to the herbicide. Resistance levels are moderate (RI < 5.0), and at high doses the plants develop toxicity symptoms and the foliage becomes necrotic. If plants remain watered, new tillers begin to emerge (at least 1 month after treatment). Initially, these tillers are almost albino, but recover and the plant matures and produces seed. This "phoenix resistance" is more difficult to diagnose by whole plant bioassays, but explains the increasing colonization by the grass in the glyphosate-treated fields.

14.4.4 *Echinochloa colona* in Argentina and Venezuela

There is suspicion that *Echinochloa colona* has evolved resistance to glyphosate in rice in Venezuela (A. Anzalone, pers. comm.) and in RR soybeans in Tucuman, Argentina (J. Delucchi, pers. comm.). Studies to confirm resistance are being conducted by local researchers. A GR biotype of this species has already been found in Australia (Preston et al. 2008).

14.5 PERSPECTIVES

Resistance to glyphosate is an emerging problem in Latin America. Nine weed species have been confirmed resistant to this herbicide, but resistance is still not well characterized. Most GR populations appeared in areas subjected to no-till agriculture and where RR crops have been grown for several years. As areas planted with these technologies continue to grow in Latin America, it is expected that additional species and populations of already resistant species will evolve resistance in the near future. The increased use of glyphosate in several countries, especially in Brazil where the RR soybean area is also increasing, will most likely result in additional cases of resistance. Systematic research on GR weeds in the region is scarce; Brazil and Chile are the only countries that have devoted more effort to characterize and study possible means to control resistant populations. A few resistance cases are also associated with perennial crops. Farmers, in many instances, are not familiar with herbicide resistance or are reluctant to report it because of uncertainty about its consequences. Similarly, the agrochemical industry is very cautious about accepting and dealing with resistance cases, as they are probably concerned about reduced sales and bad publicity. Governments, with the exception of Argentina that faces a major agricultural problem with the evolution of GR SORHA, and more recently, *L. multiflorum*, do very little to prevent and help control resistant populations. By compiling the available information about GR weeds in this region as thoroughly as possible, the author hopes to contribute to increased awareness about GR weeds and to stimulate researchers, industry representatives, and governmental authorities to promote agricultural practices that will prevent or delay resistance evolution and to provide farmers with adequate information and tools to cope with resistance once it appears in their fields.

REFERENCES

Adegas, F. S., E. Voll, and D. Gazziero. 2008. Controle químico de buva resistente ao Glyphosate, com herbicidas aplicados na operação de manejo, em pré-semeadura da cultura da soja. In D. Karam, M. H. Tabim, and J. Baptista, eds., *Resumos. XXVI Congresso Brasileiro da Ciência das Plantas Daninhas and XVIII Congreso de la Asociación Latinoamericana de Malezas.* May 4–8, Ouro Preto, MG, Brazil. Sete Lagoas, Brazil: Sociedade Brasileira da Ciência das Plantas Daninhas, p. 76.

Alonso, M. 2005. Determinación de la resistencia al herbicida glifosato, en poblaciones de *Parthenium hysterophorus* L. Thesis (Ingeniero Agrónomo), Facultad de Agronomía, Universidad Nacional de Colombia. Abstract available: http://168.176. 162.23/F/FXTUAXITDXELDVCN7AVR8G97AMFXV9LPQ388DSJ9P2YCTKY6RV-03100?func=full-set-set&set_number=729963&set_entry=000001&format=999 (accessed October 11, 2009).

Anónimo. 2009. El regreso de las malas hierbas. *El Federal* 268:39–40. http://www. infocampo.com.ar/img/supef/edicion-268.pdf (accessed October 11, 2009).

Associação Brasileira da Indústria Química (ABIQUIM). 2009. The Brazilian Chemical Industry in 2008. http://www.abiquim.org.br/english/content.asp?princ=bci&pag= outlook (accessed October 11, 2009).

Binimelis, R., W. Pengue, and I. Monterroso. 2009. "Transgenic treadmill": responses to the emergence and spread of glyphosate-resistant johnsongrass in Argentina. *Geoforum* 40:623–633.

Christoffoleti, P. J., S. J. P. Carvalho, M. Nicolai, and V. C. B. Cardinali. 2008a. Custo das medidas pró ativa de manejo da resistência de Buva (*Conyza* spp.) ao herbicida glifosato na cultura do citros. In D. Karam, M. H. Tabim, and J. Baptista, eds., *Resumos. XXVI Congresso Brasileiro da Ciência das Plantas Daninhas and XVIII Congreso de la Asociación Latinoamericana de Malezas.* May 4–8, Ouro Preto, MG, Brazil. Sete Lagoas, Brazil: Sociedade Brasileira da Ciência das Plantas Daninhas, p. 313.

Christoffoleti, P. J., A. J. B. Galli, S. J. P. Carvalho, M. S. Moreira, M. Nicolai, L. L. Foloni, B. A. B. Martins, and D. N. Ribeiro. 2008b. Glyphosate sustainability in South American cropping systems. *Pest Management Science* 64:422–427.

Christoffoleti, P. J., R. Trentin, S. Tocchetto, A. Marochi, A. J. Batista Galli, R. F. López-Ovejero, and M. Nicolai. 2005. Alternative herbicides to manage Italian ryegrass (*Lolium multiflorum* Lam) resistant to glyphosate at different phenological stages. *Journal of Environmental Science and Health, Part B* 40:59–67.

Concenço, G., E. A. Ferreira, A. A. Silva, F. A. Ferreira, R. G. Viana, L. D'Antonino, L. Vargas, and C. M. T. Fialho. 2007. Uso da água em biótipos de azevém (*Lolium multiflorum*) em condição de competição. *Planta Daninha* 25:449–455.

De la Vega, M. H., D. Fadda, A. Alonso, M. Argañaraz, J. Y. Sánchez Loria, and A. García. 2006. Curvas dose-resposta em duas populações de *Sorghum halepense* ao herbicida glyphosate no Norte Argentino. *Resumos do XXV Congresso Brasileiro da Ciência das Plantas Dañinhas*. May 29–June 2, 2006, Brasilia, Brazil.

De la Vega, M., A. Mamani, D. Fadda, R. A. Vidal, M. Argañaraz, and M. M. Vila-Aiub. 2007. Mechanism of glyphosate resistance in johnsongrass: shikimate accumulation. *Abstracts of the Weed Science Society of America* 47:114.

Diez de Ulzurrun, P. 2008. Identificación de los mecanismos de resistencia a glifosato de *Lolium multiflorum* en la provincia de Buenos Aires. Avance de Investigación. Instituto Nacional de Tecnología Agropecuaria, Argentina. http://www.inta.gov.ar/balcarce/ResumenesPG/PGPV2008/julio/Avance_DIEZdeULZURRUM.doc (accessed October 11, 2009).

Espinoza, N., C. Cerda, J. Díaz, and M. Mera. 2003. Primer biotipo de ballica (*Lolium multiflorum* Lam) chileno con resistencia múltiple a herbicidas. In G. Mondragón, J. A. Domínguez, G. Martínez, and R. A. Ocampo, eds., *Memoria XVI Congreso Latinoamericano de Malezas (ALAM)/XXIV Congreso Nacional de la Asociación Mexicana de la Ciencia de la Maleza (ASOMECIMA)*. November 10–12, 2003, Manzanillo, Mexico: Asociación Latinoamericana de Malezas, p. 393.

Espinoza, N., J. Díaz, and R. de Prado. 2005. Ballica (*Lolium multiflorum* Lam) con resistencia a glifosato, glifosato-trimesium, iodosulfuron y flucarbazone-Na. In *Resúmenes XVII Congreso de la Asociación Latinoamericana de Malezas (ALAM)*. November 8–11, 2005, Varadero, Cuba: Asociación Latinoamericana de Malezas, p. 324.

Farm Chemicals International. 2009. Report Confirms Brazil as World's Top Ag Chemical User. Market Updates, April 21, 2009. http://www.farmchemicalsinternational.com/news/marketupdates/?storyid=1666&style=1 (accessed October 11, 2009).

Fernández, L., A. J. Distéfano, M. C. Martínez, D. Tosto, and H. E. Hopp. 2009. Epidemiología molecular del sorgo de Alepo resistente a glifosato en Argentina. *VII Simposio Nacional de Biotecnología REDBIO Argentina; II Congreso Internacional REDBIO Argentina*. April 20–24, 2009, Rosario, Argentina.

Ferreira, E. A., L. Galon, I. Aspiazú, A. A. Silva, G. Concenço, A. F. Silva, J. A. Oliveira, and L. Vargas. 2008a. Glyphosate translocation in hairy fleabane (*Conyza bonariensis*) biotypes. *Planta Daninha* 26:637–643.

Ferreira, E. A., A. A. Silva, L. Vargas, M. Rodrigues dos Reis, G. Concenço, I. Aspiazu, L. Galon, and A. C. França. 2008b. Potencial competitivo de biótipos de azevém resistente e suscetível ao glyphosate. In D. Karam, M. H. Tabim, and J. Baptista, eds., *Resumos. XXVI Congresso Brasileiro da Ciência das Plantas Daninhas and XVIII Congreso de la Asociación Latinoamericana de Malezas*. May 4–8, Ouro Preto, MG, Brazil. Sete Lagoas, Brazil: Sociedade Brasileira da Ciência das Plantas Daninhas, p. 153.

Ferreira, E. A., A. A. Silva, M. R. Reis, J. B. Santos, J. A. Oliveira, L. Vargas, K. R. Khouri, and A. A. Guimarães. 2008c. Distribuição de glyphosate e acúmulo de nutrientes em biótipos de azevém. *Planta Daninha* 26:165–173.

Ferreira, E. A., J. B. Santos, A. A. Silva, J. A. Oliveira, and L. Vargas. 2006. Translocação do glyphosate em biótipos de azevém (*Lolium multiflorum*). *Planta Daninha* 24:365–370.

Formoso, F. A., A. Ríos, and G. Fernández. 2008. Evaluación de la susceptibilidad de raigras espontáneo (*Lolium multiflorum* Lam.) a glifosato en sistemas de siembra directa del litoral agrícola. *Seminario Internacional "Viabilidad del glifosato en sistemas productivos sustentables."* Instituto Nacional de Investigación Agropecuaria, Serie de Actividades de Difusión 554, Uruguay, pp. 113–125. http://webs.chasque.net/~rapaluy1/glifosato/Viabilidad_Glifosato.pdf (accessed October 10, 2009).

Fornarolli, D. A., B. N. Rodrigues, E. Sawada, M. Faco, J. F. Nietzke, F. F. Nieztke, and I. O. Santos. 2008. Alternativas de Manejo da Buva (*Conyza bonariensis*) na Região Oeste do Estado do Paraná na Cultura da Soja (*Glycine max*). In D. Karam, M. H. Tabim, and J. Baptista, eds., *Resumos. XXVI Congresso Brasileiro da Ciência das Plantas Daninhas and XVIII Congreso de la Asociación Latinoamericana de Malezas.* May 4–8, Ouro Preto, MG, Brazil. Sete Lagoas, Brazil: Sociedade Brasileira da Ciência das Plantas Daninhas, p. 533.

Franco, P. 2008. Malezas resistentes y/o tolerantes a glifosato y los mecanismos para su control. Fundación de Desarrollo Agrícola de Santa Cruz (FUNDACRUZ). *Manual de Difusión Técnica de Soya* 13:137–143. http://www.fundacruz.org.bo/documents/informacion/13_Malezas_Resis_Tolerantes_a_Glifosato_y_Mecanismos_de_Control.pdf (accessed October 10, 2009).

Galli, A. J. B., A. I. Marochi, P. J. Christoffoleti, R. Trentin, and S. Tochetto. 2005. Ocorrência de *Lolium multiflorum* Lam resistente a glyphosate no Brasil. Ponencias. *Seminario-Taller Iberoamericano Resistencia a Herbicidas y Cultivos Transgénicos.* December 6–8, 2005, Centro Politécnico del Cono Sur, Colonia del Sacramento, Uruguay. http://www.inia.org.uy/estaciones/la_estanzuela/webseminariomalezas/index.htm (accessed October 10, 2009).

Gazziero, D. L. P., F. S. Adegas, C. E. C. Prete, R. Ralisch, M. F. Guimarães. 2001. *As plantas daninhas e a semeadura direta.* Circular Técnica 33. Londrina, PR, Brazil: Embrapa Soja.

Gazziero, D. L. P., F. S. Adegas, and E. Voll. 2008. *Glifosate e a soja transgênica.* Circular Tecnica 60. Londrina, PR, Brazil: Embrapa.

Gazziero, D. L. P., A. M. Brighenti, and E. Voll. 2006. Resistência cruzada da losna-branca (*Parthenium hysterophorus*) aos herbicidas inibidores da enzima acetolactato sintase. *Planta Daninha* 24:157–162.

Gelmini, G. A., R. Victória Filho, M. C. S. S. Novo, and M. L. Adoryan. 2005. Resistência de *Euphorbia heterophylla* L. aos herbicidas inibidores da ALS na cultura da soja. *Scientia Agricola* 62:452–457.

Gigón, R., F. Labarthe, L. E. Lageyre, M. R. Vigna, R. L. López, M. F. Vergara, and P. E. Varela. 2008. Comunidades de malezas en cultivos de trigo en el Sur y Sudoeste de la provincia de Buenos Aires. *VII Congreso Nacional de Trigo*, La Pampa. http://www.inta.gov.ar/bordenave/contactos/autores/ramon/com_maleza.pdf (accessed October 10, 2009).

Gómez, C. and C. L. Fuentes. 2008. *Parthenium hysterophorus* L PARTHYS (Asteraceae): un caso de resistencia a glifosato en el área frutícola del Valle del Cauca, Colombia. *Horticultura Argentina* 27:109.

Green, J. M. 2009. Evolution of glyphosate-resistant crop technology. *Weed Science* 57:108–117.

Guimarães, A. A., E. A. Ferreira, L. Vargas, A. A. Silva, R. G. Viana, A. J. Demuner, G. Concenço, I. Aspiazu, L. Galon, M. R. Reis, and A. F. Silva. 2009. Composição química da cera epicuticular de biótipos de azevém resistente e suscetível ao glyphosate. *Planta Daninha* 27:149–154.

Heap, I. 2010. The International Survey of HR Weeds. http://www.weedscience.com (accessed April 8, 2010).

Holm, L. R., J. Doll, E. Holm, J. V. Pancho, and J. P. Herberger. 1997. *World Weeds: Natural Histories and Distribution.* New York: John Wiley & Sons, Inc.

James, C. 2008. *Global Status of Commercialized Biotech/GM Crops: 2008.* Ithaca, NY: International Service for the Acquisition of Agri-Biotech Applications.

Kleffmann & Partner SRL. 2009. Mercado argentino de productos fitosanitarios 2008. Kleffmann Group. http://www.casafe.org.ar/estad/m2008/m2008.htm (accessed October 10, 2009).

Kogan, M. and A. Pérez. 2002. Resistencia de malezas a glifosato. In *Memorias XXXII Congreso Annual COMALFI: Manejo Integrado de Arvenses.* March 20–22, Santa Marta, Colombia: Sociedad Colombiana de Control de Malezas y Fisiología Vegetal, pp. 154–163.

Lacerda, A. L. S. and R. Victoria Filho. 2004. Curvas dose-resposta em espécies de plantas daninhas com o uso do herbicida glyphosate. *Bragantia* 63:73–79.

Lamego, F. P. and R. A. Vidal. 2008. Resistência ao glyphosate em biótipos de *Conyza bonariensis* e *Conyza canadensis* no estado do Rio Grande do Sul, Brasil. *Planta Daninha* 26:467–471.

Lazaroto, C. A., N. G. Fleck, and R. A. Vidal. 2008. Biologia e ecofisiologia de buva (*Conyza bonariensis* e *Conyza canadensis*). *Ciência Rural* 38:852–860.

Leguizamón, E. and J. M. Canullo. 2008. Mapas de área de infestación de malezas en la provincia de Córdoba. *Agromensajes* 26. http://www.fcagr.unr.edu.ar/agromensajes.htm (accessed October 10, 2009).

López, R. L., M. R. Vigna, and R. Gigon. 2008. Evaluación de herbicidas para el control de *Lolium multiflorum* lam. en barbecho para cereales de invierno. *Congresso Brasileiro da Ciência das Plantas Daninhas (26.); Congreso de la Asociación Latinoamericana de Malezas (18., 2008, Ouro Preto, MG, Brazil).* Atas. Ouro Preto, SBCPD. http://www.inta.gov.ar/bordenave/contactos/autores/vigna/lolium_CI.pdf (accessed October 10, 2009).

Machado, A. F. L., L. R. Ferreira, F. A. Ferreira, C. M. T. Fialho, L. S. Tuffi Santos, and M. S. Machado. 2006. Análise de crescimento de *Digitaria insularis*. *Planta Daninha* 24:641–647.

Magro, T. D., J. Paula, L. Vargas, and D. Agostinetto. 2008. Dose de glyphosate necessária para reduzir 50% da produção de massa seca (GR_{50}) de *Conyza* sp. resistente e sensível a glyphosate. In D. Karam, M. H. Tabim, and J. Baptista, eds., *Resumos. XXVI Congresso Brasileiro da Ciência das Plantas Daninhas and XVIII Congreso de la Asociación Latinoamericana de Malezas.* May 4–8, Ouro Preto, MG, Brazil. Sete Lagoas, Brazil: Sociedade Brasileira da Ciência das Plantas Daninhas, p. 463.

Marochi, A. I., A. J. B. Galli, R. Trentin, and S. Tochetto. 2008. Estudo de doses de glifosato em pós-emergência da soja Roundup Ready, em áreas consideradas com problemas de controle de *Euphorbia heterophylla* no Rio Grande do Sul. In D.

Karam, M. H. Tabim, and J. Baptista, eds., *Resumos. XXVI Congresso Brasileiro da Ciência das Plantas Daninhas and XVIII Congreso de la Asociación Latinoamericana de Malezas.* May 4–8, Ouro Preto, MG, Brazil. Sete Lagoas, Brazil: Sociedade Brasileira da Ciência das Plantas Daninhas, p. 377.

Melo, M. S. C., H. T. Filho, M. S. Moreira, M. Nicolai, S. J. P. Carvalho, and P. J. Christoffoleti. 2008. Alternativas de controle para buva (*Conyza canadensis* e *Conyza bonariensis*) resistente ao glyphosate. In D. Karam, M. H. Tabim, and J. Baptista, eds., *Resumos. XXVI Congresso Brasileiro da Ciência das Plantas Daninhas and XVIII Congreso de la Asociación Latinoamericana de Malezas.* May 4–8, Ouro Preto, MG, Brazil. Sete Lagoas, Brazil: Sociedade Brasileira da Ciência das Plantas Daninhas, p. 399.

Menza Franco, H. D. and L. F. Salazar Gutiérrez. 2006a. Evaluación de la resistencia al glifosato de biotipos de *Erigeron bonariensis* provenientes de cafetales de la zona cafetera central colombiana. *Cenicafé* 57:220–231.

Menza Franco, H. D. and L. F. Salazar Gutiérrez. 2006b. Resistencia de *Eleusine indica* al glifosato en cafetales de la zona cafetera central de Colombia. *Cenicafé* 57:146–157.

Menza Franco, H. D. and L. F. Salazar Gutiérrez. 2007. Alternativas de control químico para la prevención y manejo de la resistencia de arvenses al glifosato. *Cenicafé* 58:91–98.

Michitte, P., R. De Prado, N. Espinoza, J. P. Ruiz-Santaella, and C. Gauvrit. 2007. Mechanisms of resistance to glyphosate in a ryegrass (*Lolium multiflorum*) biotype from Chile. *Weed Science* 55:435–440.

Moreira, M. S. 2008. Detecção, crescimento e manejo químico alternativo de biótipos das espécies de buva *Conyza canadensis* e *C. bonariensis* resistentes ao herbicida glyphosate. MS thesis, Escola Superior de Agricultura Luiz de Queiroz, Piracicaba, SP, Brazil.

Moreira, M. S., M. S. C. Melo, F. Tersi, P. Sperandio, M. Nicolai, and P. J. Christoffoleti. 2007a. Alternativas de controle para buva (*Conyza canadensis* e *Conyza bonariensis*) resistente ao glyphosate. In C. A. Carbonari, D. K. Meschede, and E. D. Velini, eds., *I Simpósio Internacional sobre Glyphosate.* October 15–19, Botucatu, SP, Brazil. Botucatu, Brazil: Facultade de Ciências Agronômicas, Universidad Estadual Paulista (UNESP), pp. 103–107.

Moreira, M. S., M. Nicolai, S. J. P. Carvalho and P. J. Christoffoleti. 2007b. Resistência de *Conyza canadensis* e *C. bonariensis* ao herbicida glyphosate. *Planta Daninha* 25:157–164.

Olea, I. L. 2007. Glifosato: distribución e importancia del problema de especies tolerantes y sorgo de alepo resistente en Argentina. Elementos fundamentales para el uso de fitoterápicos: dosis modo de acción y prevención de deriva. Sociedad Rural de Tucuman, Argentina, p. 23. http://www.eeaoc.org.ar/informes/Catalogo_exposiciones_SRT.pdf (accessed October 4, 2009).

Olea, I. and S. Sabaté. 2008. Mucho más que un problema regional. *Revista CREA (Argentina)* 36:64–68, 70.

Olea, I., S. Sabaté, and F. Vinciguerra. 2008. Sorgo de alepo resistente a glifosato. Avances para su manejo en el cultivo de soja en el NOA. *Revista CREA (Argentina)* 36:55–56, 58–60, 62.

Papa, J. C., D. H. Tuesca, and L. A. Nisensohn. 2008. Avances sobre el sorgo de alepo [*Sorghum halepense* (L.) Pers] resistente a glifosato. *Soja—Para Mejorar la Producción INTA EEA Oliveros* 39:98–102.

Paula, J., T. D. Magro, L. Vargas and D. Agostinetto. 2008a. Características morfofisiológicas de biótipo de buva (*Conyza* sp.) resistente e sensível a glyphosate. In D. Karam, M. H. Tabim, and J. Baptista, eds., *Resumos. XXVI Congresso Brasileiro da Ciência das Plantas Daninhas and XVIII Congreso de la Asociación Latinoamericana de Malezas*. May 4–8, Ouro Preto, MG, Brazil. Sete Lagoas, Brazil: Sociedade Brasileira da Ciência das Plantas Daninhas, p. 264.

Paula, J., T. D. Magro, L. Vargas and D. Agostinetto. 2008b. Eficiência de herbicidas aplicados em pré-semeadura da soja, para o controle de buva (*Conyza* sp.) In D. Karam, M. H. Tabim, and J. Baptista, eds., *Resumos. XXVI Congresso Brasileiro da Ciência das Plantas Daninhas and XVIII Congreso de la Asociación Latinoamericana de Malezas*. May 4–8, Ouro Preto, MG, Brazil. Sete Lagoas, Brazil: Sociedade Brasileira da Ciência das Plantas Daninhas, p. 265.

Pérez, A., C. Alister, and M. Kogan. 2004. Absorption, translocation and allocation of glyphosate in resistant and susceptible Chilean biotypes of *Lolium multiflorum*. *Weed Biology and Management* 4:56–58.

Pérez-Jones A., K. W. Park, N. Polge, J. Colquhoun, and C. Mallory-Smith. 2007. Investigating the mechanisms of glyphosate resistance in *Lolium multiflorum*. *Planta* 226:395–404.

Pérez, A. and M. Kogan. 2001. Resistencia de malezas a herbicidas. *Agronomia y Forestal UC (Chile)* 4:4–9.

Pérez, A. and M. Kogan. 2003. Glyphosate-resistant *Lolium multiflorum* in Chilean orchards. *Weed Research* 43:12–19.

Preston, C., P. Boutsalis, J. Malone, F. Dolman, and A. Storrie. 2008. Resistance to glyphosate in *Echinochloa colona* in Australia. In *Proceedings of 5th International Weed Science Congress*. June 23–27, Vancouver, Canada: International Weed Science Society, pp. 226–227.

Ribeiro, D. N. 2008. Caracterização da resistência ao herbicida glyphosate em biótipos da planta daninha *Lolium multiflorum* (Lam.). MS thesis, Universidade de São Paulo, Piracicaba, SP, Brazil.

Rizzardi, M. A., T. D. Lamb, L. B. Johann and W. M. Wolff. 2008. Os herbicidas aplicados em pré-semeadura afetam o controle de plantas daninhas em pós-emergência da soja. In D. Karam, M. H. Tabim, and J. Baptista, eds., *Resumos. XXVI Congresso Brasileiro da Ciência das Plantas Daninhas and XVIII Congreso de la Asociación Latinoamericana de Malezas*. May 4–8, Ouro Preto, MG, Brazil. Sete Lagoas, Brazil: Sociedade Brasileira da Ciência das Plantas Daninhas, p. 159.

Rizzardi, M. A., L. Vargas, and M. A. Bianchi. 2007. Um belo problema. *FMC Square* 3:5–9.

Roman, E. S., L. Vargas, M. A. Rizzardi, and R. W. Mattei. 2004. Resistência de azevém (*Lolium multiflorum*) ao herbicida glyphosate. *Planta Daninha* 22:301–306.

Romig, S. 2007. Argentina pampas crops threatened by HR weed. *Dow Jones Newswires*. http://www.gmwatch.org/latest-listing/1-news-items/7522-argentina-pampas-crops-threatened-by-HR-weed-2792007?format=pdf (accessed September 20, 2009).

Rosario, J. M. 2005. Resistencia de *Parthenium hysterophorus* L. al herbicida glifosato: un nuevo caso de resistencia a herbicidas en Colombia. MS thesis,

Escuela de Posgrado, Facultad de Agronomía, Universidad Nacional de Colombia, CD-ROM. Abstract available: http://168.176.162.23/F/ 5RPM8DG8EHSG1K7FGQ34SXMDLSHHXVCR2T81EGP4MKHLQ71LQI- 24686?func=full-set-set&set_number=730135&set_entry=000002&format=999 (accessed October 10, 2009).

Rosario, J. M. and C. L. Fuentes. 2005. Resistencia de *Parthenium hysterophorus* L. al herbicida glifosato: un nuevo caso de resistencia a herbicidas en Colombia. *Resúmenes 2do Congreso Bianual SODIAF*, November 24–25, 2005, Santo Domingo, Dominican Republic, p. 9. http://www.sodiaf.org.do/publica/congreso/01Resumenes %20Final.pdf (accessed August 3, 2009).

Sabaté, S., I. Olea, F. Vinciguerra, and J. Raimondo. 2007. Variedades de soja tolerantes a herbicidas del grupo químico de las sulfonilúreas—Una nueva herramienta para el manejo del sorgo de alepo resistente a glifosato en el NOA. In *El cultivo de soja en el noroeste argentino. Campaña 2007/2008*. Tucuman: Estación Experimental Agroindustrial "Obispo Colombres," pp. 189–197.

Scursoni, J. and R. Gigón. 2007. ¿Qué está pasando con las malezas en nuestra región? *Revista Desafío 21* 29:10–12.

Secretaría de Agricultura, Ganadería, Pesca y Alimentos (SAGPyA). 2009. Resultados Definitivos del Censo Nacional Agropecuario 2002. Resumen Ejecutivo. Subsecretaría de Economía Agropecuaria, Dirección de Economía Agraria. http:// www.sagpya.mecon.gov.ar/new/0-0/programas/economia_agraria/index/censo/ Parte_I.pdf (accessed October 10, 2009).

Sembrando Satelital. 2008. Sorgo de alepo: Avanza ley para prevención de plaga resistente al glifosato. http://www.sembrando.com.ar/index.php?menu=noticias&noticia =nota&sub=6423 (accessed October 10, 2009).

Trezzi, M. M., C. L. Felippi, D. Mattei, H. L. Silva, A. L. Nunes, C. Debastiani, R. A. Vidal, and A. Marques. 2005. Multiple resistance of acetolactate synthase and protoporphyrinogen oxidase inhibitors in *Euphorbia heterophylla* biotypes. *Journal of Environmental Science and Health, Part B* 40:101–109.

Trezzi, M. M., R. A. Vidal, N. D. Kruse, and A. L. Nunes. 2006. Bioensaios para identificação de biótipos de *Euphorbia heterophylla* com resistência múltipla a inibidores da ALS e da Protox. *Planta Daninha* 24:563–571.

Valverde, B. E. 2007. Status and management of grass-weed herbicide resistance in Latin America. *Weed Technology* 21:310–332.

Valverde, B. E., L. Chaves, J. González, and I. Garita. 1993. Field-evolved imazapyr resistance in *Ixophorus unisetus* and *Eleusine indica* in Costa Rica. In *Proceedings of the Brighton Crop Protection Conference—Weeds*. Surrey, UK: British Crop Protection Council, pp. 1189–1194.

Valverde, B. E. and J. Gressel. 2006. Dealing with the evolution and spread of *Sorghum halepense* glyphosate resistance in Argentina, A consultancy report to SENASA. Servicio Nacional de Sanidad y Calidad Agroalimentaria, Argentina. http://www. sinavimo.gov.ar/files/senasareport2006.pdf and http://www.weedscience.org/paper/ Johnsongrass%20Glyphosate%20Report.pdf Also available in Spanish at http:// www.sinavimo.gov.ar/files/informesensa.pdf (all sites accessed October 10, 2009).

Valverde, B. E., J. Gressel, S. Passalacqua, and J. C. Rodríguez. 2007. The emerging problem of glyphosate-resistant johnsongrass (*Sorghum halepense*) in Argentina: an account of detection, initial spread and collaborative action for its prevention and management. *Abstracts of the Weed Science Society of America* 47:183.

Vargas, L., D. Agostinetto, R. E. Toledo, and J. M. De Paula. 2007a. Herbicidas alternativos para manejo de buva resistente ao glyphosate. In C. A. Carbonari, D. K. Meschede, and E. D. Velini, eds., *I Simpósio Internacional sobre Glyphosate*. October 15–19, Botucatu, SP, Brazil. Botucatu, Brazil: Facultade de Ciências Agronômicas, Universidad Estadual Paulista (UNESP), pp. 90–92.

Vargas, L., M. A. Bianchi, M. A. Rizzardi, D. Agostinetto, and T. Dal Magro. 2007b. Buva (*Conyza bonariensis*) resistente ao glyphosate na região sul do Brasil. *Planta Daninha* 25:573–578.

Vargas, L. and D. Gazziero. 2008. Manejo de plantas daninhas tolerantes e resistentes ao glyphosate no Brasil. *Seminario Internacional "Viabilidad del glifosato en sistemas productivos sustentables."* Instituto Nacional de Investigación Agropecuaria, Serie de Actividades de Difusión 554, Uruguay, pp. 70–74. http://webs.chasque.net/~rapaluy1/glifosato/Viabilidad_Glifosato.pdf (accessed October 10, 2009).

Vargas, L., R. M. A. Moraes, and C. M. Berto. 2007c. Herança da resistência de azevém (*Lolium multiflorum*) ao glyphosate. *Planta Daninha* 25:567–571.

Vargas, L., E. S. Roman, M. A. Rizzardi, and V. C. Silva. 2004. Identificação de biótipos de azevém (*Lolium multiflorum*) resistentes ao herbicida glyphosate em pomares de maçã. *Planta Daninha* 22:617–622.

Vargas, L., E. S. Roman, M. A. Rizzardi, and V. C. Silva. 2005. Alteração das características biológicas dos biótipos de azevém (*Lolium multiflorum*) ocasionada pela resistência ao herbicida glyphosate. *Planta Daninha* 23:153–160.

Vargas, L., E. S. Roman, M. A. Rizzardi, and R. E. B. Toledo. 2006. Manejo de azevém resistente ao glyphosate em pomares de maçã com herbicida select (clethodim). *Revista Brasileira de Herbicidas* 1:30–36.

Vidal, R. A., E. S. Portes, F. P. Lamego, and M. M. Trezzi. 2006. Resistência de *Eleusine indica* aos inibidores de ACCase. *Planta Daninha* 24:163–171.

Vidal, R. A., M. M. Trezzi, R. De Prado, J. P. Ruiz-Santaella, and M. Vila-Aiub. 2007. Glyphosate resistant biotypes of wild poinsettia (*Euphorbia heterophylla* L.) and its risk analysis on glyphosate-tolerant soybeans. *Journal of Food and Agricultural Environment* 5:265–269.

Vigna, M. R., R. L. López, and R. Gigón. 2008a. Efecto de glifosato sobre cultivares de raigras en el SO de Buenos Aires. In *Congresso Brasileiro da Ciência das Plantas Daninhas (26.); Congreso de la Asociación Latinoamericana de Malezas (18., 2008, Ouro Preto, MG, Brazil).* Atas. Ouro Preto, SBCPD. http://www.inta.gov.ar/bordenave/contactos/autores/vigna/raigras.pdf (accessed October 10, 2009).

Vigna, M., R. Lopez, R. Gigon, P. Diez de Ulzurrun, and M. I. Leaden. 2008b. Raigras anual (*Lolium multiflorum*) resistente a glifosato en el sudoeste de la provincia de Buenos Aires. *Seminario Internacional "Viabilidad del glifosato en sistemas productivos sustentables."* Instituto Nacional de Investigación Agropecuaria, Serie de Actividades de Difusión 554, Uruguay, pp. 55–60. http://webs.chasque.net/~rapaluy1/glifosato/Viabilidad_Glifosato.pdf (accessed October 10, 2009).

Vigna, M. R., R. L. López, R. Gigón, and J. Mendoza. 2008c. Estudios de curvas dosisrespuesta de poblaciones de *Lolium multiflorum* a glifosato en el SO de Buenos Aires, Argentina. *Congresso Brasileiro da Ciência das Plantas Daninhas (26.); Congreso de la Asociación Latinoamericana de Malezas (18., 2008, Ouro Preto, MG,*

Brazil). Atas. Ouro Preto, SBCPD. http://www.inta.gov.ar/bordenave/contactos/autores/vigna/dosis_lolium.pdf (accessed October 10, 2009).

Vila-Aiub, M. M., M. C. Balbi, P. E. Gundel, C. M. Ghersa, and S. B. Powles. 2007. Evolution of glyphosate-resistant johnsongrass (*Sorghum halepense*) in glyphosate-resistant soybean. *Weed Science* 55:566–571.

Vila-Aiub, M. M., M. C. Balbi, P. E. Gundel, Q. Yu and S. B. Powles. 2008. Ecophysiological studies on glyphosate resisistant *Sorghum halepense* (johnsongrass). In *Seminario Internacional "Viabilidad del glifosato en sistemas productivos sustentables."* Serie de Actividades de Difusión 554, Instituto Nacional de Investigación Agropecuaria, Colonia, Uruguay, pp. 9–53. http://webs.chasque.net/~rapaluy1/glifosato/Viabilidad_Glifosato.pdf (accessed October 10, 2009).

15

STRATEGIES FOR MANAGING GLYPHOSATE RESISTANCE— AN EXTENSION PERSPECTIVE

KEN SMITH

15.1 INTRODUCTION

It was once thought that weeds could not develop resistance to herbicides. J. L. Harper (1956) first proposed the concept of weed evolution to herbicide resistance. Many scientists and practitioners at the time failed to take this theory seriously. Herbicide technology was in its infancy, and this new method of killing weeds seemed infallible. However, Harper's premise proved to be correct 12 years later when the first weed was confirmed resistant to an herbicide (Ryan 1970). Since then, extension education programs have targeted this problem with varying levels of success. During the 1970s and 1980s, over 20 different companies conducted very active new herbicide discovery and synthesis programs. It was common to have multiple new herbicides introduced each year. If a weed was discovered to be resistant to a particular herbicide mode of action, education programs focused on identifying the best alternative chemistry and convincing farmers to substitute herbicide B for herbicide A. An example of this switch technique was when barnyardgrass developed resistance to propanil, rice farmers were encouraged to switch to quinclorac, and when quinclorac resistance was discovered, farmers were again encouraged to switch to clomazone for preemergence barnyardgrass control. Switching from one herbicide to an alternative one for the most part required few, if any, changes to management programs and was relatively easy for

Glyphosate Resistance in Crops and Weeds: History, Development, and Management
Edited by Vijay K. Nandula
Copyright © 2010 John Wiley & Sons, Inc.

TABLE 15.1. Tank-Mix Combinations with Glyphosate for Annual Morning Glory and Prickly Sida Control in Soybean (Smith 2004)

Tank-Mix Partner	Annual Morning Glory			Prickly Sida		
	22 oz	32 oz	44 oz	22 oz	32 oz	44 oz
None	93	95	97	93	95	97
Classic	90	98		90	98	
Flexstar	93	88		91	92	
Resource	90	97		90	88	
Firstrate	93	98		90	98	

farmers to adopt. Because it was so easy to accomplish, many were willing to adopt new chemistry "just in case" in a proactive approach.

Glyphosate combined with Roundup Ready® (Monsanto Co., St. Louis, MO) glyphosate-resistant (GR) crops dramatically altered production practices. Weed control became easier and much more effective. This technology was adopted very quickly and conservation tillage increased while acres farmed by individual farmers increased very rapidly. The size of individual farms decreased in every size category less than 2000 ac from 1998 to 2007, while the number of acres in farms over 2000 ac continued to increase (USDA NASS 2007b). This technology permitted farmers to farm more acres with less equipment and labor than ever before. Although farm size increased following the introduction of GR crops, the number of tractors per acre declined dramatically (USDA NASS 2007a). Cultivators and other tillage equipment were quickly abandoned. A common adage was "the best tank-mix with glyphosate is more glyphosate." Although this was contrary to good resistance management and proper stewardship of the herbicide, it was very accurate from an efficacy standpoint (Table 15.1). In many similar studies, adding more glyphosate would increase efficacy on susceptible to slightly tolerant species as much as a tank-mix partner. As glyphosate price decreased with the introduction of generic formulations, increasing the rate was not only efficacious, but also economical and convenient. This created a perception by some that weed control would never again be a major problem. One farmer relayed his thought after first seeing the control of Palmer amaranth with glyphosate in Roundup Ready cotton as "what a gift." Recent history reminds us this was a false hypothesis. The development of GR weeds in this changed agriculture structure creates some unique challenges in extension education.

15.2 EXTENSION EDUCATOR ROLL

Extension was founded as a crisis-solving agency by visionary agricultural leaders such as Seaman Knapp, Washington Carver, and Walter Porter. At the beginning of the twentieth century, cotton was the only cash crop for much of southern agriculture. The Mexican boll weevil seemed to appear suddenly and quickly began to spread across cotton country creating havoc in its path. This pest devastated the economy and hearts of communities as well as the cotton crops. Dr. Knapp had become recognized as a leader and strong advocate for agriculture. He was asked to go to Texas in the heart of cotton and boll weevil country and teach farmers techniques to manage this devastating insect. Dr. Knapp, the scientist and educator, teamed up with Walter Porter, the innovative farmer, to "demonstrate" the best management practices for growing cotton in boll weevil-infested areas. A field on the Porter Farm was selected as the "demonstration field," and best management practices were outlined and integrated into the farming practices of the day. By selecting early maturing varieties and modifying tillage techniques, they showed that cotton plants could escape many of the ravages imposed by the boll weevil. It is reported that Dr. Knapp visited the demonstration farm at least once every 2 weeks to monitor the crop, make recommendations, and keep neighboring farmers interested in the project (Bailey 1945). At the end of the year, farm receipts were tallied and shared. The $700 profit from this field quickly captured the attention of other farmers and business owners in the area. Rapid adoption of the improved practices not only changed a way of life and stabilized agriculture in the region, but laid the foundation for the extension service. It is easy to recognize that many of the techniques used by these well-informed and dedicated educators and innovative farmers over 100 years ago are still valid today. But more important than the techniques employed is that the roll of extension in agriculture is as relevant today as it was in the early 1900s when the boll weevil was destroying crops and threatening lifestyles throughout the South. Extension leaders have always had a vision and worked through both conventional and unconventional channels to insure that the vision becomes a reality.

Keep up to date on new research information, develop strong county agent training programs to teach the latest developments, assist county agents conduct effective result demonstrations to showcase new technology, and wear a tie to every meeting—this was the extension philosophy for many years and was an extremely effective foundation for many successful extension careers. Strong county educational programs continue to be the backbone of extension agricultural programs. However, the roll of extension educators has changed dramatically over the past several years. They no longer have the luxury of waiting for 3 years of research on a particular problem to be summarized before a recommendation is made. Rapidly changing agriculture and quick adoption of new technology by leading farmers require making the best recommendations as information is acquired. This may mean that initial recommendations could be antiquated and necessitate revision within a few months. Prior to glyphosate resistance, the standard recommendation for controlling horseweed (*Conyza canadensis* (L.) Cronq.) in early-season burndown was 0.84 kg a.e. ha^{-1} glyphosate. When this weed first demonstrated resistance to glyphosate, there were no data to indicate the effectiveness of adding $0.57\,kg\,a.e.ha^{-1}$ 2,4-D to the glyphosate treatment. This combination had proven to be very efficacious on many other glyphosate-tolerant weeds such as cutleaf evening primrose (*Oenothera laciniata* Hill.) and henbit (*Lamium amplexicaule* L.). The low rate of 2,4-D provided an additive effect with both herbicides contributing to the control of the weed. Only after the GR biotype was identified could research be initiated to test effective herbicide combinations. It became evident after the first growing season and several hundred plots that glyphosate was adding little to no additional control to that provided by 2,4-D alone, and this low rate of 2,4-D was insufficient to control horseweed. Higher rates of 2,4-D or $0.57\,kg\,a.e.ha^{-1}$ dicamba were required to give acceptable control of this new biotype. In the meantime, farmers were abandoning conservation tillage programs and even disking and replanting existing crops due to failed control programs (Fig. 15.1). Immediate answers were not only needed but also expected from university extension professionals. This also emphasizes the need for extension educators to be flexible enough to devote large amounts of time in developing the best management practices for new resistant biotypes. Never is extension needed or appreciated more than when a crisis occurs. Lack of response and leadership during this time due to real and/or perceived time or budget restraints is a disservice to the farming clientele and will result in a loss of confidence in extension programs. During the first 3 years after GR horseweed was discovered in Tennessee, Dr. Bob Hayes, extension weed scientist, laments that greater than 50% of his time was consumed identifying dispersal patterns, control techniques, emergence patterns, and conducting educational programs on this new biotype (R. Hayes, pers. comm.). This author has also devoted nearly 75% of his time and resources attempting to address the herbicide resistance issues in Arkansas.

Many university workers at the regional or state level have joint extension and research appointments and conduct fairly extensive applied research programs along with their educational rolls. These applied research responsibilities are very effective at addressing herbicide resistance issues at the farmer level. There is no lag time in moving information from the field plots to the farmer (Fig. 15.2).

15.3 STRATEGY

Identify the problem, develop workable solutions, and obtain adoption of those solutions is a simply stated approach to most any obstacle. When applied to techniques required to manage GR weeds, these challenges have caused more than one extension worker to lose sleep at night. In order to teach, one must first understand. An understanding of the ecological, biological, and physiological characteristics of the weed as well as the psychological viewpoints of major stakeholders is critical to successful attempts to address the glyphosate resistance issue.

Figure 15.1. Glyphosate-resistant horseweed dominating a newly planted cotton field.

15.4 IDENTIFYING THE PROBLEM

The first questions that must be answered concerning any new or threatening herbicide resistant weed are "how widespread is the problem now" and "how rapidly it is spreading." No herbicide program provides 100% control of weeds in every situation. "Escapes" or susceptible weeds surviving normal herbicide applications are undesirable but not uncommon on most farms. Herbicide-resistant biotypes most often occur in small numbers during the first and

Figure 15.2. Dr. Larry Steckel has research and extension responsibilities at the University of Tennessee. Often, research plots work extremely well as educational tools for farmer tours.

second years of infestation. Many times, these are not recognized as resistant biotypes, but simply called escapes. By the time they are recognized as resistant, soil seedbanks have already been established. Some suggest that 30% of a population must be resistant before it is recognized at the field level. This is certainly too high for weeds species that produce large numbers of seed and typically have germination levels of several plants per meter square. Palmer amaranth (*Amaranthus palmeri* S. Wats.) plants have been collected and sent for screening at the University of Arkansas laboratory because they were suspected of being resistant when fewer than 10 plants per acre survived the herbicide application. These plants represented less than 1% of the total number of Palmer that germinated in the respective fields. Typically, an aggressive screening program is the most effective technique to determine the magnitude of the problem. There are multiple approaches to screening programs that will provide information about the distribution of the resistant biotype. One is to develop a sampling grid and attempt to sample a given number of populations within each segment of the grid. This is a very labor-intensive system and very dependent on large sample sizes to identify resistant individuals. It is used when a more exact number is desired to represent percent of a population that is resistant. In a survey of weeds in Mississippi soybean fields, Rankins et al. (2005) based sample intensity on a number of hectares of soybean planted in the county. Using the state agricultural statistics reports, they sampled one random soybean field in every county that had at least 2000 ha of soybean planted and an additional field for each additional 4000 ha. No samples were collected from fields in counties that had less than

2000 ha planted. Fields were randomly selected, and five subsamples were taken per field with predetermined defined area in each subsample. Weed frequency was determined and reported. These techniques are necessary in order to obtain valid data representing percent of fields infested or frequency of a particular species or biotype within fields. Another more realistic approach for determining extent of distribution is to only sample escapes in fields known to have been treated with at least one application of glyphosate. This can be accomplished very effectively by enlisting the help of extension agents throughout the state. Extension agents are alerted to be on the lookout for escapes in fields known to be planted to GR crops. When escapes are noted, seed heads are clipped and sent to the laboratory for screening. Utilizing this technique, University of Arkansas scientists were able to determine that from a single confirmed resistant location in 2005, eight additional counties in the state had Palmer amaranth plants that tested positive for resistance in 2006. By using this technique, over half of the seed samples received contained greater than 10% resistant offspring when screened. The shortcomings of this technique are that percentage of fields within an area or frequency of occurrence within field estimates cannot be extrapolated. Although this is often desired information, it is more critical to locate as many resistant weeds as possible when attempting to determine how widespread the problem has become in early resistant detection. This can best be accomplished by selecting suspect escapes when and where they are located without regard for random sampling techniques.

Unfortunately, the spread of GR horseweed was so rapid, that screening techniques were somewhat inadequate. Also, horseweed germinates in the fall and spring and is normally destroyed prior to planting warm season crops. Practically, by the time a farmer had sprayed horseweed in a burndown program prior to planting and at least two more times in crop before it produced seed, there was little doubt it was resistant and screening programs would have been of little value. The best information gathered on the distribution of this biotype was to closely monitor failures when glyphosate alone or glyphosate and low rates of other herbicides were used. Utilizing this technique, it was possible to determine that GR horseweed spread throughout the cropping area of Arkansas within 3 years after the initial confirmation (Fig. 15.3). This technique was also used to monitor the spread of GR Palmer amaranth in Arkansas (Fig. 15.4).

15.5 ANSWERING FARMER QUESTIONS

Until GR horseweed became a problem, there was no information defining the dispersal pattern of this species. Only by observing the spread of control failures did it become apparent that seed movement in excess of 50 mi was not uncommon. In later work, it was estimated that horseweed seed could travel up to 300 mi (Shields et al. 2006). Underestimating this movement caused a false sense of security in many extension workers and farmers during the first

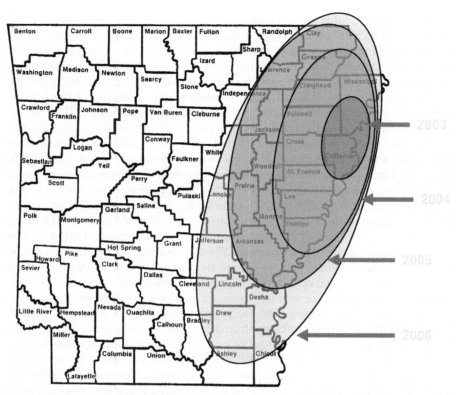

Figure 15.3. Glyphosate-resistant (GR) horseweed spread (2003–2006) in Arkansas, USA, as determined by spray failures.

few years after the resistant biotype was discovered. A key requirement in understanding and being able to provide reliable information about ecological characteristics of a new biotype is a tremendous amount of time and a very coordinated effort from county extension staff as well as state faculty. Everybody at every level must be informed and involved. As stated earlier, extensive field research on a resistant biotype is only possible after the biotype is established in the area. Unfortunately, answers are needed immediately. In-depth symposia are an excellent way to collect and share knowledge about the ecology and physiology of a weed. A four-state horseweed symposium was conducted in 2004 to bring together the collective information from agriculture industry representatives, university scientists, and private consultants in Missouri, Tennessee, Arkansas, and Mississippi. Ample time was allowed for questions and discussion after each speaker. Industry representatives presented data on their respective products that had activity on horseweed, and university scientists presented data and literature reviews available at the time. Although no one seemed to have all the needed information, it was a

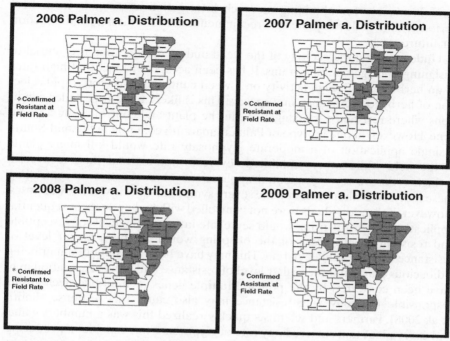

Figure 15.4. Glyphosate-resistant Palmer amaranth infestation was confirmed in only one Arkansas (USA) county in 2005, but spread to 22 counties by 2009.

very valuable meeting in that everyone was as up-to-date as possible and everyone had the same information from which to make decisions.

When Palmer amaranth was first discovered to be resistant to glyphosate in Georgia, one of the first projects initiated by Dr. Stanley Culpepper, extension weed scientist, was to attempt to measure dispersal of the resistant gene by pollen movement (see Chapter 11). Also, a GR Palmer roundtable was conducted in Little Rock, AR, to pull together the collective knowledge of scientists interested in this new pest (McClelland 2007). By design, this was a very in-depth discussion that allowed all participants to leave with the joint knowledge of the group. This is extremely valuable to extension scientists who are attempting to advise farmers on a problem of which there is often limited information available. When ryegrass began to be recognized as resistant or difficult to control with glyphosate in the mid-South, Dr. Bob Scott, extension weed scientist at the University of Arkansas Division of Agriculture, quickly organized a meeting of all industry and university scientists in the area to compile all known information and to develop a short-term management approach while more in-depth studies and surveys could be conducted to build longer-term strategies. Although many papers were presented at professional society meetings on each of these topics, these in-depth and focused meetings were invaluable for obtaining as much information as possible in a short period of time to develop the best recommendations for farmers and direction for future study.

Understanding the ecology of the weed under consideration is essential to designing management programs. It has been accepted as fact that high rates of an herbicide with high activity on a weed contributes to more rapid selection of herbicide resistance in a species. This is likely to be true under conditions where resistance is a single site in the plant and controlled by a single gene. However, in one biotype of Palmer amaranth common in the mid-South, a single application of a moderate glyphosate rate would kill many plants in the field while others exhibited severe symptoms and then recovered. A higher dose rate or sequential applications were effective in controlling these individuals that exhibited the severe symptoms, but survived lower doses. However, if the individuals were not controlled with higher doses or sequential applications, their offspring would segregate into varying levels of susceptible and resistant plants. Some of the offspring would exhibit a higher level of resistance than the female parent. This may have been a result of Palmer being a dioecious plant, and the higher level of resistance and great segregation may have been conferred by the pollen. Multiple genes or multiple mechanisms responsible for conveying resistance may also cause this response (Smith et al. 2008). Farmers and scientists quickly realized this was a numbers game, and every plant that survived any rate of glyphosate was capable of producing resistant offspring. More survivors resulted in greater numbers of resistant plants in subsequent generations. A common adage was that "dead weeds don't produce resistant offspring." Obviously, this was a short-term solution to an inevitable problem. Although it was a numbers game, the number of resistant survivors increased each year. Reduced glyphosate rates or fewer applications would have allowed these numbers to increase at a much faster rate. This phenomenon was later confirmed in rigid ryegrass (Busi and Powles 2009). After three to four cycles of glyphosate selection in two distinct environments, the progenies of the initially susceptible ryegrass populations were shifted toward glyphosate resistance. They concluded that the contribution of minor genes endowing substantial plant survival at sublethal herbicide doses is a potential complementary path to herbicide resistance evolution.

15.6 CREATING AWARENESS

Benjamin Franklin is credited with the adage "an ounce of prevention is worth a pound of cure." This has been the philosophy behind many resistance management programs gone awry. All too often when awareness programs are designed to prevent the development of glyphosate resistance, they are not taken seriously by many producers until it occurs on their farm. This "not my problem until it occurs on my farm" philosophy is often frustrating to extension workers and crop consultants. However, a better perspective of this may be gained by closely listening to farmers. When one leading farmer that had sat through several presentations on the horrors of herbicide resistance was presented with the question, "Now that you have heard all this information and understand the threats of GR weeds, what are you going to do about it on your farm?" His reply was "I am going to put it on my list of things to worry about." His answer reflected the large number of items farmers must consider on a daily basis. He indicated that high fuel prices, marketing decisions, labor costs and availability, equipment costs and maintenance, government regulations, other production inputs, and a whole host of other decisions that managers of large farms must make every day can lower the priority on anything that is not an immediate crisis. This should not discourage extension workers from developing awareness programs and developing recommendations for managing herbicide-resistant weeds. Some farmers will proactively adopt a number of the recommendations, while others may wait until resistance occurs on their farm before modifying cultural practices.

Several different techniques are effective for creating awareness, and astute extension educators recognize those effective for their particular area. Most often, more than one technique is required. Slogan programs, field days, production meetings, mass media, and one-on-one discussions are tried and tested techniques that have been used by extension educators. Slogan programs are a form of advertisement that repeats the same slogan many times that hopefully prompts a conscious or subconscious response from the listener. "Horseweed, Wanted Dead Not Alive" (Fig. 15.5) was a slogan

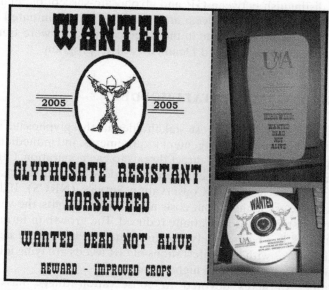

Figure 15.5. One thousand posters were distributed across a three-county area to draw attention to the glyphosate-resistant horseweed issue.

chosen to create awareness of GR horseweed in Arkansas. A small competitive educational grant was secured from extension administration by county agents in a three-county area first affected by this pest. This grant was matched with industry funds utilized to pay for program supplies and postage. The agents, working with state faculty, developed the slogan and initiated an aggressive educational program disseminating the best information available at the time. The slogan was printed on posters, hand towels, portfolios, CDs, and mail-out material. Posters were distributed to all local agricultural businesses including chemical dealerships, equipment dealers, and banks. A portfolio bearing the slogan was passed out at every county educational meeting as well as provided to agriculture leaders in the counties. Hand towels were provided to crop consultants to keep in their trucks, reminding them that resistance was an issue each time they wiped their face. The CD included some light-hearted information including "An ode to the horseweed" and "An arrest of outlaw horseweed by resistance marshal." Also included in this CD were producer quotes and best management practices as outlined by extension scientists.

The objectives of the awareness program were clearly defined as (1) the importance of controlling GR horseweed, (2) the impact GR horseweed will have on conservation tillage, (3) the economic impact of managing GR horseweed, and (4) the magnitude of the GR horseweed issue in the area. The educational component of the program sought to teach correct product selection, most economical control without sacrificing resistance management, techniques to preserve conservation tillage benefits when horseweed is present, and how to distinguish between GR and glyphosate-susceptible plants.

A grower survey conducted 1 year after the program was initiated indicated that nearly 90% of the 350,000 ac in the three-county area were treated with strategies defined in the "Wanted Dead Not Alive" program.

15.7   ENGAGING MAJOR STAKEHOLDERS

Obviously, farmers are the largest stakeholder in the glyphosate resistance issue. Lower profits, greater management requirements, and immediate changes in farming practices are real and direct threats to each operation. Other major stakeholders include landlords, community leaders, and other organizations such as the Natural Resource Conservation Service (NRCS). If herbicide-resistant weeds cause higher input costs and/or lower profits, the value of the land for rent or sale is also potentially reduced. The growth in farm size over the past 10 years has been largely due to farmers renting more acres. Some more marginal acres with high infestations of GR weeds are lying idle because it is not feasible to pay rent and higher input costs.

The number of acres devoted to conservation tillage has increased to greater than 70% of all cotton, soybean, and corn acreage since the introduction of GR crops. This increase in conservation tillage has saved 1 billion t of soil from erosion annually. Less than 5% of the cotton and soybean in the southern United States receive cultivation for weed control. These are numbers the NRCS would like to see preserved. However, current agriculture practices and the great advances in conservation are being compromised by the development of GR weeds. In a 2004 survey, county extension agents said GR horseweed had reduced conservation tillage farming in Tennessee by 18%. Even more telling, the survey showed the percentage of farms using conservation tillage in the largest cotton counties in Tennessee had dropped from 80% to 40%. Arkansas weed scientists estimate a 15% reduction in conservation tillage in their state due to glyphosate resistance. Similar trends have been reported in Georgia, Mississippi, and the Bootheel of Missouri.

GR biotypes of troublesome weed species have been confirmed in nine states since 2005 and will likely continue to infest cropland throughout the United States. These weeds have been effectively controlled with glyphosate in GR crops; however, many farmers are reverting to deep plowing and multiple cultivations for control of the GR biotypes of these species. There are fewer effective herbicide options for controlling GR weeds in cotton than in either soybean or field corn.

Although abandonment of GR technology may be a perceived solution, this is not feasible for many producers because this technology has allowed a substantial increase in their planted acreage due to increased herbicide application and timing flexibility associated with this technology. Those that choose to switch from the technology will supplement the weed control with multiple tillage passes and alternative herbicide programs. Research has identified some resistant management strategies that will help farmers manage resistant Palmer amaranth where it already exists and to help avoid infestations in areas where it is not currently a problem. Many of these practices are more costly than what has been done over the past several years. An incentive plan to help defray some of the increased cost and reward those practicing good conservation techniques would be beneficial in helping to preserve much of the conservation tillage expansion we have experienced in the past 10 years.

University extension and the NRCS have a record of promoting conservation tillage over the past 25 years and recognize that GR weeds have the potential to destroy much of the gains made. With additional tillage, the danger of increased erosion and off-site sedimentation is real. Water quality and quantity will also be reduced as more tillage is utilized to manage these pests (Fig. 15.6). Alternative programs may increase the use of higher-risk herbicides. The land-grant university systems are recognized as leading agencies to provide sound technical data for effective weed control and agricultural production. These agencies working together can develop sound management practices that encourage conservation tillage while providing weed control that promotes sustainable crop production.

Figure 15.6. Sediment in runoff water from no-till (left) and conventional till (right) areas following a single rainfall event. Photo courtesy of Bill Teeter.

15.8 DEVELOPING BEST MANAGEMENT PRACTICES—RECOMMENDATIONS

This usually requires an extensive applied research program that evaluates alternative treatments and techniques. Although much of the applied research may be conducted on university research stations, the results must be verified on farms with farm equipment and management. Best management strategies or recommendations must be clearly stated and updated as new information becomes available.

15.9 WHERE TO FROM HERE?

No doubt, additional species will evolve resistance to glyphosate and each will bring additional challenges to farmers, and thus, to extension educators. Farmers want answers and depend on extension scientists to provide best management information. They are not different from any other segment of business in that they prefer easy answers. However, they are not strangers to adversity and will adopt new technology or new ways of controlling weeds when necessary. After sitting through a long, multiday discussion of research on GR Palmer amaranth, a farmer was asked for his assessment of the discussion and what it meant to him. His response was "we are farmers, we farm, that is what we do. We know we will not be able to farm the same as we have in the past. We expect you (meaning university researchers) to find solutions and give us our options." This places a great deal of responsibility on extension scientists and clearly requests they assume leadership roles in this dilemma.

Instant and almost constant communication technology, the Internet, rapid means of transportation, university research stations, and scientific weed science societies are valuable tools to assist extension educators do a better job. However, none of these will replace walking the fields and understanding the problems first hand. Just as Dr. Knapp rode a train for 2 days every 2 weeks to make field visits and make recommendations in early extension history, we will continue a hands-on approach with a sincere desire to help farmers produce food and fiber in a sustainable agricultural system.

REFERENCES

Bailey, J. C. 1945. *Seaman A. Knapp School Master of American Agriculture*. New York: Columbia University Press.

Busi, R. and S. B. Powles. 2009. Evolution of glyphosate resistance in a *Lolium rigidum* population by glyphosate selection at sublethal doses. *Heredity* 103:318–325.

Harper, J. L. 1956. The evolution of weeds in relation to resistance to herbicides. *Proceedings of the Brighton Conference, Weeds* 3:179.

McClelland, M. 2007. Managing Glyphosate-Resistant Palmer Amaranth Roundtable Presentations. http://www.cottoninc.com/2007-Glyphosate-Resistant-Palmer-Amaranth (accessed July 15, 2009).

Rankins, A. Jr., J. D. Byrd, Jr., D. B. Mask, J. W. Barnett, and P. D. Gerard. 2005. Survey of soybean weeds in Mississippi. *Weed Technology* 19:492–498.

Ryan, G. F. 1970. Resistance of common groundsel to simazine and atrazine. *Weed Science* 18:614–616.

Shields, E. J., J. T. Dauer, M. J. VanGessel, and G. Neumann. 2006. Horseweed (*Conyza canadensis*) seed collected in the planetary boundary layer. *Weed Science* 54:1063–1067.

Smith, K. L. 2004. *Weed Control Demonstration and Research Trial Results*. Little Rock, AR: University of Arkansas Division of Agriculture Southeast Research and Extension Center, pp. 460–466.

Smith, K. L., J. Norsworthy, R. Scott, and N. Burgos. 2008. Field evidence of multiple glyphosate resistance mechanisms in *Amaranthus palmeri*. In *Proceedings of the 5th International Weed Science Congress*. Vancouver, Canada: International Weed Science Society, p. 50.

United States Department of Agriculture National Agricultural Statistics Service (USDA NASS). 2007a. Chapter 2: state level data. Table 8. In *Census Report*, Vol. 1. Washington, DC: US Department of Agriculture, http://www.agcensus.usda.gov/Publications/2007/Full_Report/usv1.pdf (accessed April 8, 2010).

United States Department of Agriculture National Agricultural Statistics Service (USDA NASS). 2007b. Chapter 2: state level data. Table 9. In *Census Report*, Vol. 1. Washington, DC: US Department of Agriculture, http://www.agcensus.usda.gov/Publications/2007/Full_Report/usv1.pdf (accessed April 8, 2010).

16

ECONOMIC IMPACT OF GLYPHOSATE-RESISTANT WEEDS

Janet E. Carpenter and Leonard P. Gianessi

16.1 INTRODUCTION

Genetically engineered herbicide-resistant (HR) crops have been widely planted in the United States and globally since their introduction in 1995, totaling 106 million ha planted worldwide in 2008 (James 2008a). The popularity of the technology is primarily due to the simplicity and flexibility of weed control programs that rely on herbicides with efficacy against a broad spectrum of weeds without crop injury or crop rotation restrictions (Carpenter and Gianessi 1999). Growers have been able to reduce their managerial intensity as they reduced the number of herbicides needed to control a wide range of weed species and take advantage of a wider window of application to treat larger weeds while maintaining efficacy (Gianessi 2008). In addition, HR crops fit into established trends toward postemergence weed control, adoption of conservation tillage practices, and narrow-row spacing. Further, the use of glyphosate in glyphosate-resistant (GR) crops has allowed management of weeds that have developed resistance to other classes of herbicides. Growers have realized significant cost savings through the adoption of GR crops (Fernandez-Cornejo and McBride 2000; Marra et al. 2004).

Over the past few years, an increasing number of cases of weed populations with confirmed resistance to glyphosate have been reported. The first confirmed case of glyphosate resistance in an area growing GR crops in the United States was in horseweed in Delaware in 2000 (VanGessel 2001).

Glyphosate Resistance in Crops and Weeds: History, Development, and Management
Edited by Vijay K. Nandula
Copyright © 2010 John Wiley & Sons, Inc.

The field had been treated with glyphosate only for weed control in continuous GR soybean. Since then, glyphosate resistance has been confirmed in populations of 10 weed species across several states. Most of these cases have been reported where GR crops are commonly grown. However, GR weeds have also been reported in California in almonds and roadsides, in orchards in Oregon, and in nurseries in Michigan (Hanson et al. 2009; Heap 2010), none of which are related to the planting of GR crops. It is difficult to estimate the acreage that is currently infested with GR weeds, particularly in the case of a rapidly evolving situation such as Palmer amaranth. Weed scientists work to confirm cases of reported glyphosate resistance, which implies a necessary delay in reporting while the confirmation work is completed. It is generally recognized that, currently, GR horseweed and Palmer amaranth are by far the most widespread GR weeds in the United States. In many cases, confirmed cases of GR biotypes continue to be localized for a number of years.

The development of weeds resistant to glyphosate in a particular field will likely require modification to weed control programs where practices in addition to applying glyphosate are needed to control the resistant populations. This chapter explores the economic implications to growers of the development of GR weeds and the changes recommended for weed management programs compared with common programs, which frequently have relied on glyphosate only.

16.2   CHANGES IN MANAGEMENT PRACTICES SINCE THE INTRODUCTION OF GR CROPS

Glyphosate resistance was introduced in the United States in soybean and cotton in 1996, followed by corn in 1998, canola in 1999, and sugar beet in 2007. By 2008, 92% of soybean acres planted in the United States were GR, and 68% of the cotton and 63% of corn acres planted were HR (primarily GR) (USDA NASS 2008a). It is estimated that 62% of canola acreage was planted to GR canola in 2005 (Sankula 2006). Figure 16.1 shows the trends in adoption of HR soybean, cotton, and corn acres planted from 2000 to 2008. Regulatory approval is currently pending for GR alfalfa[1] and creeping bentgrass (USDA APHIS 2009). With adoption of GR crops, growers have changed their weed management programs in several important respects, including chemical use, tillage, and other integrated weed management practices. These changes have allowed growers to reduce weed management costs while maintaining yields.

One of the most dramatic changes in weed management related to the adoption of GR crops has been an overall reduction in the number of herbicide active ingredients applied in some crops because of the broad spectrum of activity of glyphosate. This shift has been most dramatic in soybeans (Givens

[1]Glyphosate-tolerant alfalfa was deregulated in 2005 but was later returned to regulated status following a preliminary court injunction issued on March 12, 2007 (Green 2009).

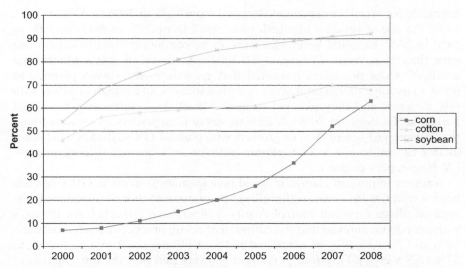

Figure 16.1. Percent of U.S. acres planted with herbicide-resistant varieties. *Source*: USDA NASS Acreage 2000–2008.

et al. 2009a). In 1995, 12 different active ingredients were used on at least 10% of soybean acres (USDA NASS 1996). By 2006, no herbicide besides glyphosate was used on more than 10% of soybean acres, and only one other herbicide (2,4-D) was used on more than 5% of soybean acres (USDA NASS 2007). Similarly, glyphosate use in cotton has increased from 9% of acres treated in 1995 to 85% in 2007. Nine herbicides were used on at least 10% of cotton acres in 1995 (USDA NASS 1996), compared with six in 2007 (USDA NASS 2008b).

Alternatively, in corn, while glyphosate use has increased from 4% of acres treated in 1997 (USDA NASS 1998a) to 31% in 2005, several other active ingredients continue to be widely used. For example, atrazine, which is still the most commonly used herbicide on corn, was used on 66% of acres in 2005, down only slightly from 69% in 1997 (USDA NASS 2006). The same number of herbicides was used on at least 10% of corn acres in 1997 and 2005 (USDA NASS 1998a, 2006). In part, these figures are the result of slower adoption of GR corn, due to restricted availability in hybrids suited to various geographic locations and trade restrictions in export markets. However, residual herbicides continue to be used in GR corn due to the earlier planting of corn and greater susceptibility to early-season weed competition (Gianessi 2008).

These trends toward reducing the number of herbicide active ingredients may already be reversing. In the first 2 years of a 4-year study, grower cooperators in six states provided information on weed management practices during the 2006 and 2007 growing seasons. The percentage of fields in continuous GR cotton, corn, and soybean that were treated with glyphosate as the sole herbicide for weed management dropped from 45% in 2006 to 33% in 2007, while the use of residual herbicides increased from 39% of fields in 2006 to 54% in 2007. Residual herbicides were more commonly used in cotton and corn than in soybean (Young et al. 2007). A survey of 400 southern and southern plains producers indicated that growers in the survey planned to plant an average of 955 total acres of cotton in 2008, an average of 779 ac with GR varieties, and they planned to treat 536 of these acres with a residual herbicide (Hurley et al. 2009). A 2009 survey of Illinois soybean growers found that between 63% and 86% of growers who planted GR soybeans used other modes of action either in tank-mix with or in sequence with glyphosate (A. Hager, pers. comm.).

Another important change in weed management systems in GR crops has been a shift toward conservation tillage practices, and therefore, reduced reliance on tillage for weed control. A survey of soybean growers from 19 states with over 200 ac showed that the adoption of no-till practices doubled between 1996 and 2001 to 49% of acres, and reduced till increased by one-fourth, to 33% (ASA 2001). Fifty-four percent of growers said that the introduction of GR soybeans had the greatest impact on their decision to adopt conservation tillage. Another survey conducted in 2001 and 2002, found that GR soybean growers, on average, make 25% fewer tillage passes each season compared with non-adopters. When growers switch to GR soybeans, they tend to switch from conventional tillage to no-till, with the amount of reduced tillage remaining fairly constant. More than 40% of GR soybean adopters used no-till, while about 30% of non-adopters used no-till. Five percent of non-adopters, 50% of partial adopters, and 77% of full adopters used reduced tillage on at least 50% of their soybean acreage (Marra et al. 2004).

A survey of 1195 growers in six states (Illinois, Indiana, Iowa, Mississippi, Nebraska, and North Carolina) found that the adoption of GR cropping systems resulted in a large increase in the percentage of growers using no-till and reduced-till systems (Givens et al. 2009b). The decline in tillage was greater in continuous GR cotton and GR soybean, with reductions of 45% and 23%, respectively, than in cropping systems that included GR corn or conventional crops.

Growers also appear to have shifted away from some other integrated weed management practices since the introduction of HR crops. In corn, 15% of acres were mapped for weed problems in 1997, compared with 11% in 2005 (USDA NASS 1998b, 2006). In soybeans, field mapping of weed problems dropped from 14% to 9% between 1997 and 2006 (USDA NASS 1998b, 2007). Professional scouting for weeds was used on only 8% of soybean acreage in 2006. Similarly, in cotton, field mapping was used on 9% of acreage in 2007, compared with 14% in 1997 (USDA NASS 1998b, 2008a, 2008b). However, in a recent survey of cotton growers in the south and southern plains, 95% of respondents indicated that they always or often adopted five or more weed resistance management practices, while more than 70% always or often adopted seven or more practices (Frisvold et al. 2009). Weed resistance management practices considered in the study were scouting before or after herbicide application, starting with a clean field, early weed control, prevention of weed escapes, cleaning equipment, buying new seed, using multiple herbicide modes of action, supplemental tillage, and using recommended herbicide application rates.

Surveys have also directly addressed grower attitudes and practices with respect to GR weeds. Indiana corn and soybean growers were surveyed during the winter in 2003/2004 about the importance of GR weeds and any management practices they employed to prevent the development of resistant populations (Johnson and Gibson 2006). Although a relatively low percentage of respondents were highly concerned about resistance (36%), many reported the use of integrated weed resistance management practices, including field scouting, the use of tank-mix partners with glyphosate for burndown and postemergence weed control, and soil-applied residual herbicides as resistance management strategies. Growers who farmed at least 800 ha were more concerned about glyphosate resistance and more likely to adopt resistance management strategies than smaller growers.

A related point about changes in management practices since the introduction of GR crops is that overall weed pressure may be decreasing in some fields. Another major survey showed that overall growers were experiencing reduced weed pressure after the adoption of GR crops. Between 36% and 70% of surveyed corn, cotton, and soybean growers in six states felt that weed pressure had decreased since they had planted GR crops (Kruger et al. 2009).

16.3   COSTS OF ADDRESSING GR WEEDS IN WEED MANAGEMENT PROGRAMS

Since the emergence of GR weed biotypes, weed scientists have been working to develop recommended modifications to glyphosate-based weed control programs. Here we gather the available recommendations and cost estimates for programs recommended for growers to control GR weeds. Table 16.1 summarizes these estimates by state and crop, which are described in greater detail below. It should be noted that in several cases, a range of estimates is provided that reflects the different approaches that growers might choose, depending on the level of infestation, timing, and other factors that would influence herbicide choice, such as presence of other weeds. In general, it is not possible to say where the majority of growers fall in the range of cost estimates, though growers will try to minimize costs where possible.

16.3.1   Corn

GR weeds in corn have been reported in four states: common waterhemp in Illinois and Missouri, horseweed in Mississippi, and Palmer amaranth in North Carolina (Heap 2010). However, it is likely that GR weeds reported in other

TABLE 16.1. Estimates of Increased Costs Associated with Control of Glyphosate-Resistant Weeds

State	Reference	Crop	Weed	Increased Cost/Acre
Arkansas	K. Smith (pers. comm.)	Cotton	Palmer amaranth	$14.07
	Bryant (2007)	Cotton	Palmer amaranth	$35
Delaware	M. VanGessel (pers. comm.)	Soybean	Horseweed	$3–12
Georgia	S. Culpepper (pers. comm.)	Cotton	Palmer amaranth	$3–100
Illinois	Mueller et al. (2005)	Soybean	Common waterhemp	$35.82
	Mueller et al. (2005)	Corn	Common waterhemp	$0
Minnesota	J. Gunsolus (pers. comm.)	Corn	Common waterhemp and giant ragweed	No change
		Soybean	Common waterhemp and giant ragweed	Equal or slightly lower
Mississippi	D. Shaw (pers. comm.)	Soybean	Horseweed	$8.40–15.50
			Italian ryegrass	$10.85–20.45
			Palmer amaranth	$6.01–11.00
		Corn	Horseweed	$1.82–16.00
			Italian ryegrass	$4.20–21.96
			Palmer amaranth	$1.82–35.02
		Cotton	Horseweed	$5.44–15.41
			Italian ryegrass	$14.52–26.50
			Palmer amaranth	$6.19–20.44
Missouri	K. Bradley (pers. comm.)	Soybean	Common ragweed	$20–25
		Soybean	Common waterhemp	$20–25
		Corn	Common waterhemp	$0–15
		Cotton	Horseweed	$5
New Jersey	M. VanGessel (pers. comm.)	Soybean	Horseweed	$3–12
North Carolina	A. York (pers. comm.)	Cotton	Horseweed	$10
		Corn	Palmer amaranth	$13
		Cotton	Palmer amaranth	$15–40[a]
		Soybean	Palmer amaranth	$19[b]
		Soybean	Horseweed	$10
South Carolina	M. Marshall (pers. comm.)	Cotton	Palmer amaranth	$25–50

TABLE 16.1. *Continued*

State	Reference	Crop	Weed	Increased Cost/Acre
Tennessee	L. Steckel (pers. comm.)	Cotton	Giant ragweed	$16
		Soybean	Giant ragweed	$30
		Cotton	Horseweed	$20
		Soybean	Horseweed	$13–23
		Cotton	Palmer amaranth	$30–33
		Soybean	Palmer amaranth	$32–42
	Mueller et al. (2005)	Cotton	Horseweed	$25.49
		Soybean	Horseweed	$11.51
		Corn	Horseweed	$0

Note: Costs include application costs where appropriate if herbicides are applied separately from glyphosate application and may also include costs of cultivation or hand weeding.
[a]Yields may decrease by as much as 15% on infested acreage (A. York, pers. comm.).
[b]Yields may decrease by 1–2% on infested acreage (A. York, pers. comm.).

crops also infest corn but are effectively controlled with current weed management programs and therefore unreported. In North Carolina, corn growers with GR Palmer amaranth have several options, including treatment with atrazine in formulations or tank-mixes either pre- or postemergence (A. York, pers. comm.).

Rotation to corn is considered to be beneficial in the management of GR weeds due to the availability of a range of herbicides that are effective against weeds with reported resistance to glyphosate (e.g., Davis et al. 2008). University of Missouri Extension suggests that one of the most effective and economical method to prevent or manage GR waterhemp is to rotate to corn and use alternative herbicides (Bradley et al. 2008). However, growers in areas without irrigation, with low rainfall, and/or sandy soils are limited in their ability to rotate to corn, which requires relatively high amounts of soil moisture to achieve high yields. During the early years of GR corn, it was speculated that some growers will shift toward more glyphosate use and away from the use of alternative herbicides (Owen 2006). However, whether this expectation is borne out is uncertain, as more and more growers become aware of the development of GR weeds.

Extension weed scientists estimate that corn growers with GR weeds may incur increased weed control costs in some severe cases of up to $35 per acre compared with commonly used glyphosate-based programs, primarily due to applying herbicides with an additional mode of action (Table 16.1). However, growers may be able to control GR weeds in corn without increasing costs, due to the availability of low-cost herbicides with efficacy against GR weeds such as waterhemp and giant ragweed in Minnesota (J. Gunsolus, pers. comm.) and horseweed in Tennessee (Mueller et al. 2005).

16.3.2   Cotton

Five species of weeds have developed GR biotypes that infest cotton. GR horseweed has been reported in six states (Arkansas, Kansas, Mississippi, Missouri, North Carolina, and Tennessee), and Palmer amaranth is reported in four states (Georgia, North Carolina, Tennessee, and Mississippi). Giant ragweed and Italian ryegrass are each reported in one state, Tennessee and Mississippi, respectively (Heap 2010). GR horseweed is the most widely established across the United States currently, while Palmer amaranth has become a troublesome issue particularly in cotton-growing regions (B. Nichols, pers. comm.).

Cotton growers have several options for managing GR horseweed. In Tennessee, growers might add dicamba and flumioxazin to glyphosate in a preplant burndown application, followed by diuron and monosodium methylarsonate (MSMA) in a postdirected in-crop application (L. Steckel, pers. comm.). In North Carolina, growers can add 2,4-D and flumioxazin to the burndown to control GR horseweed (A. York, pers. comm.). Mississippi cotton growers may apply flumioxazin or trifloxysulfuron-sodium in the fall, may tank-mix dicamba, flumioxazin, or glufosinate in a burndown application with or without glyphosate, followed by trifloxysulfuron-sodium postemergence (D. Shaw, pers. comm.). The increased cost of these programs can range from $5 to 25 per acre (Table 16.1).

Managing GR Palmer amaranth can be more challenging for cotton growers than other currently known GR weeds. While there are a number of herbicides labeled for preemergence use in cotton that will control GR Palmer amaranth (such as dicamba, diuron, fomesafen, fluomioxazin, fluometuron, metolachlor, MSMA, pendimethalin), their efficacy depends on timely rain or irrigation for their activation. With timely rainfall or irrigation to activate the preemergence options, control is possible with minimum added cost (M. Marshall, D. Shaw, K. Smith, and L. Steckel, pers. comm.). Options for postemergence control of Palmer amaranth are limited to pyrithiobac sodium, S-metolachlor, MSMA, and trifloxysulfuron-sodium. In Mississippi, recommendations include postdirected treatments of diuron, flumioxazin, or MSMA (D. Shaw, pers. comm.). However, all of these herbicides must be used on relatively small Palmer amaranth to be effective.

Planting glufosinate-tolerant cotton is another option, allowing the postemergence use of glufosinate. In one study in dryland cotton production, where residual herbicides are not activated by rainfall or irrigation, only glufosinate-tolerant cotton programs with timely applications of glufosinate in combination with cultivation provided adequate control of GR Palmer amaranth (Culpepper et al. 2008). However, glufosinate is only effective on Palmer amaranth up to 4 in. (*Ignite® 280 SL Herbicide Product Label* 2008). Another significant limitation to expanded use of glufosinate-tolerant cotton is the lack of availability of the trait in varieties suited for a range of growing areas, which implies a yield penalty that may be substantial. However, this should be resolved over time as glufosinate tolerance is bred into more varieties. Once GR Palmer amaranth grows above the cotton canopy, the only options are tillage or hand weeding.

In Georgia, cotton growers are considered to be in one of three categories with respect to GR Palmer amaranth. Extension agents from 20 cotton-producing counties were surveyed in 2008 to estimate the adoption of various weed control practices. Growers with low levels of infestation are adding alternative herbicide modes of action to their weed management programs, have increased hand weeding, while at the same time have increased adoption of conservation tillage. Growers with moderate infestation are also adding modes of action and spot hand weeding, and have increased their use of cultivation. Growers who have a severe level of infestation (with 100 or more plants per square yard) may be implementing a range of practices. It was estimated that 20% of growers are cultivating, 45% are hand weeding, and 88% are adding two or more residual herbicides when there is a severe level of infestation. Adoption of glufosinate-tolerant varieties has increased between 2004 and 2008 from 0% to 26% for growers with severe infestations, despite the lower yields associated with switching to varieties that are less suited to their geographic conditions. Most Georgia cotton growers have low or no GR Palmer amaranth, a few have light to moderate infestations and about 100,000 ac are estimated to have severe infestations (S. Culpepper, pers. comm.).

Across the states where GR Palmer amaranth has been confirmed, weed management costs are estimated to have increased from $0 to 100 per acre (Table 16.1). At the high end of the range are growers with severe levels of infestation who are using cultivation and hand weeding to achieve satisfactory control (S. Culpepper, pers. comm.). In addition, growers with GR Palmer amaranth face an increased risk of crop failure, due to the chance that pre-emergence herbicides are not activated by timely rainfall and the lack of options for controlling weeds once they grow over the crop canopy, which is not captured in the increased weed management costs (K. Smith, pers. comm.). However, this risk exists for conventional growers as well because uncontrolled GR Palmer amaranth is as difficult to control in conventional cotton as it is in GR cotton.

GR giant ragweed has been confirmed in cotton in Tennessee (Heap 2010). There are no preemergence herbicides for the control of giant ragweed. A postemergence program could include an application of trifloxysulfuron-sodium, followed by a postdirected application of glufosinate, at an increased cost of $16 per acre compared with current weed management programs using GR cotton (L. Steckel, pers. comm.).

In Mississippi, the only state where Italian ryegrass with resistance to glyphosate has been confirmed in an area growing GR crops, cotton growers may use clomazone or S-metolachlor in a fall treatment, or clethodim and/or paraquat in a burndown treatment at a cost of $15–27 per acre (D. Shaw, pers. comm.).

16.3.3   Soybeans

GR biotypes of seven weed species have been confirmed in soybeans in 16 states of the United States. GR horseweed is the most widespread, confirmed in 11 states: Delaware, Illinois, Indiana, Kansas, Kentucky, Maryland, Mississippi, New Jersey, Pennsylvania, Tennessee, and Ohio (Heap 2010). Growers can control GR horseweed by adding herbicides such as 2,4-D, dicamba, paraquat, glufosinate, or fluomoxazin to the burndown, with follow-up application of cloransulam-methyl or metribuzin/chlorimuron-ethyl (L. Steckel, M. VanGessel, and A. York, pers. comm.; Mellendorf et al. 2007).

There is one confirmed case of a horseweed population resistant to both glyphosate and acetolactate synthase (ALS)-inhibiting herbicides (e.g. chlorimuron-ethyl and cloransulam-methyl) in Ohio (Heap 2010). In those cases, growers can use 2,4-D or paraquat preplant, and have no other postemergence soybean herbicide options (Johnson et al. 2008).

The increased costs associated with managing GR horseweed in soybean are estimated between $3 and 23 per acre, depending on the choice of herbicides, and whether growers choose to add postemergence treatments in addition to preemergence treatments (Table 16.1). In Delaware, where GR horseweed was first reported, surveyed growers reported the most frequent change in soybean weed management was to apply another herbicide with a different mode of action before planting. Forty-eight percent of growers with resistance on-farm reported a $2–7 per acre increase in weed management costs, and 28% reported an increase greater than $7 per acre (Scott and VanGessel 2007). A benefit of adding preemergence herbicides back into soybean weed management programs could be an increase in yields due to improved early-season weed control (M. VanGessel and A. York, pers. comm.). This added benefit can counterbalance the added cost of using a preemergence product for resistant weed management.

Common ragweed populations with glyphosate resistance have been confirmed in three states: Arkansas, Kansas, and Missouri (Heap 2010). In addition, studies in Ohio have confirmed the presence of common ragweed biotypes with multiple herbicide resistance, to glyphosate and ALS inhibitors, or to protoporphyrinogen oxidase (PPO) and ALS inhibitors (Stachler and Loux 2007). GR common ragweed can be managed using preemergence and/ or alternative postemergence herbicides. The increased cost of managing GR common ragweed in Missouri is estimated at $20–25 per acre (K. Bradley, pers. comm.).

Common waterhemp with resistance to glyphosate has been confirmed in four states: Missouri, Kansas, Minnesota, and Illinois (Heap 2010). Multiple resistance has been confirmed to ALS inhibitors in Illinois and to both ALS and PPO inhibitors in Missouri. Growers with GR common waterhemp are encouraged to use a preemergence residual herbicide with good activity on waterhemp. If there is no resistance to ALS- or PPO-inhibiting herbicides, chlorimuron, fomesafen, or acifluorfen are registered for postermergence use in soybeans. For waterhemp populations that are resistant to both glyphosate and PPO inhibitors, between-row cultivation is the only option for controlling waterhemp escapes in crop, unless growers plant glufosinate-tolerant soybeans and apply glufosinate to waterhemp that are less than 4 in. (Bradley et al. 2008). Increased costs of weed management for GR common waterhemp in soybean is estimated at $20–25 per acre in Missouri (K. Bradley, pers. comm.). In Missouri weed control trials conducted in 2006 and 2007, preemergence applications of S-metolachlor plus metribuzin provided one of the highest net incomes in both years, between $110 and 138 per acre higher than glyphosate-only treatment (Legleiter et al. 2009).

Giant ragweed with glyphosate resistance has been confirmed in six states: Arkansas, Indiana, Kansas, Minnesota, Ohio, and Tennessee (Heap 2010). In addition, giant ragweed with multiple resistance, to glyphosate and ALS inhibitors has been confirmed in Ohio (Loux and Stachler 2007). As in cotton, there are no preemergence herbicides that provide reliable control of giant ragweed (L. Steckel, pers. comm.). Soybean growers in Tennessee that have GR giant ragweed can use cloransulam-methyl for $30 per acre increased cost (Table 16.1).

GR Palmer amaranth in soybean has been confirmed in three states: Arkansas, Georgia, and North Carolina (Heap 2010). As for the case of waterhemp, growers are encouraged to use a preemergence herbicide. In most cases, the same preemergence herbicide can be used for control of waterhemp and Palmer amaranth. In Tennessee, soybean growers can add dicamba and flumioxazin in a burndown application, with chlorimuron and metribuzin at planting and maybe fomesafen postemergence (L. Steckel, pers. comm.). In Mississippi, recommendations include tank-mixes of flumioxazin, S-metolachlor, and/or fomesafen with glyphosate in postemergence treatments (D. Shaw, pers. comm.). The estimated additional cost of managing GR Palmer amaranth in soybean can be $6–42 per acre (Table 16.1).

GR populations of two other weed species have been confirmed in soybeans: Italian ryegrass in Mississippi and johnsongrass in Arkansas (Heap 2010). Both of these cases are isolated at this point (Nandula et al. 2008; J. Norsworthy, pers. comm.). However, soybean growers in Mississippi with GR Italian ryegrass may use paraquat in a burndown, or treat with S-metolachlor or clethodim postemergence at an additional cost of $11–20 per acre (Table 16.1) (D. Shaw, pers. comm.).

The estimated change in the cost of weed management associated with GR weeds in soybean is estimated to be slightly less than a glyphosate-only program or up to $42 per acre, depending on the particular weed, level of infestation, and choice of herbicides (Table 16.1).

16.4 ECONOMICS OF PROACTIVE VERSUS REACTIVE RESISTANCE MANAGEMENT

Several studies have explored the costs to growers of taking a proactive approach to preventing the development of GR weeds compared with the costs of a reactive approach of changing practices only once a problem has become apparent in the field. A proactive approach assumes that growers incur increased costs associated with implementing integrated weed management practices to avert the development of resistance immediately. Growers who choose to wait until herbicide resistance to develop save on costs initially, waiting to implement integrated weed management practices until after they observe HR weeds in the field.

In a model of the development of GR common waterhemp in a corn–soybean rotation in Illinois, researchers estimated the annual cost of a proactive approach to be $1.83 per acre compared with the cost of a reactive approach at $17.91 per acre. Given these costs, and using an 8% discount rate, it was estimated that growers would benefit from implementing the proactive approach if resistance developed in less than 29 years (Mueller et al. 2005).

In another study, researchers compared net returns of a proactive approach to practices that relied mainly on the use of glyphosate. Researchers selected 156 farm sites in six states (Illinois, Indiana, Iowa, Nebraska, North Carolina, and Mississippi) that were divided into two halves with one half managed by the farmer using standard GR corn, cotton, or soybean weed management practices and the other half managed according to university recommendations based on weed resistance management principles, which included the use of other modes of action in addition to glyphosate. All costs were kept constant between the farmer and researcher systems except for the cost and application of the different herbicides. In 2006–2007, yields and net returns were similar between the farmer and university systems for all rotations, even though the herbicide costs were higher for the university system (Weirich et al. 2007).

Mid-South cotton growers are estimated to face an increase in cost of $13 per acre for a program that incorporates residual herbicides that would delay the onset of glyphosate resistance in Palmer amaranth in Roundup Ready Flex® (Monsanto Co., St. Louis, MO) cotton compared with an increase of $35 per acre for a program that would control an established problem with GR Palmer amaranth (Bryant 2007).

One preemptive strategy that has been investigated is known as "double knockdown," which involves a follow-up application of another nonselective herbicide, paraquat, after glyphosate to eliminate any weeds that survive glyphosate application. Analysis of the development of GR rigid ryegrass in Australia showed that this approach can drive genes for glyphosate resistance to local extinction, potentially allowing continued use of glyphosate into the indefinite future (Weersink et al. 2005).

16.5   LOOKING AHEAD TO NEW TECHNOLOGIES

The rotation of herbicide modes of action is an important component of weed resistance management. Several new HR crop technologies are under development that will facilitate the rotation of herbicides while increasing the flexibility of timing of herbicide application. Glufosinate-tolerant soybeans were recently approved for importation by the European Commission, which clears the way for commercialization in 2009 (James 2008b). Stacked glyphosate/ALS inhibitor-tolerant soybean was recently granted nonregulated status by the United States Department of Agriculture (USDA) and is expected to be commercialized in 2009. An application for nonregulated status is pending for stacked glyphosate and imidazolinone-tolerant corn (USDA APHIS 2009). Cotton with tolerance to both glufosinate and glyphosate is also planned for commercial release in 2009 (Holloway et al. 2008). Field trials have been conducted using soybeans with stacked glyphosate/dicamba tolerance (Steckel and Montgomery 2007, 2008). Other HR transgenes are at various stages of development for resistance to ALS inhibitors, acetyl coenzyme A carboxylase (ACCase) and auxin herbicides, 4-hydroxyphenylpyruvate dioxygenase (HPPD)-inhibiting herbicides, and PPO-inhibiting herbicides, among others (Green 2009). The availability of new technology has the potential of reducing costs by expanding weed management options for growers.

The development of HR weeds is not unique to glyphosate. The first confirmed report of a weed population expressing tolerance to an herbicide was in 1964, where field bindweed in Kansas was found to be resistant to 2,4-D. To date, over 125 HR weeds have been confirmed in the United States (Heap 2010). Glyphosate continues to be effective on 90 different weed species, including many that have developed resistance to other herbicides. Growers continue to value the GR crops (Aultman et al. 2009; Hurley et al. 2009). Maintaining the efficacy of glyphosate and other herbicides is critical to controlling weed control costs (Powles 2008; Robinson 2009). Growers will modify weed management programs to address the development of GR weeds, either before or after they become a problem in the field.

REFERENCES

American Soybean Association (ASA). 2001. Conservation Tillage Study. http://www.soygrowers.com/ctstudy/Default.htm (accessed March 11, 2009).

Aultman, S., T. Hurley, P. Mitchell, and G. Frisvold. 2009. Valuing the Roundup Ready corn and soybean weed management program. Agricultural and Applied Economics Association Annual Meeting, Milwaukee, WI.

Bradley, K., R. Smeda, and R. Massey. 2008. *Management of Glyphosate-Resistant Waterhemp in Corn and Soybean*. Columbia, MO: University of Missouri Extension IPM1030.

Bryant, K. 2007. What will glyphosate-resistant pigweed cost mid-South cotton? *Delta Farm Press*, May 11. http://deltafarmpress.com/mag/farming_glyphosateresistant_pigweed_cost/index.html (accessed April 1, 2010).

Carpenter, J. E. and L. P. Gianessi. 1999. Herbicide tolerant soybeans: why growers are adopting Roundup Ready varieties. *AgBioForum* 2:65–72.

Culpepper, A. S., A. C. York, J. M. Kichler, and A. W. MacRae. 2008. Glyphosate-resistant Palmer amaranth response to weed management programs in Roundup

Ready and Liberty Link cotton. In *Proceedings of the Beltwide Cotton Conferences*. pp. 1689–1690.

Davis, V. M., K. D. Gibson, and W. G. Johnson. 2008. Crop rotation and herbicide use influence population dynamics of glyphosate-resistant horseweed (*Conyza canadensis*) in no-till crop management systems. In *Proceedings of the North Central Weed Science Society*. 63:98.

Fernandez-Cornejo, J. and W. D. McBride. 2000. Genetically engineered crops for pest management in U.S. Agriculture: farm-level effects. Agricultural Economic Report No. 786, US Department of Agriculture Economic Research Service, Washington, DC.

Frisvold, G., T. Hurley, and P. Mitchell. 2009. Cotton grower adoption of weed resistance management practices. *Proceedings of the Beltwide Cotton Conferences* 306–313.

Gianessi, L. P. 2008. Economic impacts of glyphosate-resistant crops. *Pest Management Science* 64:346–352.

Givens, W. A., D. R. Shaw, W. G. Johnson, S. C. Weller, B. G. Young, R. G. Wilson, M. D. K. Owen, and D. Jordan. 2009a. A grower survey of herbicide use patterns in glyphosate-resistant cropping systems. *Weed Technology* 23:156–161.

Givens, W. A., D. R. Shaw, G. R. Kruger, W. G. Johnson, S. C. Weller, B. G. Young, R. G. Wilson, M. D. K. Owen, and D. Jordan. 2009b. Survey of tillage trends following the adoption of glyphosate-resistant crops. *Weed Technology* 23:150–155.

Green, J. M. 2009. Evolution of glyphosate-resistant crop technology. *Weed Science* 57:108–117.

Hanson, B. D., A. Shrestha, and D. L. Shaner. 2009. Distribution of glyphosate-resistant horseweed (*Conyza canadensis*) and relationship to cropping systems in the Central Valley of California. *Weed Science* 57:48–53.

Heap, I. 2010. International Survey of Herbicide Resistant Weeds. http://www. weedscience.com (accessed April 8, 2010).

Holloway, J., L. Trolinder, J. M. Ellis, and S. Baker. 2008. New herbicide tolerance technology for glyphosate resistant weed management in cotton. In *Proceedings of the Beltwide Cotton Conferences* 1726–1730.

Hurley, T., P. Mitchell, and G. Frisvold. 2009. Adoption, residual herbicide use and farmer values for Roundup Ready cotton. *Proceedings of the Beltwide Cotton Conferences* 314–325.

Ignite® 280 SL Herbicide Product Label. 2008. Research Triangle Park, NC: Bayer CropScience.

James, C. 2008a. *Global Status of Commercialized Biotech/GM Crops: 2008*. Ithaca, NY: International Service for the Acquisition of Agri-Biotech Applications.

James, L. 2008b. LibertyLink promising weapon against resistant weed problems. *Delta Farm Press*, October 10, p. 14.

Johnson, B., V. Davis, and A. Westhoven. 2008. Glyphosate-resistant weeds: yes it can happen, even when you think you know how to kill weeds. In *2008 Illinois Crop Protection Technology Conference Proceedings*. Urbana, IL: University of Illinois Extension, pp. 38–39.

Johnson, W. G. and K. D. Gibson. 2006. Glyphosate-resistant weeds and resistance management strategies: an Indiana grower perspective. *Weed Technology* 20:768–772.

Kruger, G. R., W. G. Johnson, S. C. Weller, M. D. K. Owen, D. R. Shaw, J. W. Wilcut, D. L. Jordan, R. G. Wilson, M. L. Bernards, and B. G. Young. 2009. US grower views on problematic weeds and changes in weed pressure in glyphosate-resistant corn, cotton, and soybean cropping systems. *Weed Technology* 23:162–166.

Legleiter, T. R., K. W. Bradley, and R. E. Massey. 2009. Glyphosate-resistant waterhemp (*Amaranthus rudis*) control and economic returns with herbicide programs in soybean. *Weed Technology* 23:54–61.

Loux, M. M. and J. M. Stachler. 2007. Giant ragweed biotypes with resistance to glyphosate and ALS inhibitors. In *Proceedings of the North Central Weed Science Society*. 62:179.

Marra, M. C., N. E. Piggott, and G. A. Carlson. 2004. *The Net Benefits, Including Convenience, of Roundup Ready Soybeans: Results from a National Survey. NSF Center for IPM Technical Bulletin 2004-3*. Raleigh, NC: North Carolina State University.

Mellendorf, T. G., B. G. Young, and J. L. Matthews. 2007. Management of glyphosate-resistant horseweed for southern Illinois. In *Proceedings of the North Central Weed Science Society*. 62:177.

Mueller, T. C., P. D. Mitchell, B. G. Young, and A. S. Culpepper. 2005. Proactive versus reactive management of glyphosate-resistant or -tolerant weeds. *Weed Technology* 19:924–933.

Nandula, V. K., K. N. Reddy, D. H. Poston, A. M. Rimando, and S. O. Duke. 2008. Glyphosate tolerance mechanism in Italian ryegrass (*Lolium multiflorum*) from Mississippi. *Weed Science* 56:344–349.

Owen, M. D. K. 2006. Herbicide resistance, weed populations shifts, and weed management stewardship: is anything new? In *Proceedings of the 18th Annual Integrated Crop Management Conference*. Ames, IA: Iowa State University Extension Agribusiness Education Program, pp. 135–140.

Powles, S. B. 2008. Evolution in action: glyphosate-resistant weeds threaten world crops. *Outlooks on Pest Management* 19:256–259.

Robinson, E. 2009. Fewer new chemicals chasing resistant weeds. *Southwest Farm Press*, January 8, p. 14.

Sankula, S. 2006. *Quantification of the Impacts on US Agriculture of Biotechnology-Derived Crops Planted in 2005*. Washington, DC: National Center for Food and Agricultural Policy.

Scott, B. A. and M. J. VanGessel. 2007. Delaware soybean grower survey on glyphosate-resistant horseweed (*Conyza canadensis*). *Weed Technology* 21:270–274.

Stachler, J. M. and M. M. Loux. 2007. Multiple herbicide resistance in common ragweed. In *Proceedings of the North Central Weed Science Society*. 62:180.

Steckel, L. E. and R. F. Montgomery. 2007. Glyphosate-resistant horseweed control in soybean tolerant to both dicamba and glyphosate. In *Proceedings of the North Central Weed Science Society*. 62:178.

Steckel, L. E. and R. F. Montgomery. 2008. Glyphosate-tolerant horseweed control in dicamba glyphosate resistant soybeans. In *Proceedings of the North Central Weed Science Society*. 63:92.

United States Department of Agriculture Animal and Plant Health Inspection Service (USDA APHIS). 2009. Biotechnology Regulatory Service. Petitions of Nonregulated Status Granted or Pending by APHIS as of February 5, 2009. http://www.aphis.usda.gov/brs/not_reg.html (accessed March 11, 2009).

United States Department of Agriculture National Agricultural Statistics Service (USDA NASS). 1996. *Agricultural Chemical Usage 1995 Field Crops Summary.* Washington, DC: US Department of Agriculture.

United States Department of Agriculture National Agricultural Statistics Service (USDA NASS). 1998a. *Agricultural Chemical Usage 1997 Field Crops Summary.* Washington, DC: US Department of Agriculture.

United States Department of Agriculture National Agricultural Statistics Service (USDA NASS). 1998b. *Pest Management Practices.* Washington, DC: US Department of Agriculture.

United States Department of Agriculture National Agricultural Statistics Service (USDA NASS). 2006. *Agricultural Chemical Usage 2005 Field Crops Summary.* Washington, DC: US Department of Agriculture.

United States Department of Agriculture National Agricultural Statistics Service (USDA NASS). 2007. *Agricultural Chemical Usage 2006 Field Crops Summary.* Washington, DC: US Department of Agriculture.

United States Department of Agriculture National Agricultural Statistics Service (USDA NASS). 2008a. *Acreage.* Washington, DC: US Department of Agriculture.

United States Department of Agriculture National Agricultural Statistics Service (USDA NASS). 2008b. *Agricultural Chemical Usage 2007 Field Crops Summary.* Washington, DC: US Department of Agriculture.

VanGessel, M. J. 2001. Glyphosate-resistant horseweed in Delaware. *Weed Science* 49:703–705.

Weersink, A., R. S. Llewellyn, and D. J. Pannell. 2005. Economics of pre-emptive management to avoid weed resistance to glyphosate in Australia. *Crop Protection* 24:659–665.

Weirich, J. W., D. R. Shaw, W. A. Givens, J. A. Huff, R. G. Wilson, and W.G. Johnson. 2007. Assessing long-term viability of glyphosate-resistant technology as a foundation for cropping systems—on-farm economic comparisons of management systems. *Abstracts of Weed Science Society of America* 47:318.

Young, B. G., R. G. Wilson, W. G. Johnson, S. C. Weller, M. D. K. Owen, D. R. Shaw, and J. W. Wilcut. 2007. Grower-implemented herbicide strategies for weed management in glyphosate-resistant crops. In *Proceedings of the North Central Weed Science Society.* 62:198.

INDEX

ABC transporter genes, 156
ABC transporters, 133, 153, 156
Absorption, 48, 106, 109, 122, 124, 129, 202
  reduced, 37
Accession, 37
Acid, 10–11, 21, 105
  transaconitic, 105–6
Active ingredients, 12, 22, 25–26, 37, 187, 298–99
Adjuvants, 24–25, 94, 110
Agrobacterium-mediated transformation, 55, 75–77, 80
Agroecosystem, 215, 222–23, 225, 227
Ala, 127, 128, 130, 131
Alanine, 74, 75, 80, 127, 130, 144, 150
Alkyl ester pathways, 17–18
Alleles, 37–38, 145, 219
ALS, 56–57, 69–70, 80, 93, 106, 125, 129, 205–6, 214, 219, 236, 255–56, 306
ALS herbicides, 69, 80, 206, 255–56, 263
ALS-inhibiting herbicides, 73–74, 80, 83, 205, 270
  resistance, 80

ALS inhibitors, 127, 221, 225, 306–7, 309
ALS-resistant corn, 69
Alteration, 103, 106–8, 122, 133, 150–51, 224
Amaranthus palmeri, 80, 96, 122, 145, 151, 195
Amaranthus retroflexus, 198
Amaranthus species, 173
Amaranthus spp., 129
Ammonium sulfate, 25, 94
AMPA, 1, 6, 14–15, 46, 48, 72, 75, 77, 121
Amphibians, 12
Annual ryegrass, 174
Answering farmer questions, 287
Apoplast, 4, 121, 132, 189
Applicator exposure, 15–17
Applications, 1, 4, 14–17, 24, 48, 60, 82, 94, 96, 103, 190, 204–7, 305–6, 308–9
Argentina, 7, 120, 128, 176, 249–50, 256–60, 263–66, 269, 271
Arkansas, 124, 176–78, 196, 201, 220, 284, 287–89, 292–93, 304, 306–7
Asian soybean rust (ASR). *See* ASR

Glyphosate Resistance in Crops and Weeds: History, Development, and Management
Edited by Vijay K. Nandula
Copyright © 2010 John Wiley & Sons, Inc.

Printed and bound by CPI Group (UK) Ltd, Croydon, CR0 4YY

28/06/2023

03231151-0005